Graph Theory
with Applications to Engineering & Computer Science

NARSINGH DEO
Millican Chair Professor, Dept. of Computer Science
Director, Center for Parallel Computation,
University of Central Florida

DOVER PUBLICATIONS, INC.
Mineola, New York

Copyright

Copyright © 1974 by Narsingh Deo
All rights reserved.

Bibliographical Note

This Dover edition, first published in 2016, is an unabridged republication of the work originally published in 1974 by Prentice-Hall, Inc., Englewood Cliffs, New Jersey.

Library of Congress Cataloging-in-Publication Data

Names: Deo, Narsingh 1936–
Title: Graph theory with applications to engineering and computer science / Narsingh Deo.
Description: Dover edition. | Mineola, New York : Dover Publications, 2016. | Originally published: Englewood Cliffs, New Jersey : Prentice-Hall, Inc., 1974. | Includes bibliographical references and index.
Identifiers: LCCN 2016008025 | ISBN 9780486807935 | ISBN 0486807932
Subjects: LCSH: Graphy theory. | Engineering mathematics.

Classification: LCC TA338.G7 D46 2016 | DDC 511/.5—dc23 LC record available at http://lccn.loc.gov/2016008025

Manufactured in the United States by LSC Communications
80793203   2019
www.doverpublications.com

*To the memory of my father,
 who did not live to realize his greatest ambition—
 that of witnessing his eldest son matriculate.*

CONTENTS

PREFACE xiii

1 INTRODUCTION 1

1-1 What is a Graph? 1
1-2 Application of Graphs 3
1-3 Finite and Infinite Graphs 6
1-4 Incidence and Degree 7
1-5 Isolated Vertex, Pendant Vertex, and Null Graph 8
1-6 Brief History of Graph Theory 10
Summary 11
References 11
Problems 12

2 PATHS AND CIRCUITS 14

2-1 Isomorphism 14
2-2 Subgraphs 16
2-3 A Puzzle With Multicolored Cubes 17
2-4 Walks, Paths, and Circuits 19
2-5 Connected Graphs, Disconnected Graphs, and Components 21
2-6 Euler Graphs 23
2-7 Operations On Graphs 26
2-8 More on Euler Graphs 28
2-9 Hamiltonian Paths and Circuits 30
2-10 The Traveling Salesman Problem 34
Summary 35
References 35
Problems 36

3 TREES AND FUNDAMENTAL CIRCUITS 39

- 3-1 Trees 39
- 3-2 Some Properties of Trees 41
- 3-3 Pendant Vertices in a Tree 43
- 3-4 Distance and Centers in a Tree 45
- 3-5 Rooted and Binary Trees 48
- 3-6 On Counting Trees 52
- 3-7 Spanning Trees 55
- 3-8 Fundamental Circuits 57
- 3-9 Finding All Spanning Trees of a Graph 58
- 3-10 Spanning Trees in a Weighted Graph 61
  - Summary 64
  - References 64
  - Problems 65

4 CUT-SETS AND CUT-VERTICES 68

- 4-1 Cut-Sets 68
- 4-2 Some Properties of a Cut-Set 69
- 4-3 All Cut-Sets in a Graph 71
- 4-4 Fundamental Circuits and Cut-Sets 73
- 4-5 Connectivity and Separability 75
- 4-6 Network Flows 79
- 4-7 1-Isomorphism 80
- 4-8 2-Isomorphism 82
  - Summary 85
  - References 86
  - Problems 86

5 PLANAR AND DUAL GRAPHS 88

- 5-1 Combinatorial Vs. Geometric Graphs 88
- 5-2 Planar Graphs 90
- 5-3 Kuratowski's Two Graphs 90
- 5-4 Different Representations of a Planar Graph 93
- 5-5 Detection of Planarity 99
- 5-6 Geometric Dual 102
- 5-7 Combinatorial Dual 104
- 5-8 More on Criteria of Planarity 107
- 5-9 Thickness and Crossings 108
  - Summary 109
  - References 110
  - Problems 110

6 VECTOR SPACES OF A GRAPH   112

6-1 Sets with One Operation   112
6-2 Sets with Two Operations   116
6-3 Modular Arithmetic and Galois Fields   118
6-4 Vectors and Vector Spaces   120
6-5 Vector Space Associated with a Graph   121
6-6 Basis Vectors of a Graph   123
6-7 Circuit and Cut-Set Subspaces   125
6-8 Orthogonal Vectors and Spaces   129
6-9 Intersection and Join of W and W_S 131
Summary   134
References   135
Problems   135

7 MATRIX REPRESENTATION OF GRAPHS   137

7-1 Incidence Matrix   137
7-2 Submatrices of $A(G)$ 141
7-3 Circuit Matrix   142
7-4 Fundamental Circuit Matrix and Rank of B 144
7-5 An Application to a Switching Network   146
7-6 Cut-Set Matrix   151
7-7 Relationships among A_f, B_f, and C_f 153
7-8 Path Matrix   156
7-9 Adjacency Matrix   157
Summary   162
References   162
Problems   162

8 COLORING, COVERING, AND PARTITIONING   165

8-1 Chromatic Number   165
8-2 Chromatic Partitioning   169
8-3 Chromatic Polynomial   174
8-4 Matchings   177
8-5 Coverings   182
8-6 The Four Color Problem   186
Summary   190
References   190
Problems   192

9 DIRECTED GRAPHS   194

- 9-1 What Is a Directed Graph?   194
- 9-2 Some Types of Digraphs   197
- 9-3 Digraphs and Binary Relations   198
- 9-4 Directed Paths and Connectedness   201
- 9-5 Euler Digraphs   203
- 9-6 Trees with Directed Edges   206
- 9-7 Fundamental Circuits in Digraphs   212
- 9-8 Matrices A, B, and C of Digraphs   213
- 9-9 Adjacency Matrix of a Digraph   220
- 9-10 Paired Comparisons and Tournaments   227
- 9-11 Acyclic Digraphs and Decyclization   230
    - Summary   233
    - References   234
    - Problems   234

10 ENUMERATION OF GRAPHS   238

- 10-1 Types of Enumeration   238
- 10-2 Counting Labeled Trees   240
- 10-3 Counting Unlabeled Trees   241
- 10-4 Pólya's Counting Theorem   250
- 10-5 Graph Enumeration With Pólya's Theorem   260
    - Summary   264
    - References   264
    - Problems   265

11 GRAPH THEORETIC ALGORITHMS AND COMPUTER PROGRAMS   268

- 11-1 Algorithms   269
- 11-2 Input: Computer Representation of a Graph   270
- 11-3 The Output   273
- 11-4 Some Basic Algorithms   274
    - Algorithm 1: Connectedness and Components   274
    - Algorithm 2: A Spanning Tree   277
    - Algorithm 3: A Set of Fundamental Circuits   280
    - Algorithm 4: Cut-Vertices and Separability   284
    - Algorithm 5: Directed Circuits   287
- 11-5 Shortest-Path Algorithms   290
    - Algorithm 6: Shortest Path from a Specified Vertex to Another Specified Vertex   292
    - Algorithm 7: Shortest Path between All Pairs of Vertices   297
- 11-6 Depth-First Search on a Graph   301
    - Algorithm 8: Planarity Testing   304
- 11-7 Algorithm 9: Isomorphism   310

11-8 Other Graph-Theoretic Algorithms 312
11-9 Performance of Graph-Theoretic Algorithms 314
11-10 Graph-Theoretic Computer Languages 316
Summary 317
References 318
Problems 321
Appendix of Programs 323

12 GRAPHS IN SWITCHING AND CODING THEORY 328

12-1 Contact Networks 329
12-2 Analysis of Contact Networks 330
12-3 Synthesis of Contact Networks 334
12-4 Sequential Switching Networks 342
12-5 Unit Cube and Its Graph 348
12-6 Graphs in Coding Theory 351
Summary 354
References 354

13 ELECTRICAL NETWORK ANALYSIS BY GRAPH THEORY 356

13-1 What Is an Electrical Network? 357
13-2 Kirchhoff's Current and Voltage Laws 358
13-3 Loop Currents and Node Voltages 359
13-4 RLC Networks with Independent Sources: Nodal Analysis 362
13-5 RLC Networks with Independent Sources: Loop Analysis 371
13-6 General Lumped, Linear, Fixed Networks 373
Summary 379
References 381
Problems 381

14 GRAPH THEORY IN OPERATIONS RESEARCH 384

14-1 Transport Networks 384
14-2 Extensions of Max-Flow Min-Cut Theorem 390
14-3 Minimal Cost Flows 393
14-4 The Multicommodity Flow 395
14-5 Further Applications 396
14-6 More on Flow Problems 397
14-7 Activity Networks in Project Planning 400
14-8 Analysis of an Activity Network 402
14-9 Further Comments on Activity Networks 408
14-10 Graphs in Game Theory 409
Summary 414
References 414

15 SURVEY OF OTHER APPLICATIONS 416

15-1 Signal-Flow Graphs 416
15-2 Graphs in Markov Processes 424
15-3 Graphs in Computer Programming 439
15-4 Graphs in Chemistry 449
15-5 Miscellaneous Applications 454

Appendix A BINET–CAUCHY THEOREM 458

Appendix B NULLITY OF A MATRIX AND SYLVESTER'S LAW 460

INDEX 463

PREFACE

The last two decades have witnessed an upsurge of interest and activity in graph theory, particularly among applied mathematicians and engineers. Clear evidence of this is to be found in an unprecedented growth in the number of papers and books being published in the field. In 1957 there was exactly one book on the subject (namely, König's *Théorie der Endlichen und Unendlichen Graphen*). Now, sixteen years later, there are over two dozen textbooks on graph theory, and almost an equal number of proceedings of various seminars and conferences.

Each book has its own strength and points of emphasis, depending on the axe (or the pen) the author has to grind. I have emphasized the computational and algorithmic aspects of graphs. This emphasis arises from the experience and conviction that whenever graph theory is applied to solving any practical problem (be it in electrical network analysis, in circuit layout, in data structures, in operations research, or in social sciences), it almost always leads to large graphs—graphs that are virtually impossible to analyze without the aid of the computer. An engineer often finds that those real-life problems that can be modeled into graphs small enough to be worked on by hand are also small enough to be solved by means other than graph theory. (In this respect graph theory is different from college algebra, elementary calculus, or complex variables.) In fact, the high-speed digital computer is one of the reasons for the recent growth of interest in graph theory.

Convinced that a student of applied graph theory must learn to enlist the help of a digital computer for handling large graphs, I have emphasized algorithms and their efficiencies. In proving theorems, constructive proofs have been given preference over nonconstructive existence proofs. Chapter 11, the largest in the book, is devoted entirely to computational aspects of graph theory, including graph-theoretic algorithms and samples of several tested computer programs for solving problems on graphs. I believe this approach has not been used in any of the earlier books on graph theory. The material covered in Chapter 11 and in many sections from other chapters is appearing for the first time in any textbook.

Yet the applied and algorithmic aspect of this book has not been allowed to spoil the rigor and mathematical elegance of graph theory. Indeed, the book contains enough material for a course in "pure" graph theory. The book has been made as much self-contained as was possible.

The level of presentation is appropriate for advanced undergraduate and first-year graduate students in all disciplines requiring graph theory. The book is organized so that the first half (Chapters 1 through 9) serves as essential and introductory material on graph theory. This portion requires no special background, except some elementary concepts from set theory and matrix algebra and, of course, a certain amount of mathematical maturity. Although the illustrations of applications are interwoven with the theory even in this portion, the examples selected are short and mostly of the nature of puzzles and games. This is done so that a student in almost any field can read and grasp the first half.

The second half of the book is more advanced, and different chapters require different backgrounds as they deal with applications to nontrivial, real-world, complex problems in different fields. Keeping this in mind, Chapters 10 through 15 have been made independent of each other. One could study a later chapter without going through the earlier ones, provided the first nine chapters have been covered.

Since there is more material here than what can be covered in a one-semester course, it is suggested that the contents be tailored to suit the requirements of the students in different disciplines, for example:

1. Electrical Engineering: Chapters 1–9, and 11, 12, and 13.
2. Computer Science: Chapters 1–9, 11, 12, and parts of 10 and 15.
3. Operations Research: Chapters 1–9, and 11, 14, and parts of 15.
4. Applied Mathematics: Chapters 1–11 and parts of 15.
5. Introductory "pure" graph theory: Chapters 1–10.

In fact, the book grew out of a number of such courses and lecture-series given by the author at the Jet Propulsion Laboratory, California State University at Los Angeles, the Indian Institute of Technology at Kanpur, and the University of Illinois at Urbana-Champaign.

ACKNOWLEDGMENTS

It is a pleasure to acknowledge the help I have received from different individuals and institutions. I am particularly indebted to Mr. David K. Rubin, a dear friend and former colleague at the Jet Propulsion Laboratory; Mr. Mateti Prabhaker, a former graduate student of mine at the Indian

Institute of Technology, Kanpur; and Professor Jurg Nievergelt of the University of Illinois at Urbana-Champaign for having read the entire manuscript and made numerous suggestions for improvements throughout the book.

Other friends, colleagues, and students who read parts of the manuscript and made helpful suggestions are: Professor Harry Lass and Mr. Marvin Perlman of the Jet Propulsion Laboratory, Professor Nandlal Jhunjhunwala of California State University at Los Angeles, Dr. George Shuraym of Texas Instruments, Mr. Jean A. DeBeule of Xerox Data Systems, Mr. Nicholas Karpov of Bell & Howell, Professor C. L. Liu of the University of Illinois at Urbana-Champaign, Messrs. M. S. Krishnamoorthy, K. G. Ramakrishnan, and Professors C. R. Muthukrishnan and S. K. Basu of the Indian Institute of Technology at Kanpur.

I am also grateful to the late Professor George E. Forsythe of Stanford University for his encouragement at the very outset of this project.

Support in writing this book was received from the Jet Propulsion Laboratory, the Indian Institute of Technology at Kanpur, and the Computer Science Department of the University of Illinois at Urbana-Champaign.

Just as one does not thank himself, expressing gratitude to one's wife in public is not a Hindu custom. For the wife is considered a part of the husband, and her coauthorship is tacitly assumed in any book her husband writes. There is little doubt that without Kiran's help this book would not have been possible.

NARSINGH DEO

Kanpur

Graph Theory
with Applications to Engineering
& Computer Science

1 INTRODUCTION

1-1. WHAT IS A GRAPH?

A *linear*† *graph* (or simply a *graph*) $G = (V, E)$ consists of a set of objects $V = \{v_1, v_2, \ldots\}$ called *vertices*, and another set $E = \{e_1, e_2, \ldots\}$, whose elements are called *edges*, such that each edge e_k is identified with an unordered pair (v_i, v_j) of vertices. The vertices v_i, v_j associated with edge e_k are called the *end vertices* of e_k. The most common representation of a graph is by means of a diagram, in which the vertices are represented as points and each edge as a line segment joining its end vertices. Often this diagram itself is referred to as the graph. The object shown in Fig. 1-1, for instance, is a graph.

Observe that this definition permits an edge to be associated with a vertex pair (v_i, v_i). Such an edge having the same vertex as both its end vertices is called a *self-loop* (or simply a *loop*. The word loop, however, has a different meaning in electrical network theory; we shall therefore use the term self-loop to avoid confusion). Edge e_1 in Fig. 1-1 is a self-loop. Also note that

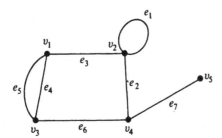

Fig. 1-1 Graph with five vertices and seven edges.

†The adjective "linear" is dropped as redundant in our discussions, because by a graph we always mean a linear graph. There is no such thing as a nonlinear graph

the definition allows more than one edge associated with a given pair of vertices, for example, edges e_4 and e_5 in Fig. 1-1. Such edges are referred to as *parallel edges*.

A graph that has neither self-loops nor parallel edges is called a *simple graph*. In some graph-theory literature, a graph is defined to be only a simple graph, but in most engineering applications it is necessary that parallel edges and self-loops be allowed; this is why our definition includes graphs with self-loops and/or parallel edges. Some authors use the term *general graph* to emphasize that parallel edges and self-loops are allowed.

It should also be noted that, in drawing a graph, it is immaterial whether the lines are drawn straight or curved, long or short: what is important is the incidence between the edges and vertices. For example, the two graphs drawn in Figs. 1-2(a) and (b) are the same, because incidence between edges and vertices is the same in both cases.

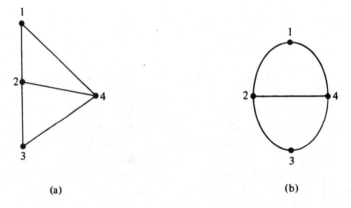

Fig. 1-2 Same graph drawn differently.

In a diagram of a graph, sometimes two edges may seem to intersect at a point that does not represent a vertex, for example, edges e and f in Fig. 1-3. Such edges should be thought of as being in different planes and thus having no common point. (Some authors break one of the two edges at such a crossing to emphasize this fact.)

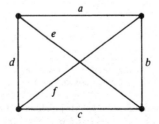

Fig. 1-3 Edges e and f have no common point.

A graph is also called a *linear complex*, a *1-complex*, or a *one-dimensional complex*. A vertex is also referred to as a *node*, a *junction*, a *point*, *0-cell*, or an *0-simplex*. Other terms used for an edge are a *branch*, a *line*, an *element*, a *1-cell*, an *arc*, and a *1-simplex*. In this book we shall generally use the terms graph, vertex, and edge.

1-2. APPLICATIONS OF GRAPHS

Because of its inherent simplicity, graph theory has a very wide range of applications in engineering, in physical, social, and biological sciences, in linguistics, and in numerous other areas. A graph can be used to represent almost any physical situation involving discrete objects and a relationship among them. The following are four examples from among hundreds of such applications.

Königsberg Bridge Problem: The Königsberg bridge problem is perhaps the best-known example in graph theory. It was a long-standing problem until solved by Leonhard Euler (1707-1783) in 1736, by means of a graph. Euler wrote the first paper ever in graph theory and thus became the originator of the theory of graphs as well as of the rest of topology. The problem is depicted in Fig. 1-4.

Two islands, *C* and *D*, formed by the Pregel River in Königsberg (then the capital of East Prussia but now renamed Kaliningrad and in West Soviet Russia) were connected to each other and to the banks *A* and *B* with seven bridges, as shown in Fig. 1-4. The problem was to start at any of the four land areas of the city, *A*, *B*, *C*, or *D*, walk over each of the seven bridges *exactly once*, and return to the starting point (without swimming across the river, of course).

Euler represented this situation by means of a graph, as shown in Fig. 1-5. The vertices represent the land areas and the edges represent the bridges.

As we shall see in Chapter 2, Euler proved that a solution for this problem does not exist.

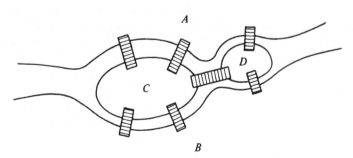

Fig. 1-4 Königsberg bridge problem.

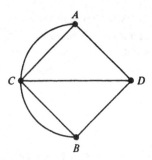

Fig. 1-5 Graph of Königsberg bridge problem.

The Königsberg bridge problem is the same as the problem of drawing figures without lifting the pen from the paper and without retracing a line (Problems 2-1 and 2-2). We all have been confronted with such problems at one time or another.

Utilities Problem: There are three houses (Fig. 1-6) H_1, H_2, and H_3, each to be connected to each of the three utilities—water (W), gas (G), and elec-

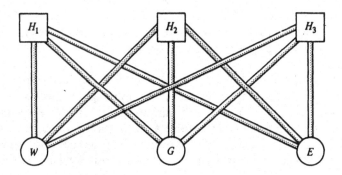

Fig. 1-6 Three-utilities problem.

tricity (E)—by means of conduits. Is it possible to make such connections without any crossovers of the conduits?

Figure 1-7 shows how this problem can be represented by a graph—the conduits are shown as edges while the houses and utility supply centers are vertices. As we shall see in Chapter 5, the graph in Fig. 1-7 cannot be drawn in the plane without edges crossing over. Thus the answer to the problem is no.

Electrical Network Problems: Properties (such as transfer function and input impedance) of an electrical network are functions of only two factors:

1. The nature and value of the elements forming the network, such as resistors, inductors, transistors, and so forth.

2. The way these elements are connected together, that is, the topology of the network.

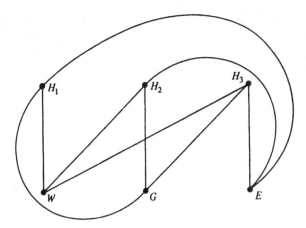

Fig. 1-7 Graph of three-utilities problem.

Since there are only a few different types of electrical elements, the variations in networks are chiefly due to the variations in topology. Thus electrical network analysis and synthesis are mainly the study of network topology. In the topological study of electrical networks, factor 2 is separated from 1 and is studied independently. The advantage of this approach will be clearer in Chapter 13, a chapter devoted solely to applying graph theory to electrical networks.

The topology of a network is studied by means of its graph. In drawing a graph of an electrical network the junctions are represented by vertices, and branches (which consist of electrical elements) are represented by edges, regardless of the nature and size of the electrical elements. An electrical network and its graph are shown in Fig. 1-8.

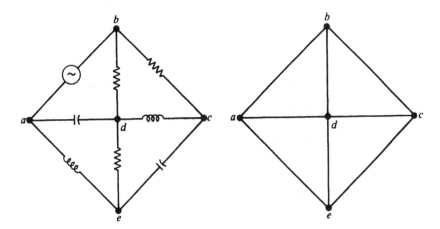

Fig. 1-8 Electrical network and its graph.

Seating Problem: Nine members of a new club meet each day for lunch at a round table. They decide to sit such that every member has different neighbors at each lunch. How many days can this arrangement last?

This situation can be represented by a graph with nine vertices such that each vertex represents a member, and an edge joining two vertices represents the relationship of sitting next to each other. Figure 1-9 shows two possible

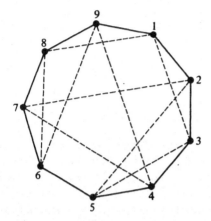

Fig. 1-9 Arrangements at a dinner table.

seating arrangements—these are 1 2 3 4 5 6 7 8 9 1 (solid lines), and 1 3 5 2 7 4 9 6 8 1 (dashed lines). It can be shown by graph-theoretic considerations that there are only two more arrangements possible. They are 1 5 7 3 9 2 8 4 6 1 and 1 7 9 5 8 3 6 2 4 1. In general it can be shown that for n people the number of such possible arrangements is

$$\frac{n-1}{2}, \quad \text{if } n \text{ is odd,}$$

and

$$\frac{n-2}{2}, \quad \text{if } n \text{ is even.}$$

The reader has probably noticed that three of the four examples of applications above are puzzles and not engineering problems. This was done to avoid introducing at this stage background material not pertinent to graph theory. More substantive applications will follow, particularly in the last four chapters.

1-3. FINITE AND INFINITE GRAPHS

Although in our definition of a graph neither the vertex set V nor the edge set E need be finite, in most of the theory and almost all applications these sets are finite. A graph with a finite number of vertices as well as a finite number of edges is called a *finite graph;* otherwise, it is an *infinite graph*. The graphs in Figs. 1-1, 1-2, 1-5, 1-7, and 1-8 are all examples of finite graphs. Portions of two infinite graphs are shown in Fig. 1-10.

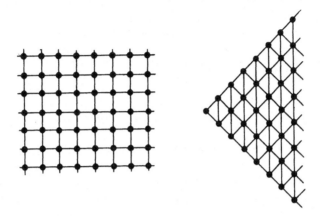

Fig. 1-10 Portions of two infinite graphs.

In this book we shall confine ourselves to the study of finite graphs, and unless otherwise stated the term "graph" will always mean a finite graph.

1-4. INCIDENCE AND DEGREE

When a vertex v_i is an end vertex of some edge e_j, v_i and e_j are said to be *incident with (on or to)* each other. In Fig. 1-1, for example, edges e_2, e_6, and e_7 are incident with vertex v_4. Two nonparallel edges are said to be *adjacent* if they are incident on a common vertex. For example, e_2 and e_7 in Fig. 1-1 are adjacent. Similarly, two vertices are said to be adjacent if they are the end vertices of the same edge. In Fig. 1-1, v_4 and v_5 are adjacent, but v_1 and v_4 are not.

The number of edges incident on a vertex v_i, with self-loops counted twice, is called the *degree*, $d(v_i)$, of vertex v_i. In Fig. 1-1, for example, $d(v_1) =$

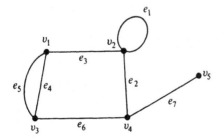

Fig. 1-1 A graph with five vertices and seven edges.

$d(v_3) = d(v_4) = 3$, $d(v_2) = 4$, and $d(v_5) = 1$. The degree of a vertex is sometimes also referred to as its *valency*.

Let us now consider a graph G with e edges and n vertices $v_1, v_2, \ldots, v_n$. Since each edge contributes two degrees, the sum of the degrees of all vertices in G is twice the number of edges in G. That is,

$$\sum_{i=1}^{n} d(v_i) = 2e. \tag{1-1}$$

Taking Fig. 1-1 as an example, once more,

$$d(v_1) + d(v_2) + d(v_3) + d(v_4) + d(v_5)$$
$$= 3 + 4 + 3 + 3 + 1 = 14 = \text{twice the number of edges.}$$

From Eq. (1-1) we shall derive the following interesting result.

THEOREM 1-1

The number of vertices of odd degree in a graph is always even.

Proof: If we consider the vertices with odd and even degrees separately, the quantity in the left side of Eq. (1-1) can be expressed as the sum of two sums, each taken over vertices of even and odd degrees, respectively, as follows:

$$\sum_{i=1}^{n} d(v_i) = \sum_{\text{even}} d(v_j) + \sum_{\text{odd}} d(v_k). \tag{1-2}$$

Since the left-hand side in Eq. (1-2) is even, and the first expression on the right-hand side is even (being a sum of even numbers), the second expression must also be even:

$$\sum_{\text{odd}} d(v_k) = \text{an even number.} \tag{1-3}$$

Because in Eq. (1-3) each $d(v_k)$ is odd, the total number of terms in the sum must be even to make the sum an even number. Hence the theorem. ■

A graph in which all vertices are of equal degree is called a *regular graph* (or simply a *regular*). The graph of three utilities shown in Fig. 1-7 is a regular of degree three.

1-5. ISOLATED VERTEX, PENDANT VERTEX, AND NULL GRAPH

A vertex having no incident edge is called an *isolated vertex*. In other words, isolated vertices are vertices with zero degree. Vertices v_4 and v_7 in Fig. 1-11, for example, are isolated vertices. A vertex of degree one is called

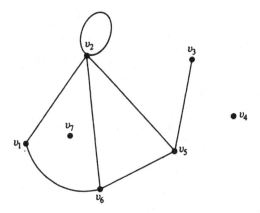

Fig. 1-11 Graph containing isolated vertices, series edges, and a pendant vertex.

a *pendant vertex* or an *end vertex*. Vertex v_3 in Fig. 1-11 is a pendant vertex. Two adjacent edges are said to be in *series* if their common vertex is of degree two. In Fig. 1-11, the two edges incident on v_1 are in series.

In the definition of a graph $G = (V, E)$, it is possible for the edge set E to be empty. Such a graph, without any edges, is called a *null graph*. In other words, every vertex in a null graph is an isolated vertex. A null graph of six vertices is shown in Fig. 1-12. Although the edge set E may be empty, the

Fig. 1-12 Null graph of six vertices.

vertex set V must not be empty; otherwise, there is no graph. In other words, by definition, a graph must have at least one vertex.†

---

†Some authors (see, for example, [2-9], p. 1, or [15-58], p. 17) do allow the case in which the vertex set V is also empty. This they call the null graph, and they call a graph with $E = \emptyset$ and $V \neq \emptyset$ a *vertex graph*. For our purposes this distinction is of no consequence. For a lively discussion on paradoxes arising out of different definitions of the null graph, see pp. 40-41 in *Theory of Graphs: a Basis for Network Theory*, by L. M. Maxwell and M. B. Reed (Pergamon Press, N. Y. 1971).

1-6. A BRIEF HISTORY OF GRAPH THEORY

As mentioned before, graph theory was born in 1736 with Euler's paper in which he solved the Königsberg bridge problem [1-4].† For the next 100 years nothing more was done in the field.

In 1847, G. R. Kirchhoff (1824–1887) developed the theory of trees for their applications in electrical networks [1-6]. Ten years later, A. Cayley (1821–1895) discovered trees while he was trying to enumerate the isomers of saturated hydrocarbons $C_n H_{2n+2}$ [1-3].

About the time of Kirchhoff and Cayley, two other milestones in graph theory were laid. One was the *four-color conjecture*, which states that four colors are sufficient for coloring any atlas (a map on a plane) such that the countries with common boundaries have different colors.

It is believed that A. F. Möbius (1790–1868) first presented the four-color problem in one of his lectures in 1840. About 10 years later, A. De Morgan (1806–1871) discussed this problem with his fellow mathematicians in London. De Morgan's letter is the first authenticated reference to the four-color problem. The problem became well known after Cayley published it in 1879 in the first volume of the *Proceedings of the Royal Geographic Society*. To this day, the four-color conjecture is by far the most famous unsolved problem in graph theory; it has stimulated an enormous amount of research in the field [1-11].

The other milestone is due to Sir W. R. Hamilton (1805–1865). In the year 1859 he invented a puzzle and sold it for 25 guineas to a game manufacturer in Dublin. The puzzle consisted of a wooden, regular dodecahedron (a polyhedron with 12 faces and 20 corners, each face being a regular pentagon and three edges meeting at each corner; see Fig. 2-21). The corners were marked with the names of 20 important cities: London, NewYork, Delhi, Paris, and so on. The object in the puzzle was to find a route along the edges of the dodecahedron, passing through each of the 20 cities exactly once [1-12].

Although the solution of this specific problem is easy to obtain (as we shall see in Chapter 2), to date no one has found a necessary and sufficient condition for the existence of such a route (called Hamiltonian circuit) in an arbitrary graph.

This fertile period was followed by half a century of relative inactivity. Then a resurgence of interest in graphs started during the 1920s. One of the pioneers in this period was D. König. He organized the work of other mathematicians and his own and wrote the first book on the subject, which was published in 1936 [1-7].

The past 30 years has been a period of intense activity in graph theory—both pure and applied. A great deal of research has been done and is being

---

†Bracketed numbers refer to references at the end of chapters.

done in this area. Thousands of papers have been published and more than a dozen books written during the past decade. Among the current leaders in the field are Claude Berge, Oystein Ore (recently deceased), Paul Erdös, William Tutte, and Frank Harary.

SUMMARY

In this chapter some basic concepts of graph theory have been introduced, and some elementary results have been obtained. An attempt has also been made to show that graphs can be used to represent almost any problem involving discrete arrangements of objects, where concern is not with the internal properties of these objects but with the relationships among them.

REFERENCES

As an elementary text on graph theory, Ore's book [1-10] is recommended. Busacker and Saaty [1-2] is a good intermediate-level book. Seshu and Reed [1-13] is specially suited for electrical engineers. Berge [1-1] and Ore [1-9] are good general texts, but are somewhat advanced. Harary's book [1-5] contains an excellent treatment of the subject. It is compact and clear, but it contains no applications and is written for an advanced student of graph theory. For relating graph theory to the rest of topology one should read [1-8], a well-written elementary book on important aspects of topology. The entertaining book of Rouse Ball [1-12] contains a variety of puzzles and games to which graphs have been applied.

1-1. BERGE, C., *The Theory of Graphs and Its Applications*, John Wiley & Sons, Inc., New York, 1962. English translation of the original book in French: *Théorie des graphes et ses applications*, Dunod Editeur, Paris, 1958.
1-2. BUSACKER, R. G., and T. L. SAATY, *Finite Graphs and Networks: An Introduction with Applications*, McGraw-Hill Book Company, New York, 1965.
1-3. CAYLEY, A., "On the Theory of Analytical Forms Called Trees," *Phil. Mag.*, Vol. 13, 1857, 172–176.
1-4. EULER, L., "Solutio Problematis ad Geometriam Situs Pertinantis," *Academimae Petropolitanae* (St. Petersburg Academy), Vol. 8, 1736, 128–140. English translation in *Sci. Am.*, July 1953, 66–70.
1-5. HARARY, F., *Graph Theory*, Addison-Wesley Publishing Company, Inc., Reading, Mass., 1969.
1-6. KIRCHHOFF, G., "Über die Auflösung der Gleichungen, auf welche man bei der Untersuchungen der Linearen Verteilung Galvanisher Ströme geführt wird," *Poggendorf Ann. Physik*, Vol. 72, 1847, 497–508. English translation, *IRE Trans. Circuit Theory*, Vol. CT-5, March 1958, 4–7.
1-7. KÖNIG, D., *Theorie der endlichen und unendlichen Graphen*, Leipzig, 1936; Chelsea, New York, 1950.
1-8. LIETZMANN, W., *Visual Topology*, American Elsevier Publishing Company, Inc., New York, 1965. English translation of the German book *Anschauliche Topologie*, R. Oldenbourg K. G., Munich, 1955.
1-9. ORE, O., *Theory of Graphs*, American Mathematical Society, Providence, R. I., 1962.
1-10. ORE, O., *Graphs and Their Uses*, Random House, Inc., New York, 1963.
1-11. ORE, O., *The Four Color Problem*, Academic Press, Inc., New York, 1967.

1-12. Rouse Ball, W., *Mathematical Recreations and Essays*, London and New York, 1892; and The Macmillan Company, New York, 1962.
1-13. Seshu, S., and M. B. Reed, *Linear Graphs and Electrical Networks*, Addison-Wesley Publishing Company, Inc., Reading, Mass., 1961.

PROBLEMS

1-1. Draw all simple graphs of one, two, three, and four vertices.

1-2. Draw graphs representing problems of (a) two houses and three utilities; (b) four houses and four utilities, say, water, gas, electricity, and telephone.

1-3. Name 10 situations (games, activities, real-life problems, etc.) that can be represented by means of graphs. Explain what the vertices and the edges denote.

1-4. Draw the graph of the Wheatstone bridge circuit.

1-5. Draw graphs of the following chemical compounds: (a) CH_4, (b) C_2H_6, (c) C_6H_6, (d) N_2O_3. (*Hint:* Represent atoms by vertices and chemical bonds between them by edges.)

1-6. Draw a graph with 64 vertices representing the squares of a chessboard. Join these vertices appropriately by edges, each representing a move of the knight. You will see that in this graph every vertex is of degree two, three, four, six, or eight. How many vertices are of each type?

1-7. Given a maze as shown in Fig. 1-13, represent this maze by means of a graph such that a vertex denotes either a corridor or a dead end (as numbered). An edge represents a possible path between two vertices. (This is one of numerous mazes that were drawn or built by the Hindus, Greeks, Romans, Vikings, Arabs, etc.)

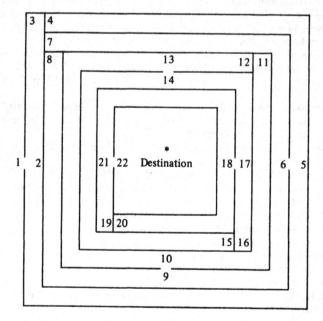

Fig. 1-13   A maze.

1-8. *Decanting problem.* You are given three vessels A, B, and C of capacities 8, 5, and 3 gallons, respectively. A is filled, while B and C are empty. Divide the liquid in A into two equal quantities. [*Hint:* Let a, b, and c be the amounts of liquid in A, B, and C, respectively. We have $a + b + c = 8$ at all times. Since at least one of the vessels is always empty or full, at least one of the following equations must always be satisfied: $a = 0$, $a = 8$; $b = 0$, $b = 5$; $c = 0$, $c = 3$. You will find that with these constraints there are 16 possible states (situations) in this process. Represent this problem by means of a 16-vertex graph. Each vertex stands for a state and each edge for a permissible change of states between its two end vertices. Now when you look at this graph it will be clear to you how to go from state (8, 0, 0) to state (4, 4, 0).] This is the classical decanting problem.

1-9. Convince yourself that an infinite graph with a finite number of edges (i.e., a graph with a finite number of edges and an infinite number of vertices) must have an infinite number of isolated vertices.

1-10. Show that an infinite graph with a finite number of vertices (i.e., a graph with a finite number of vertices and an infinite number of edges) will have at least one pair of vertices (or one vertex in case of parallel self-loops) joined by an infinite number of parallel edges.

1-11. Convince yourself that the maximum degree of any vertex in a simple graph with n vertices is $n - 1$.

1-12. Show that the maximum number of edges in a simple graph with n vertices is $n(n - 1)/2$.

2 PATHS AND CIRCUITS

This chapter serves two purposes. The first is to introduce additional concepts and terms in graph theory. These concepts, such as paths, circuits, and Euler graphs, deal mainly with the nature of connectivity in graphs. The degree of vertices, which is a local property of each vertex, will be shown to be related to the more global properties of the graph.

The second purpose is to illustrate with examples how to solve actual problems using graph theory. The celebrated Königsberg bridge problem, which was introduced in Chapter 1, will be solved. The solution of the seating arrangement problem, also introduced in Chapter 1, will be obtained by means of Hamiltonian circuits. A third problem, which involves stacking four multicolored cubes, will also be solved. These three unrelated problems will demonstrate the problem-solving power of graph theory. The reader may attempt to solve these problems without using graphs; the difficulty of such an approach will quickly convince him of the value of graph theory.

2-1. ISOMORPHISM

In geometry two figures are thought of as equivalent (and called congruent) if they have identical behavior in terms of geometric properties. Likewise, two graphs are thought of as equivalent (and called *isomorphic*) if they have identical behavior in terms of graph-theoretic properties. More precisely: Two graphs G and G' are said to be isomorphic (to each other) if there is a one-to-one correspondence between their vertices and between their edges such that the incidence relationship is preserved. In other words, suppose that edge e is incident on vertices v_1 and v_2 in G; then the corresponding edge e' in G' must be incident on the vertices v'_1 and v'_2 that correspond to

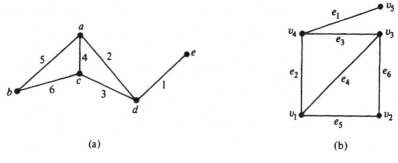

Fig. 2-1 Isomorphic graphs.

v_1 and v_2, respectively. For example, one can verify that the two graphs in Fig. 2-1 are isomorphic. The correspondence between the two graphs is as follows: The vertices a, b, c, d, and e correspond to v_1, v_2, v_3, v_4, and v_5, respectively. The edges 1, 2, 3, 4, 5, and 6 correspond to e_1, e_2, e_3, e_4, e_5, and e_6, respectively.

Except for the labels (i.e., names) of their vertices and edges, isomorphic graphs are the same graph, perhaps drawn differently. As indicated in Chapter 1, a given graph can be drawn in many different ways. For example, Fig. 2-2 shows two different ways of drawing the same graph.

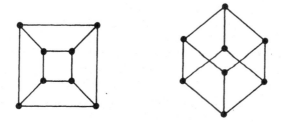

Fig. 2-2 Isomorphic graphs.

It is not always an easy task to determine whether or not two given graphs are isomorphic. For instance, the three graphs shown in Fig. 2-3 are all isomorphic, but just by looking at them you cannot tell that. It is left as an exercise for the reader to show, by redrawing and labeling the vertices and edges, that the three graphs in Fig. 2-3 are isomorphic (see Problem 2-3).

It is immediately apparent by the definition of isomorphism that two isomorphic graphs must have

1. The same number of vertices.
2. The same number of edges.
3. An equal number of vertices with a given degree.

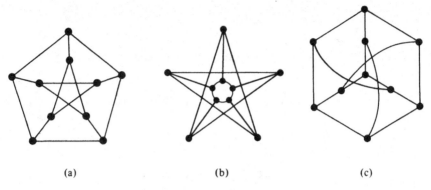

Fig. 2-3 Isomorphic graphs.

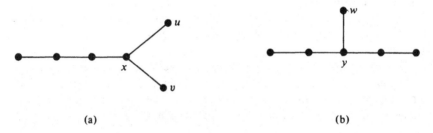

Fig. 2-4 Two graphs that are not isomorphic.

However, these conditions are by no means sufficient. For instance, the two graphs shown in Fig. 2-4 satisfy all three conditions, yet they are not isomorphic. That the graphs in Figs. 2-4(a) and (b) are not isomorphic can be shown as follows: If the graph in Fig. 2-4(a) were to be isomorphic to the one in (b), vertex x must correspond to y, because there are no other vertices of degree three. Now in (b) there is only one pendant vertex, w, adjacent to y, while in (a) there are two pendant vertices, u and v, adjacent to x.

Finding a simple and efficient criterion for detection of isomorphism is still actively pursued and is an important unsolved problem in graph theory. In Chapter 11 we shall discuss various proposed algorithms and their programs for automatic detection of isomorphism by means of a computer. For now, we move to a different topic.

2-2. SUBGRAPHS

A graph g is said to be a *subgraph* of a graph G if all the vertices and all the edges of g are in G, and each edge of g has the same end vertices in g as in G. For instance, the graph in Fig. 2-5(b) is a subgraph of the one in Fig. 2-5(a). (Obviously, when considering a subgraph, the original graph must

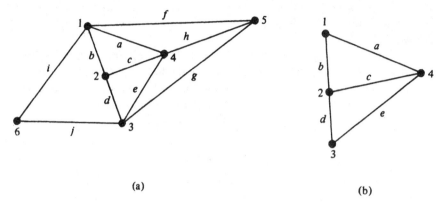

Fig. 2-5 Graph (a) and one of its subgraphs (b).

not be altered by identifying two distinct vertices, or by adding new edges or vertices.) The concept of subgraph is akin to the concept of subset in set theory. A subgraph can be thought of as being contained in (or a part of) another graph. The symbol from set theory, $g \subset G$, is used in stating "g is a subgraph of G."

The following observations can be made immediately:

1. Every graph is its own subgraph.
2. A subgraph of a subgraph of G is a subgraph of G.
3. A single vertex in a graph G is a subgraph of G.
4. A single edge in G, together with its end vertices, is also a subgraph of G.

Edge-Disjoint Subgraphs: Two (or more) subgraphs g_1 and g_2 of a graph G are said to be *edge disjoint* if g_1 and g_2 do not have any edges in common. For example, the two graphs in Figs. 2-7(a) and (b) are edge-disjoint subgraphs of the graph in Fig. 2-6. Note that although edge-disjoint graphs do not have any edge in common, they may have vertices in common. Subgraphs that do not even have vertices in common are said to be *vertex disjoint*. (Obviously, graphs that have no vertices in common cannot possibly have edges in common.)

2-3. A PUZZLE WITH MULTICOLORED CUBES

Now we shall take a brief pause to illustrate, with an example, how a problem can be solved by using graphs. Two steps are involved here: First, the physical problem is converted into a problem of graph theory. Second, the graph-theory problem is then solved. Let us consider the following problem, a well-known puzzle available in toy stores (under the name *Instant Insanity*).

Problem: We are given four cubes. The six faces of every cube are variously colored blue, green, red, or white. Is it possible to stack the cubes one on top of another to form a column such that no color appears twice on any of the four sides of this column? (Clearly, a trial-and-error method is unsatisfactory, because we may have to try all 41,472 ($= 3 \times 24 \times 24 \times 24$) possibilities.)

Solution: Step 1: Draw a graph with four vertices B, G, R, and W—one for each color (Fig. 2-6). Pick a cube and call it cube 1; then represent its

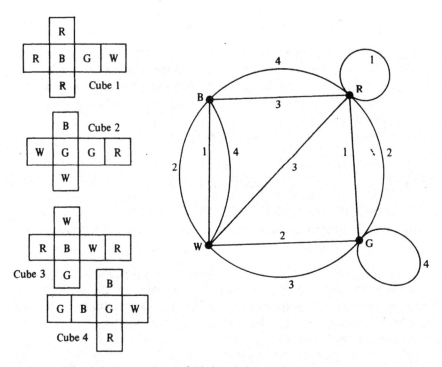

Fig. 2-6 Four cubes unfolded and the graph representing their colors.

three pairs of opposite faces by three edges, drawn between the vertices with appropriate colors. In other words, if a blue face in cube 1 has a white face opposite to it, draw an edge between vertices B and W in the graph. Do the same for the remaining two pairs of faces in cube 1. Put label 1 on all three edges resulting from cube 1. A self-loop with the edge labeled 1 at vertex R, for instance, would result if cube 1 had a pair of opposite faces both colored

SEC. 2-4 WALKS, PATHS, AND CIRCUITS 19

red. Repeat the procedure for the other three cubes one by one on the same graph until we have a graph with four vertices and 12 edges. A particular set of four colored cubes and their graph are shown in Fig. 2-6.

Step 2: Consider the graph resulting from this representation. The degree of each vertex is the total number of faces with the corresponding color. For the cubes of Fig. 2-6, we have five blue faces, six green, seven red, and six white.

Consider two opposite vertical sides of the desired column of four cubes, say facing north and south. A subgraph (with four edges) will represent these eight faces—four facing south and four north. Each of the four edges in this subgraph will have a different label—1, 2, 3, and 4. Moreover, no color occurs twice on either the north side or south side of the column if and only if every vertex in this subgraph is of degree two.

Exactly the same argument applies to the other two sides, east and west, of the column.

Thus the four cubes can be arranged (to form a column such that no color appears more than once on any side) if and only if there exist two edge-disjoint subgraphs, each with four edges, each of the edges labeled differently, and such that each vertex is of degree two. For the set of cubes shown in Fig. 2-6, this condition is satisfied, and the two subgraphs are shown in Fig. 2-7.

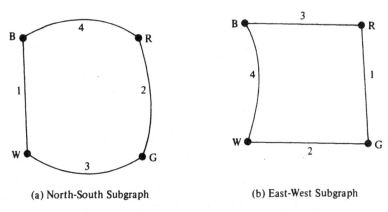

(a) North-South Subgraph (b) East-West Subgraph

Fig. 2-7 Two edge-disjoint subgraphs of the graph in Fig. 2-6.

2-4. WALKS, PATHS, AND CIRCUITS

A *walk* is defined as a finite alternating sequence of vertices and edges, beginning and ending with vertices, such that each edge is incident with the vertices preceding and following it. No edge appears (is covered or traversed) more than once in a walk. A vertex, however, may appear more than once. In Fig. 2-8(a), for instance, $v_1\, a\, v_2\, b\, v_3\, c\, v_3\, d\, v_4\, e\, v_2\, f\, v_5$ is a walk shown with

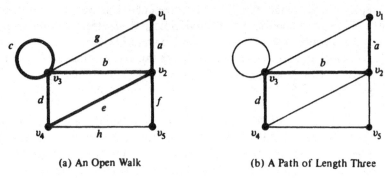

(a) An Open Walk (b) A Path of Length Three

Fig. 2-8 A walk and a path.

heavy lines. A walk is also referred to as an *edge train* or a *chain*. The set of vertices and edges constituting a given walk in a graph G is clearly a subgraph of G.

Vertices with which a walk begins and ends are called its *terminal vertices*. Vertices v_1 and v_5 are the terminal vertices of the walk shown in Fig. 2-8(a). It is possible for a walk to begin and end at the same vertex. Such a walk is called a *closed walk*. A walk that is not closed (i.e., the terminal vertices are distinct) is called an *open walk* [Fig. 2-8(a)].

An open walk in which no vertex appears more than once is called a *path* (or a *simple path* or an *elementary path*). In Fig. 2-8, v_1 a v_2 b v_3 d v_4 is a path, whereas v_1 a v_2 b v_3 c v_3 d v_4 e v_2 f v_5 is not a path. In other words, a path does not intersect itself. The number of edges in a path is called the *length of a path*. It immediately follows, then, that an edge which is not a self-loop is a path of length one. It should also be noted that a self-loop can be included in a walk but not in a path (Fig. 2-8).

The terminal vertices of a path are of degree one, and the rest of the vertices (called *intermediate vertices*) are of degree two. This degree, of course, is counted only with respect to the edges included in the path and not the entire graph in which the path may be contained.

A closed walk in which no vertex (except the initial and the final vertex) appears more than once is called a *circuit*. That is, a circuit is a closed, non-

Fig. 2-9 Three different circuits.

intersecting walk. In Fig. 2-8(a), $v_2\ b\ v_3\ d\ v_4\ e\ v_2$ is, for example, a circuit. Three different circuits are shown in Fig. 2-9. Clearly, every vertex in a circuit is of degree two; again, if the circuit is a subgraph of another graph, one must count degrees contributed by the edges in the circuit only.

A circuit is also called a *cycle, elementary cycle, circular path,* and *polygon*. In electrical engineering a circuit is sometimes referred to as a *loop*—not to be confused with self-loop. (Every self-loop is a circuit, but not every circuit is a self-loop.)

The definitions in this section are summarized in Fig. 2-10. The arrows are in the direction of increasing restriction.

You may have observed that although the concepts of a path and a circuit are very simple, the formal definition becomes involved.

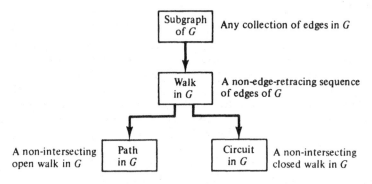

Fig. 2-10 Walks, paths, and circuits as subgraphs.

2-5. CONNECTED GRAPHS, DISCONNECTED GRAPHS, AND COMPONENTS

Intuitively, the concept of *connectedness* is obvious. A graph is connected if we can reach any vertex from any other vertex by traveling along the edges. More formally:

A graph G is said to be *connected* if there is at least one path between every pair of vertices in G. Otherwise, G is *disconnected*. For instance, the graph in Fig. 2-8(a) is connected, but the one in Fig. 2-11 is disconnected. A null graph of more than one vertex is disconnected (Fig. 1-12).

It is easy to see that a disconnected graph consists of two or more connected graphs. Each of these connected subgraphs is called a *component*. The graph in Fig. 2-11 consists of two components. Another way of looking at a component is as follows: Consider a vertex v_i in a disconnected graph G. By definition, not all vertices of G are joined by paths to v_i. Vertex v_i and all the vertices of G that have paths to v_i, together with all the edges incident on them, form a component. Obviously, a component itself is a graph.

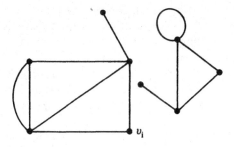

Fig. 2-11 A disconnected graph with two components.

Theorem 2-1

A graph G is disconnected if and only if its vertex set V can be partitioned into two nonempty, disjoint subsets V_1 and V_2 such that there exists no edge in G whose one end vertex is in subset V_1 and the other in subset V_2.

Proof: Suppose that such a partitioning exists. Consider two arbitrary vertices a and b of G, such that $a \in V_1$ and $b \in V_2$. No path can exist between vertices a and b; otherwise, there would be at least one edge whose one end vertex would be in V_1 and the other in V_2. Hence, if a partition exists, G is not connected.

Conversely, let G be a disconnected graph. Consider a vertex a in G. Let V_1 be the set of all vertices that are joined by paths to a. Since G is disconnected, V_1 does not include all vertices of G. The remaining vertices will form a (nonempty) set V_2. No vertex in V_1 is joined to any in V_2 by an edge. Hence the partition. ■

Two interesting and useful results involving connectedness are:

Theorem 2-2

If a graph (connected or disconnected) has exactly two vertices of odd degree, there must be a path joining these two vertices.

Proof: Let G be a graph with all even vertices† except vertices v_1 and v_2, which are odd. From Theorem 1-1, which holds for every graph and therefore for every component of a disconnected graph, no graph can have an odd number of odd vertices. Therefore, in graph G, v_1 and v_2 must belong to the same component, and hence must have a path between them. ■

Theorem 2-3

A simple graph (i.e., a graph without parallel edges or self-loops) with n vertices and k components can have at most $(n - k)(n - k + 1)/2$ edges.

Proof: Let the number of vertices in each of the k components of a graph G be $n_1, n_2, \ldots, n_k$. Thus we have

$$n_1 + n_2 + \cdots + n_k = n,$$
$$n_i \geq 1.$$

†For brevity, a vertex with odd degree is called an *odd vertex*, and a vertex with even degree an *even vertex*.

The proof of the theorem depends on an algebraic inequality†

$$\sum_{i=1}^{k} n_i^2 \leq n^2 - (k - 1)(2n - k). \qquad (2\text{-}1)$$

Now the maximum number of edges in the ith component of G (which is a simple connected graph) is $\frac{1}{2}n_i(n_i - 1)$. (See Problem 1-12.) Therefore, the maximum number of edges in G is

$$\frac{1}{2}\sum_{i=1}^{k}(n_i - 1)n_i = \frac{1}{2}\left(\sum_{i=1}^{k} n_i^2\right) - \frac{n}{2} \qquad (2\text{-}2)$$

$$\leq \frac{1}{2}[n^2 - (k - 1)(2n - k)] - \frac{n}{2}, \quad \text{from (2-1)}$$

$$= \frac{1}{2}\cdot(n - k)(n - k + 1). \quad \blacksquare \qquad (2\text{-}3)$$

It may be noted that Theorem 2-3 is a generalization of the result in Problem 1-12. The solution to Problem 1-12 is given by (2-3), where $k = 1$.

Now we are equipped to handle the Königsberg bridge problem introduced in Chapter 1.

2-6. EULER GRAPHS

As mentioned in Chapter 1, graph theory was born in 1736 with Euler's famous paper in which he solved the Königsberg bridge problem. In the same paper, Euler posed (and then solved) a more general problem: In what type of graph G is it possible to find a closed walk running through every edge of G exactly once? Such a walk is now called an *Euler line*, and a graph that consists of an Euler line is called an *Euler graph*. More formally:

If some closed walk in a graph contains all the edges of the graph, then the walk is called an *Euler line* and the graph an *Euler graph*.

By its very definition a walk is always connected. Since the Euler line (which is a walk) contains all the edges of the graph, an *Euler graph* is always connected, except for any isolated vertices the graph may have. Since isolated vertices do not contribute anything to the understanding of an Euler graph, it is hereafter assumed that Euler graphs do not have any isolated vertices and are therefore connected.

Now we shall state and prove an important theorem, which will enable us to tell immediately whether or not a given graph is an Euler graph.

† *Proof:* $\sum_{i=1}^{k}(n_i - 1) = n - k$. Squaring both sides,

$$\left(\sum_{i=1}^{k}(n_i - 1)\right)^2 = n^2 + k^2 - 2nk$$

or $\sum_{i=1}^{k}(n_i^2 - 2n_i) + k +$ nonnegative cross terms $= n^2 + k^2 - 2nk$ because $(n_i - 1) \geq 0$, for all i. Therefore, $\sum_{i=1}^{k} n_i^2 \leq n^2 + k^2 - 2nk - k + 2n = n^2 - (k - 1)(2n - k)$. $\blacksquare$

Theorem 2-4

A given connected graph G is an Euler graph if and only if all vertices of G are of even degree.

Proof: Suppose that G is an Euler graph. It therefore contains an Euler line (which is a closed walk). In tracing this walk we observe that every time the walk meets a vertex v it goes through two "new" edges incident on v—with one we "entered" v and with the other "exited." This is true not only of all intermediate vertices of the walk but also of the terminal vertex, because we "exited" and "entered" the same vertex at the beginning and end of the walk, respectively. Thus if G is an Euler graph, the degree of every vertex is even.

To prove the sufficiency of the condition, assume that all vertices of G are of even degree. Now we construct a walk starting at an arbitrary vertex v and going through the edges of G such that no edge is traced more than once. We continue tracing as far as possible. Since every vertex is of even degree, we can exit from every vertex we enter; the tracing cannot stop at any vertex but v. And since v is also of even degree, we shall eventually reach v when the tracing comes to an end. If this closed walk h we just traced includes all the edges of G, G is an Euler graph. If not, we remove from G all the edges in h and obtain a subgraph h' of G formed by the remaining edges. Since both G and h have all their vertices of even degree, the degrees of the vertices of h' are also even. Moreover, h' must touch h at least at one vertex a, because G is connected. Starting from a, we can again construct a new walk in graph h'. Since all the vertices of h' are of even degree, this walk in h' must terminate at vertex a; but this walk in h' can be combined with h to form a new walk, which starts and ends at vertex v and has more edges than h. This process can be repeated until we obtain a closed walk that traverses all the edges of G. Thus G is an Euler graph. ∎

Königsberg Bridge Problem: Now looking at the graph of the Königsberg bridges (Fig. 1-5), we find that not all its vertices are of even degree. Hence, it is not an Euler graph. Thus it is not possible to walk over each of the seven bridges exactly once and return to the starting point.

One often encounters Euler lines in various puzzles. The problem common to these puzzles is to find how a given picture can be drawn in one continuous line without retracing and without lifting the pencil from the paper. Two such pictures are shown in Fig. 2-12. The drawing in Fig. 2-12(a) is called *Mohammed's scimitars* and is believed to have come from the Arabs. The one in Fig. 2-12(b) is, of course, the *star of David*. (Equal time!)

In defining an Euler line some authors drop the requirement that the walk be closed. For example, the walk $a\ 1\ c\ 2\ d\ 3\ a\ 4\ b\ 5\ d\ 6\ e\ 7\ b$ in Fig. 2-13, which includes all the edges of the graph and does not retrace any edge, is not closed. The initial vertex is a and the final vertex is b. We shall call such an open walk that includes (or traces or covers) all edges of a graph without retracing any edge a *unicursal line* or an *open Euler line*. A (connected) graph that has a unicursal line will be called a *unicursal graph*.

SEC. 2-6   EULER GRAPHS   25

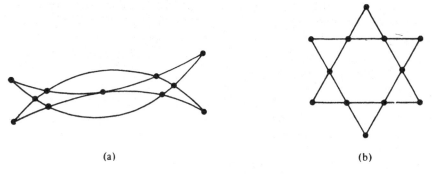

Fig. 2-12 Two Euler graphs.

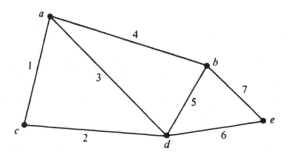

Fig. 2-13 Unicursal graph.

It is clear that by adding an edge between the initial and final vertices of a unicursal line we shall get an Euler line. Thus a connected graph is unicursal if and only if it has exactly two vertices of odd degree. This observation can be generalized as follows:

THEOREM 2-5

In a connected graph G with exactly $2k$ odd vertices, there exist k edge-disjoint subgraphs such that they together contain all edges of G and that each is a unicursal graph.

Proof: Let the odd vertices of the given graph G be named $v_1, v_2, \ldots, v_k$; $w_1, w_2, \ldots, w_k$ in any arbitrary order. Add k edges to G between the vertex pairs $(v_1, w_1), (v_2, w_2), \ldots, (v_k, w_k)$ to form a new graph G'.

Since every vertex of G' is of even degree, G' consists of an Euler line p. Now if we remove from p the k edges we just added (no two of these edges are incident on the same vertex), p will be split into k walks, each of which is a unicursal line: The first removal will leave a single unicursal line; the second removal will split that into two unicursal lines; and each successive removal will split a unicursal line into two unicursal lines, until there are k of them. Thus the theorem. ■

We shall interrupt our study of Euler graphs to define some commonly used graph-theoretic operations. One of these operations is required immediately in the next section; others will be needed later.

2-7. OPERATIONS ON GRAPHS

As is the case with most mathematical entities, it is convenient to consider a large graph as a combination of small ones and to derive its properties from those of the small ones. Since graphs are defined in terms of the sets of vertices and edges, it is natural to employ the set-theoretical terminology to define operations between graphs. In particular:

The *union* of two graphs $G_1 = (V_1, E_1)$ and $G_2 = (V_2, E_2)$ is another graph G_3 (written as $G_3 = G_1 \cup G_2$) whose vertex set $V_3 = V_1 \cup V_2$ and the edge set $E_3 = E_1 \cup E_2$. Likewise, the *intersection* $G_1 \cap G_2$ of graphs G_1 and G_2 is a graph G_4 consisting only of those vertices and edges that are in both G_1 and G_2. The *ring sum* of two graphs G_1 and G_2 (written as $G_1 \oplus G_2$) is a graph consisting of the vertex set $V_1 \cup V_2$ and of edges that are either in G_1 or G_2, but *not* in both. Two graphs and their union, intersection, and ring sum are shown in Fig. 2-14.[†]

It is obvious from their definitions that the three operations just mentioned are commutative. That is,

$$G_1 \cup G_2 = G_2 \cup G_1, \quad G_1 \cap G_2 = G_2 \cap G_1,$$
$$G_1 \oplus G_2 = G_2 \oplus G_1.$$

If G_1 and G_2 are edge disjoint, then $G_1 \cap G_2$ is a null graph, and $G_1 \oplus G_2 = G_1 \cup G_2$. If G_1 and G_2 are vertex disjoint, then $G_1 \cap G_2$ is empty.

For any graph G,

$$G \cup G = G \cap G = G,$$

and

$$G \oplus G = \text{a null graph}.$$

If g is a subgraph of G, then $G \oplus g$ is, by definition, that subgraph of G which remains after all the edges in g have been removed from G. Therefore, $G \oplus g$ is written as $G - g$, whenever $g \subseteq G$. Because of this complementary nature, $G \oplus g = G - g$ is often called the complement of g in G.

Decomposition: A graph G is said to have been *decomposed* into two subgraphs g_1 and g_2 if

$$g_1 \cup g_2 = G,$$

and

$$g_1 \cap g_2 = \text{a null graph}.$$

[†] If an edge e_i is in two graphs G_1 and G_2, its end vertices in G_1 must have the same labels as in G_2.

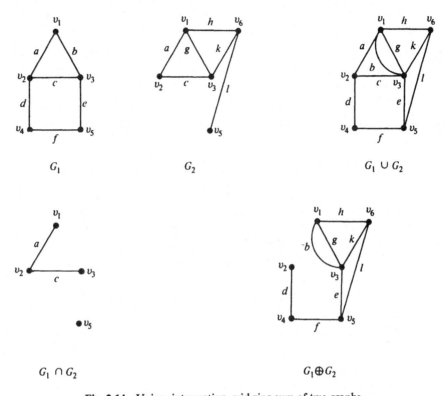

Fig. 2-14 Union, intersection, and ring sum of two graphs.

In other words, every edge of G occurs either in g_1 or in g_2, but not in both. Some of the vertices, however, may occur in both g_1 and g_2. In decomposition, isolated vertices are disregarded. A graph containing m edges $\{e_1, e_2, \ldots, e_m\}$ can be decomposed in $2^{m-1} - 1$ different ways into pairs of subgraphs g_1, g_2 (why?).

Although union, intersection, and ring sum have been defined for a pair of graphs, these definitions can be extended in an obvious way to include any finite number of graphs. Similarly, a graph G can be decomposed into more than two subgraphs—subgraphs that are (pairwise) edge disjoint and collectively include every edge in G.

Deletion: If v_i is a vertex in graph G, then $G - v_i$ denotes a subgraph of G obtained by deleting (i.e., removing) v_i from G. Deletion of a vertex always implies the deletion of all edges incident on that vertex. (See Fig. 2-15.) If e_j is an edge in G, then $G - e_j$ is a subgraph of G obtained by deleting e_j from G. Deletion of an edge does not imply deletion of its end vertices. Therefore $G - e_j = G \oplus e_j$.

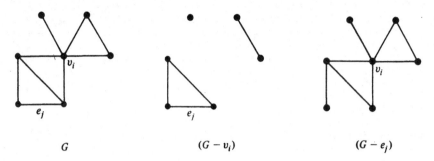

Fig. 2-15 Vertex deletion and edge deletion.

Fusion: A pair of vertices *a, b* in a graph are said to be *fused* (*merged* or *identified*) if the two vertices are replaced by a single new vertex such that every edge that was incident on either *a* or *b* or on both is incident on the new vertex. Thus fusion of two vertices does not alter the number of edges, but it reduces the number of vertices by one. See Fig. 2-16 for an example.

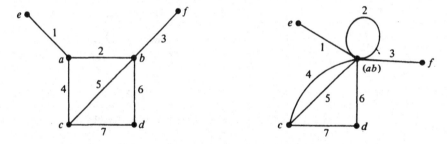

Fig. 2-16 Fusion of vertices *a* and *b*.

These are some of the elementary operations on graphs. More complex operations have been defined and are used in graph-theory literature. For a survey of such operations see the paper by Harary and Wilcox [2-10].

2-8. MORE ON EULER GRAPHS

The following are some more results on the important topic of Euler graphs.

THEOREM 2-6

A connected graph G is an Euler graph if and only if it can be decomposed into circuits.

Proof: Suppose graph G can be decomposed into circuits; that is, G is a union of edge-disjoint circuits. Since the degree of every vertex in a circuit is two, the degree of every vertex in G is even. Hence G is an Euler graph.

Conversely, let G be an Euler graph. Consider a vertex v_1. There are at least two edges incident at v_1. Let one of these edges be between v_1 and v_2. Since vertex v_2 is also of even degree, it must have at least another edge, say between v_2 and v_3. Proceeding in this fashion, we eventually arrive at a vertex that has previously been traversed, thus forming a circuit Γ. Let us remove Γ from G. All vertices in the remaining graph (not necessarily connected) must also be of even degree. From the remaining graph remove another circuit in exactly the same way as we removed Γ from G. Continue this process until no edges are left. Hence the theorem. ∎

Arbitrarily Traceable Graphs: Consider the graph in Fig. 2-17, which is an Euler graph. Suppose that we start from vertex a and trace the path $a\,b\,c$.

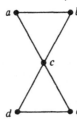

Fig. 2-17 Arbitrarily traceable graph from c.

Now at c we have the choice of going to a, d, or e. If we took the first choice, we would only trace the circuit $a\,b\,c\,a$, which is not an Euler line. Thus, starting from a, we cannot trace the entire Euler line simply by moving along any edge that has not already been traversed. This raises the following interesting question:

What property must a vertex v in an Euler graph have such that an Euler line is always obtained when one follows any walk from vertex v according to the single rule that whenever one arrives at a vertex one shall select *any* edge (which has not been previously traversed)?

Such a graph is called an *arbitrarily traceable graph from vertex v*. For instance, the Euler graph in Fig. 2-17 is an arbitrarily traceable graph from vertex c, but not from any other vertex. The Euler graph in Fig. 2-18 is not arbitrarily traceable from any vertex; the graph in Fig. 2-19 is arbitrarily

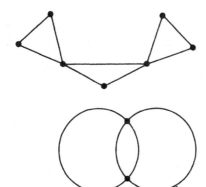

Fig. 2-18 Euler graph; not arbitrarily traceable.

Fig. 2-19 Arbitrarily traceable graph from all vertices.

traceable from all its vertices. The following interesting theorem, due to Ore [2-5], answers the question just raised.

THEOREM 2-7

An Euler graph G is arbitrarily traceable from vertex v in G if and only if every circuit in G contains v.

For a proof of the theorem the reader is referred to [2-5].

2-9. HAMILTONIAN PATHS AND CIRCUITS

An Euler line of a connected graph was characterized by the property of being a closed walk that traverses *every edge* of the graph (exactly once). A *Hamiltonian circuit* in a connected graph is defined as a closed walk that traverses *every vertex* of G exactly once, except of course the starting vertex, at which the walk also terminates. For example, in the graph of Fig. 2-20(a)

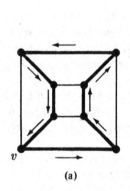

(a)　　　　　　　　　　　　　　　(b)

Fig. 2-20 Hamiltonian circuits.

starting at vertex v, if one traverses along the edges shown in heavy lines—passing through each vertex exactly once—one gets a Hamiltonian circuit. A Hamiltonian circuit for the graph in Fig. 2-20(b) is also shown by heavy lines. More formally:

A circuit in a connected graph G is said to be Hamiltonian if it includes every vertex of G. Hence a Hamiltonian circuit in a graph of n vertices consists of exactly n edges.

Obviously, not every connected graph has a Hamiltonian circuit. For example, neither of the graphs shown in Figs. 2-17 and 2-18 has a Hamiltonian circuit. This raises the question: What is a necessary and sufficient condition for a connected graph G to have a Hamiltonian circuit?

SEC. 2-9   HAMILTONIAN PATHS AND CIRCUITS   31

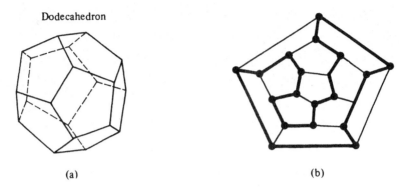

Fig. 2-21 Dodecahedron and its graph shown with a Hamiltonian circuit.

This problem, first posed by the famous Irish mathematician Sir William Rowan Hamilton in 1859, is still unsolved. As was mentioned in Chapter 1, Hamilton made a regular dodecahedron of wood [see Fig. 2-21(a)], each of whose 20 corners was marked with the name of a city. The puzzle was to start from any city and find a route along the edge of the dodecahedron that passes through every city exactly once and returns to the city of origin. The graph of the dodecahedron is given in Fig. 2-21(b), and one of many such routes (a Hamiltonian circuit) is shown by heavy lines.

The resemblance between the problem of an Euler line and that of a Hamiltonian circuit is deceptive. The latter is infinitely more complex. Although one can find Hamiltonian circuits in many specific graphs, such as those shown in Figs. 2-20 and 2-21, there is no known criterion we can apply to determine the existence of a Hamiltonian circuit in general. There are, however, certain types of graphs that always contain Hamiltonian circuits, as will be presently shown.

Hamiltonian Path: If we remove any one edge from a Hamiltonian circuit, we are left with a path. This path is called a *Hamiltonian path*. Clearly, a Hamiltonian path in a graph G traverses every vertex of G. Since a Hamiltonian path is a subgraph of a Hamiltonian circuit (which in turn is a subgraph of another graph), every graph that has a Hamiltonian circuit also has a Hamiltonian path. There are, however, many graphs with Hamiltonian paths that have no Hamiltonian circuits (Problem 2-23). The length of a Hamiltonian path (if it exists) in a connected graph of n vertices is $n - 1$.

In considering the existence of a Hamiltonian circuit (or path), we need only consider simple graphs. This is because a Hamiltonian circuit (or path) traverses every vertex exactly once. Hence it cannot include a self-loop or a set of parallel edges. Thus a general graph may be made simple by removing parallel edges and self-loops before looking for a Hamiltonian circuit in it.

It is left as an exercise for the reader to show that neither of the two graphs

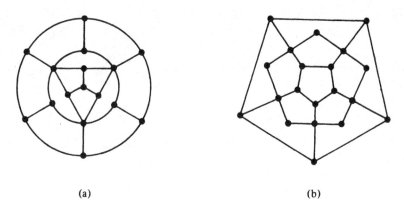

Fig. 2-22 Graphs without Hamiltonian circuits.

shown in Fig. 2-22 has a Hamiltonian circuit (or Hamiltonian path). See Problem 2-24.

What general class of graphs is guaranteed to have a Hamiltonian circuit? Complete graphs of three or more vertices constitute one such class.

Complete Graph: A simple graph in which there exists an edge between every pair of vertices is called a *complete graph*. Complete graphs of two, three, four, and five vertices are shown in Fig. 2-23. A complete graph is

Fig. 2-23 Complete graphs of two, three, four, and five vertices.

sometimes also referred to as a *universal graph* or a *clique*. Since every vertex is joined with every other vertex through one edge, the degree of every vertex is $n - 1$ in a complete graph G of n vertices. Also the total number of edges in G is $n(n - 1)/2$. See Problem 1-12.

It is easy to construct a Hamiltonian circuit in a complete graph of n vertices. Let the vertices be numbered $v_1, v_2, \ldots, v_n$. Since an edge exists between any two vertices, we can start from v_1 and traverse to v_2, and v_3, and so on to v_n, and finally from v_n to v_1. This is a Hamiltonian circuit.

Number of Hamiltonian Circuits in a Graph: A given graph may contain more than one Hamiltonian circuit. Of interest are all the edge-disjoint Hamiltonian circuits in a graph. The determination of the exact number of *edge-disjoint* Hamiltonian circuits (or paths) in a graph in general is also an unsolved problem. However, the number of edge-disjoint Hamiltonian circuits in a complete graph with odd number of vertices is given by Theorem 2-8.

THEOREM 2-8

In a complete graph with n vertices there are $(n - 1)/2$ edge-disjoint Hamiltonian circuits, if n is an odd number ≥ 3.

Proof: A complete graph G of n vertices has $n(n - 1)/2$ edges, and a Hamiltonian circuit in G consists of n edges. Therefore, the number of edge-disjoint Hamiltonian circuits in G cannot exceed $(n - 1)/2$. That there are $(n - 1)/2$ edge-disjoint Hamiltonian circuits, when n is odd, can be shown as follows:

The subgraph (of a complete graph of n vertices) in Fig. 2-24 is a Hamiltonian circuit. Keeping the vertices fixed on a circle, rotate the polygonal pattern clockwise

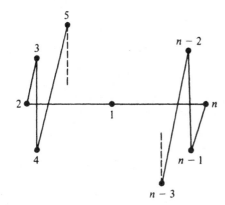

Fig. 2-24 Hamiltonian circuit; n is odd.

by $360/(n - 1), 2 \cdot 360/(n - 1), 3 \cdot 360/(n - 1), \ldots, (n - 3)/2 \cdot 360/(n - 1)$ degrees. Observe that each rotation produces a Hamiltonian circuit that has no edge in common with any of the previous ones. Thus we have $(n - 3)/2$ new Hamiltonian circuits, all edge disjoint from the one in Fig. 2-24 and also edge disjoint among themselves. Hence the theorem. ∎

This theorem enables us to solve the problem of the seating arrangement at a round table, introduced in Chapter 1, as follows:

Representing a member x by a vertex and the possibility of his sitting next to another member y by an edge between x and y, we construct a graph G. Since every member is allowed to sit next to any other member, G is a complete graph of nine vertices—nine being the number of people to be seated around the table. Every seating arrangement around the table is clearly a Hamiltonian circuit.

The first day of their meeting they can sit in any order, and it will be a Hamiltonian circuit H_1. The second day, if they are to sit such that every member must have different neighbors, we have to find another Hamiltonian circuit H_2 in G, with an entirely different set of edges from those in H_1; that is,

H_1 and H_2 are edge-disjoint Hamiltonian circuits. From Theorem 2-8 the number of edge-disjoint Hamiltonian circuits in G is four; therefore, only four such arrangements exist among nine people.

Another interesting result on the question of existence of Hamiltonian circuits in a graph, obtained by G. A. Dirac, is:

THEOREM 2-9

A sufficient (but by no means necessary) condition for a simple graph G to have a Hamiltonian circuit is that the degree of every vertex in G be at least $n/2$, where n is the number of vertices in G.

Proof: For proof the reader is referred to the original paper by Dirac [2-3].

2-10. TRAVELING-SALESMAN PROBLEM

A problem closely related to the question of Hamiltonian circuits is the *traveling-salesman problem*, stated as follows: A salesman is required to visit a number of cities during a trip. Given the distances between the cities, in what order should he travel so as to visit every city precisely once and return home, with the minimum mileage traveled?

Representing the cities by vertices and the roads between them by edges, we get a graph. In this graph, with every edge e_i there is associated a real number (the distance in miles, say), $w(e_i)$. Such a graph is called a *weighted graph*; $w(e_i)$ being the *weight* of edge e_i.

In our problem, if each of the cities has a road to every other city, we have a *complete weighted graph*. This graph has numerous Hamiltonian circuits, and we are to pick the one that has the smallest sum of distances (or weights).

The total number of different (not edge disjoint, of course) Hamiltonian circuits in a complete graph of n vertices can be shown to be $(n-1)!/2$. This follows from the fact that starting from any vertex we have $n-1$ edges to choose from the first vertex, $n-2$ from the second, $n-3$ from the third, and so on. These being independent choices, we get $(n-1)!$ possible number of choices. This number is, however, divided by 2, because each Hamiltonian circuit has been counted twice.

Theoretically, the problem of the traveling salesman can always be solved by enumerating all $(n-1)!/2$ Hamiltonian circuits, calculating the distance traveled in each, and then picking the shortest one. However, for a large value of n, the labor involved is too great even for a digital computer (try solving it for the 50 state capitals in the United States; $n = 50$).

The problem is to prescribe a manageable algorithm for finding the shortest route. No efficient algorithm for problems of arbitrary size has yet been found, although many attempts have been made. Since this problem has applications in operations research, some specific large-scale examples have been worked out (see [2-1]). There are also available several heuristic methods of solution that give a route very close to the shortest one, but do not guarantee the shortest (see [2-4] for such a method).

SUMMARY

In this chapter we discussed the subgraph—a graph that is part of another graph. Walks, paths, circuits, Euler lines, Hamiltonian paths, and Hamiltonian circuits in a graph G are its subgraphs with special properties. A given graph G can be characterized and studied in terms of the presence or absence of these subgraphs. Many physical problems can be represented by graphs and solved by observing the relevant properties of the corresponding graphs.

Various types of walks discussed in this chapter are summarized in Fig. 2-25. The arrows point in the direction of increasing restriction.

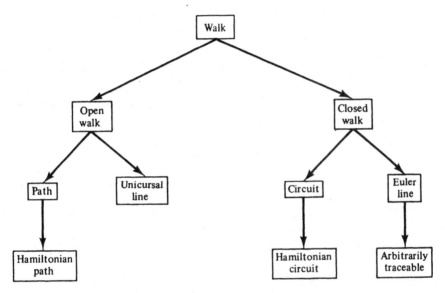

Fig. 2-25 Different types of walks.

REFERENCES

Textbooks listed in Chapter 1 are to be read for this chapter also. Specially recommended are

1. Berge [1-1], Chapters 1, 7, 11, and 17.
2. Busacker and Saaty [1-2], Chapters 1, 3, and 6.
3. Harary [1-5], Chapter 7.
4. Ore [1-9], Chapters 2 and 3.
5. Ore [1-10], Chapters 1 and 2.

6. Seshu and Reed [1-13], Chapter 2.

For arbitrarily traceable graphs, one should read Ore's paper [2-5]. Additional information on properties of Hamiltonian graphs can be found in papers by Tutte [2-8], Ore [2-6], Smith and Tutte [2-7], and Dirac [2-3]. Chapters 4 and 5 of Tutte's book [2-9] are also devoted to paths and Euler paths. On the traveling-salesman problem there are many papers. In an excellent survey Bellmore and Nemhauser [2-1] summarize and list most of these papers. Deo and Hakimi [2-2] generalized the Hamiltonian-path problem and applied it to a wiring problem in computers.

2-1. BELLMORE, M., and G. L. NEMHAUSER, "The Traveling Salesman Problem: A Survey," *Operations Res.*, Vol. 16, 1968, 538–558.
2-2. DEO, N., and S. L. HAKIMI, "The Shortest Generalized Hamiltonian Tree," *Proc. Third Annual Allerton Conf.*, University of Illinois, 1965, 879–888.
2-3. DIRAC, G. A., "Connectivity Theorems for Graphs," *Quart J. Math. Oxford*, Ser. (2), Vol. 3, 1952, 171–174.
2-4. LIN, S., "Computer Solution of the Traveling Salesman Problem," *BSTJ*, Vol. 44, 1965, 2245–2269.
2-5. ORE, O., "A Problem Regarding the Tracing of Graphs," *Rev. Elementary Math.*, Vol. 6, 1961, 49–53.
2-6. ORE, O., "Note on Hamilton Circuits," *Am. Math. Monthly*, Vol. 67, 1960, 55.
2-7. SMITH, C. A. B., and W. T. TUTTE, "On Unicursal Paths in a Network of Degree Four," *Am. Math. Monthly*, Vol. 48, 1941, 233–237.
2-8. TUTTE, W. T., "On Hamiltonian Circuits," *J. London Math. Soc.*, Vol. 21, 1946, 98–101.
2-9. TUTTE, W. T., *Connectivity in Graphs*, University of Toronto Press, Toronto, 1966.
2-10. HARARY, F., and G. W. WILCOX, "Boolean Operations on Graphs," *Math. Scand.*, Vol. 20, 1967, 41–51.

PROBLEMS

2-1. Verify that the two graphs in Fig. 2-2 are isomorphic. Label the corresponding vertices and edges.
2-2. Show by redrawing, step by step, that graphs (b) and (c) in Fig. 2-3 are isomorphic to (a).
2-3. Show that the two graphs in Figs. 2-26(a) and (b) are isomorphic.

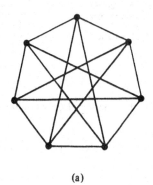
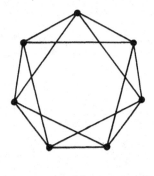

(a)　　　　　　　　　　　　　(b)

Fig. 2-26

2-4. Construct three more examples to show that conditions 1, 2, and 3 in Section 2-1 are not sufficient for isomorphism between graphs.

2-5. Prove that any two simple connected graphs with n vertices, all of degree two, are isomorphic.

2-6. Are the two graphs in Fig. 2-27 isomorphic? Why?

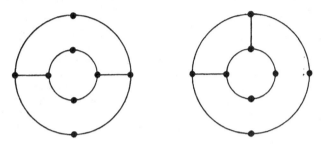

Fig. 2-27

2-7. Given the set of cubes represented by the graph in Fig. 2-6, is it possible to stack all four cubes into a column such that each side shows only one color? Explain.

2-8. Prove that a simple graph with n vertices must be connected if it has more than $[(n-1)(n-2)]/2$ edges. (*Hint:* Use Theorem 2-3.)

2-9. Prove that if a connected graph G is decomposed into two subgraphs g_1 and g_2, there must be at least one vertex common between g_1 and g_2.

2-10. Prove that a connected graph G remains connected after removing an edge e_i from G, if and only if e_i is in some circuit in G.

2-11. Draw a connected graph that becomes disconnected when any edge is removed from it.

2-12. Prove that a graph with n vertices satisfying the condition of Problem 2-11 is (a) simple, and (b) has exactly $n-1$ edges.

2-13. What is the length of the path from the entrance to the center of the maze in Problem 1-7?

2-14. List all the different paths between vertices 5 and 6 in Fig. 2-5(a). Give the length of each of these paths.

2-15. Group the paths listed in Problem 2-14 into sets of edge-disjoint paths. Demonstrate that the union of two edge-disjoint paths between a pair of vertices forms a circuit.

2-16. In a graph G let p_1 and p_2 be two different paths between two given vertices. Prove that $p_1 \oplus p_2$ is a circuit or a set of circuits in G.

2-17. Let a, b, and c be three distinct vertices in a graph. There is a path between a and b and also there is a path between b and c. Prove that there is a path between a and c.

2-18. If the intersection of two paths is a disconnected graph, show that the union of the two paths has at least one circuit.

2-19. You are given a 10-piece domino set whose titles have the following set of dots: $(1, 2)$; $(1, 3)$; $(1, 4)$; $(1, 5)$; $(2, 3)$; $(2, 4)$; $(2, 5)$; $(3, 4)$; $(3, 5)$; $(4, 5)$. Discuss the possibility of arranging the tiles in a connected series such that one number on a title always touches the same number on its neighbor. (*Hint:* Use a five-vertex complete graph and see if it is an Euler graph.)

2-20. Is it possible to move a knight on a chessboard such that it completes every permissible move exactly once? A move between two squares is counted as one regardless of the direction in which it is made. (*Hint:* Is the graph of Problem 1-6 unicursal?)

2-21. A round-robin tournament (when every player plays against every other) among n players (n being an even number) can be represented by a complete graph of n vertices. Discuss how you would schedule the tournaments to finish in the shortest possible time.

2-22. Observe that there can be no path longer than a Hamiltonian path (if it exists) in a graph.

2-23. Draw a graph that has a Hamiltonian path but does not have a Hamiltonian circuit.

2-24. Show that neither of the graphs in Fig. 2-22 has a Hamiltonian path (and therefore no Hamiltonian circuit). [*Hint:* For Fig. 2-22(a), of all the edges incident at a vertex only two can be included in a Hamiltonian circuit. Count the number of edges that have to be excluded. You will find that 13 edges must be excluded from Fig. 2-22(a). The number of remaining edges is insufficient to form a Hamiltonian circuit. For Fig. 2-22(b), first consider all vertices of degree two.]

2-25. Show that the graph of a rhombic dodecahedron (with eight vertices of degree three and six vertices of degree four) has no Hamiltonian path (and therefore no Hamiltonian circuit).

2-26. Draw a graph in which an Euler line is also a Hamiltonian circuit. What can you say about such graphs in general?

2-27. Is it possible, starting from any of the 64 squares of the chessboard, to move a knight such that it occupies every square exactly once and returns to the initial position? If so, give one such tour. (*Hint:* Look for a Hamiltonian circuit in the graph of Problem 1-6.)

2-28. Prove that a graph G with n vertices always has a Hamiltonian path if the sum of the degrees of every pair of vertices v_i, v_j in G satisfies the condition

$$d(v_i) + d(v_j) \geq n - 1.$$

(*Hint:* First show that G is connected. Then use induction on path length in G.)

2-29. Using the result of Problem 2-28, show that in a dancing ring of n children it is always possible to arrange the children so that everyone has a friend at each side if every child enjoys friendship with at least half the children.

3
TREES AND FUNDAMENTAL CIRCUITS

The concept of a *tree* is probably the most important in graph theory, especially for those interested in applications of graphs. In the first half of this chapter we shall define a *tree* and study its properties. As usual, we shall point out some of its applications to simple situations and puzzles and games, deferring the applications to more complex scientific problems till Chapter 12. Other graph-theoretic terms related to trees will also be introduced and discussed.

The second part of the chapter introduces the spanning tree—another important notion in the theory of graphs. The relationships among circuits, trees, and so on, in a graph are explored. Unavoidably, as with Chapters 1 and 2, this chapter also has a large number of definitions. In studying any new branch of mathematics, there is no way to avoid new terms and definitions.

3-1. TREES

A *tree* is a connected graph without any circuits. The graph in Fig. 3-1, for instance, is a tree. Trees with one, two, three, and four vertices are shown in Fig. 3-2. As pointed out in Chapter 1, a graph must have at least one vertex, and therefore so must a tree. Some authors allow the *null tree*, a tree without any vertices. We have excluded such an entity from being a tree. Similarly, as we are considering only finite graphs, our trees are also finite.

It follows immediately from the definition that a tree has to be a simple graph, that is, having neither a self-loop nor parallel edges (because they both form circuits).

Trees appear in numerous instances. The genealogy of a family is often

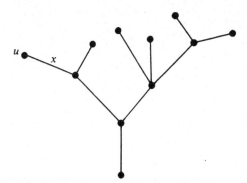

Fig. 3-1 Tree.

Fig. 3-2 Trees with one, two, three, and four vertices.

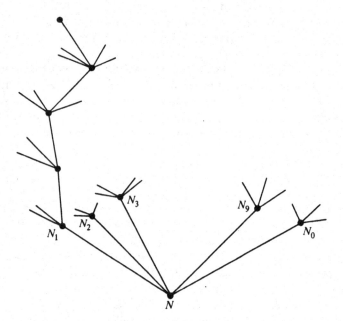

Fig. 3-3 Decision tree.

represented by means of a tree (in fact the term *tree* comes from *family tree*). A river with its tributaries and subtributaries can be represented by a tree. The sorting of mail according to zip code and the sorting of punched cards are done according to a tree (called *decision tree* or *sorting tree*).

Figure 3-3 might represent the flow of mail. All the mail arrives at some local office, vertex N. The most significant digit in the zip code is read at N, and the mail is divided into 10 piles $N_1, N_2, \ldots, N_9$, and N_0, depending on the most significant digit. Each pile is further divided into 10 piles according to the second most significant digit, and so on, till the mail is subdivided into 10^5 possible piles, each representing a unique five-digit zip code.

In many sorting problems we have only two alternatives (instead of 10 as in the preceding example) at each intermediate vertex, representing a dichotomy, such as large or small, good or bad, 0 or 1. Such a decision tree with two choices at each vertex occurs frequently in computer programming and switching theory. We shall deal with such trees and their applications in Section 3-5. Let us first obtain a few simple but important theorems on the general properties of trees.

3-2. SOME PROPERTIES OF TREES

THEOREM 3-1

There is one and only one path between every pair of vertices in a tree, T.

Proof: Since T is a connected graph, there must exist at least one path between every pair of vertices in T. Now suppose that between two vertices a and b of T there are two distinct paths. The union of these two paths will contain a circuit and T cannot be a tree. ∎

Conversely:

THEOREM 3-2

If in a graph G there is one and only one path between every pair of vertices, G is a tree.

Proof: Existence of a path between every pair of vertices assures that G is connected. A circuit in a graph (with two or more vertices) implies that there is at least one pair of vertices a, b such that there are two distinct paths between a and b. Since G has one and only one path between every pair of vertices, G can have no circuit. Therefore, G is a tree. ∎

THEOREM 3-3

A tree with n vertices has $n - 1$ edges.

Proof: The theorem will be proved by induction on the number of vertices.

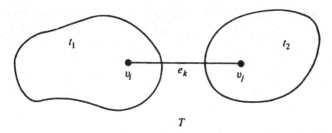

Fig. 3-4 Tree T with n vertices.

It is easy to see that the theorem is true for $n = 1, 2$, and 3 (see Fig. 3-2). Assume that the theorem holds for all trees with fewer than n vertices.

Let us now consider a tree T with n vertices. In T let e_k be an edge with end vertices v_i and v_j. According to Theorem 3-1, there is no other path between v_i and v_j except e_k. Therefore, deletion of e_k from T will disconnect the graph, as shown in Fig. 3-4. Furthermore, $T - e_k$ consists of exactly two components, and since there were no circuits in T to begin with, each of these components is a tree. Both these trees, t_1 and t_2, have fewer than n vertices each, and therefore, by the induction hypothesis, each contains one less edge than the number of vertices in it. Thus $T - e_k$ consists of $n - 2$ edges (and n vertices). Hence T has exactly $n - 1$ edges. ∎

THEOREM 3-4

Any connected graph with n vertices and $n - 1$ edges is a tree.

Proof: The proof of the theorem is left to the reader as an exercise (Problem 3-5).

You may have noticed another important feature of a tree: its vertices are connected together with the minimum number of edges. A connected graph is said to be *minimally connected* if removal of any one edge from it disconnects the graph. A minimally connected graph cannot have a circuit; otherwise, we could remove one of the edges in the circuit and still leave the graph connected. Thus a minimally connected graph is a tree. Conversely, if a connected graph G is not minimally connected, there must exist an edge e_i in G such that $G - e_i$ is connected. Therefore, e_i is in some circuit, which implies that G is not a tree. Hence the following theorem:

THEOREM 3-5

A graph is a tree if and only if it is minimally connected.

The significance of Theorem 3-5 is obvious. Intuitively, one can see that to interconnect n distinct points, the minimum number of line segments needed is $n - 1$. It requires no background in electrical engineering to realize

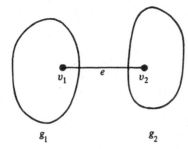

Fig. 3-5 Edge e added to $G = g_1 \cup g_2$.

that to short (electrically) n pins together, one needs at least $n - 1$ pieces of wire. The resulting structure, according to Theorem 3-5, is a tree.

We showed that a connected graph with n vertices and without any circuits has $n - 1$ edges. We can also show that a graph with n vertices which has no circuit and has $n - 1$ edges is always connected (i.e., it is a tree), in the following theorem.

THEOREM 3-6

A graph G with n vertices, $n - 1$ edges, and no circuits is connected.

Proof: Suppose there exists a circuitless graph G with n vertices and $n - 1$ edges which is disconnected. In that case G will consist of two or more circuitless components. Without loss of generality, let G consist of two components, g_1 and g_2. Add an edge e between a vertex v_1 in g_1 and v_2 in g_2 (Fig. 3-5). Since there was no path between v_1 and v_2 in G, adding e did not create a circuit. Thus $G \cup e$ is a circuitless, connected graph (i.e., a tree) of n vertices and n edges, which is not possible, because of Theorem 3-3. ■

The results of the preceding six theorems can be summarized by saying that the following are five different but equivalent definitions of a tree. That is, a graph G with n vertices is called a tree if

1. G is *connected* and is *circuitless*, or
2. G is *connected* and has $n - 1$ *edges*, or
3. G is *circuitless* and has $n - 1$ *edges*, or
4. There is *exactly one path* between every pair of vertices in G, or
5. G is a *minimally connected* graph.

3-3. PENDANT VERTICES IN A TREE

You must have observed that each of the trees shown in the figures has several pendant vertices (a pendant vertex was defined as a vertex of degree

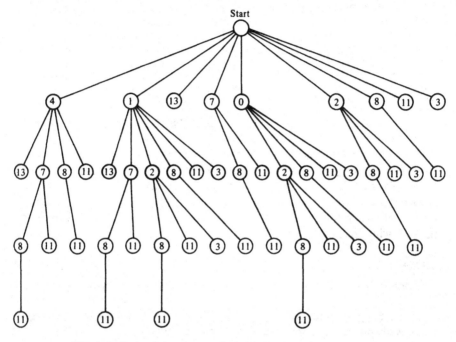

Fig. 3-6 Tree of the monotonically increasing sequences in 4, 1, 13, 7, 0, 2, 8, 11, 3.

one). The reason is that in a tree of n vertices we have $n - 1$ edges, and hence $2(n - 1)$ degrees to be divided among n vertices. Since no vertex can be of zero degree, we must have at least two vertices of degree one in a tree. This of course makes sense only if $n \geq 2$. More formally:

THEOREM 3-7

In any tree (with two or more vertices), there are at least two pendant vertices.

An Application: The following problem is used in teaching computer programming. Given a sequence of integers, no two of which are the same, find the largest monotonically increasing subsequence in it. Suppose that the sequence given to us is 4, 1, 13, 7, 0, 2, 8, 11, 3; it can be represented by a tree in which the vertices (except the start vertex) represent individual numbers in the sequence, and the path from the start vertex to a particular vertex v describes the monotonically increasing subsequence terminating in v. As shown in Fig. 3-6, this sequence contains four longest monotonically increasing subsequences, that is, (4, 7, 8, 11), (1, 7, 8, 11), (1, 2, 8, 11), and (0, 2, 8, 11). Each is of length four. Such a tree used in representing data is referred to as a data tree by computer programmers.

3-4. DISTANCE AND CENTERS IN A TREE

The tree in Fig. 3-7 has four vertices. Intuitively, it seems that vertex b is located more "centrally" than any of the other three vertices. We shall ex-

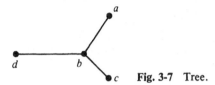

Fig. 3-7 Tree.

plore this idea further and see if in a tree there exists a "center" (or centers). Inherent in the concept of a center is the idea of "distance," so we must define distance before we can talk of a center.

In a connected graph G, the *distance* $d(v_i, v_j)$ between two of its vertices v_i and v_j is the length of the shortest path (i.e., the number of edges in the shortest path) between them.

The definition of distance between any two vertices is valid for any connected graph (not necessarily a tree). In a graph that is not a tree, there are generally several paths between a pair of vertices. We have to enumerate all these paths and find the length of the shortest one. (There may be several shortest paths.)

For instance, some of the paths between vertices v_1 and v_2 in Fig. 3-8 are (a, e), (a, c, f), (b, c, e), (b, f), (b, g, h), and (b, g, i, k). There are two shortest paths, (a, e) and (b, f), each of length two. Hence $d(v_1, v_2) = 2$.

In a tree, since there is exactly one path between any two vertices (Theorem 3-1), the determination of distance is much easier. For instance, in the tree of Fig. 3-7, $d(a, b) = 1$, $d(a, c) = 2$, $d(c, b) = 1$, and so on.

A Metric: Before we can legitimately call a function $f(x, y)$ of two variables a "distance" between them, this function must satisfy certain requirements. These are

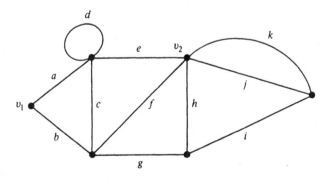

Fig. 3-8 Distance between v_1 and v_2 is two.

1. Nonnegativity: $f(x, y) \geq 0$, and $f(x, y) = 0$ if and only if $x = y$.
2. Symmetry: $f(x, y) = f(y, x)$.
3. Triangle inequality: $f(x, y) \leq f(x, z) + f(z, y)$ for any z.

A function that satisfies these three conditions is called a *metric*. That the distance $d(v_i, v_j)$ between two vertices of a connected graph satisfies conditions 1 and 2 is immediately evident. Since $d(v_i, v_j)$ is the length of the shortest path between vertices v_i and v_j, this path cannot be longer than another path between v_i and v_j, which goes through a specified vertex v_k. Hence $d(v_i, v_j) \leq d(v_i, v_k) + d(v_k, v_j)$. Therefore,

THEOREM 3-8

The distance between vertices of a connected graph is a metric.

Coming back to our original topic of relative location of different vertices in a tree, let us define another term called *eccentricity* (also referred to as *associated number* or *separation*) of a vertex in a graph.

The eccentricity $E(v)$ of a vertex v in a graph G is the distance from v to the vertex farthest from v in G; that is,

$$E(v) = \max_{v_i \in G} d(v, v_i).$$

A vertex with minimum eccentricity in graph G is called a *center* of G. The eccentricities of the four vertices in Fig. 3-7 are $E(a) = 2$, $E(b) = 1$, $E(c) = 2$, and $E(d) = 2$. Hence vertex b is the center of that tree. On the other hand, consider the tree in Fig. 3-9. The eccentricity of each of its six vertices is shown next to the vertex. This tree has two vertices having the same minimum eccentricity. Hence this tree has two centers. Some authors refer to such centers as *bicenters;* we shall call them just centers, because there will be no occasion for confusion.

The reader can easily verify that a graph, in general, has many centers. For example, in a graph that consists of just a circuit (a polygon), every vertex is a center. In the case of a tree, however, König [1-7] proved the following theorem:

THEOREM 3-9

Every tree has either one or two centers.

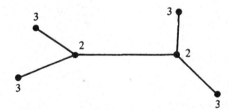

Fig 3-9 Eccentricities of the vertices of a tree.

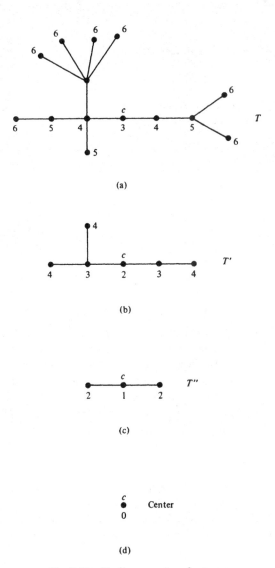

Fig. 3-10 Finding a center of a tree.

Proof: The maximum distance, max $d(v, v_i)$, from a given vertex v to any other vertex v_i occurs only when v_i is a pendant vertex. With this observation, let us start with a tree T having more than two vertices. Tree T must have two or more pendant vertices (Theorem 3-7). Delete all the pendant vertices from T. The resulting graph T' is still a tree. What about the eccentricities of the vertices in T'? A little deliberation will reveal that removal of all pendant vertices from T uniformly reduced the eccentricities of the remaining vertices (i.e., vertices in T') by one. Therefore, all vertices that T had as centers will still remain centers in T'. From T' we can again remove all pendant vertices and get another tree T''. We continue this process (which is illustrated in Fig. 3-10) until there is left either a vertex (which is the center of T) or an edge (whose end vertices are the two centers of T). Thus the theorem. ∎

COROLLARY

From the argument used in proving Theorem 3-9, we see that if a tree T has two centers, the two centers must be adjacent.

A Sociological Application: Suppose that the communication among a group of 14 persons in a society is represented by the graph in Fig. 3-10(a), where the vertices represent the persons and an edge represents the communication link between its two end vertices. Since the graph is connected, we know that all the members can be reached by any member, either directly or through some other members. But it is also important to note that the graph is a tree—minimally connected. The group cannot afford to lose any of the communication links.

The eccentricity of each vertex, $E(v)$, represents how close v is to the farthest member of the group. In Fig. 3-10(a), vertex c should be the leader of the group, if closeness of communication were the criterion for leadership.

Radius and Diameter: If a tree has a center (or two centers), does it have a radius also? Yes. The eccentricity of a center (which is the distance from the center of the tree to the farthest vertex) in a tree is defined as the *radius* of the tree. For instance, the radius of the tree in Fig. 3-10(a) is three. The *diameter* of a tree T, on the other hand, is defined as the length of the longest path in T. It is left as an exercise for the reader (Problem 3-6) to show that a radius in a tree is not necessarily half its diameter.

3-5. ROOTED AND BINARY TREES

A tree in which one vertex (called the *root*) is distinguished from all the others is called a *rooted tree*. For instance, in Fig. 3-3 vertex N, from where all the mail goes out, is distinguished from the rest of the vertices. Hence N can be considered the root of the tree, and so the tree is rooted. Similarly, in Fig. 3-6 the start vertex may be considered as the root of the tree shown. In a diagram of a rooted tree, the root is generally marked distinctly. We will show the root enclosed in a small triangle. All rooted trees with four vertices are shown in Fig. 3-11. Generally, the term *tree* means trees without any root. However, for emphasis they are sometimes called *free trees* (or *nonrooted trees*) to differentiate them from the rooted kind.

Fig. 3-11 Rooted trees with four vertices.

Binary Trees: A special class of rooted trees, called binary rooted trees, is of particular interest, since they are extensively used in the study of computer search methods, binary identification problems, and variable-length binary codes. A *binary tree* is defined as a tree in which there is exactly one vertex of degree two, and each of the remaining vertices is of degree one or three (Fig. 3-12). (Obviously, we are talking about trees with three or more vertices.) Since the vertex of degree two is distinct from all other vertices, this vertex serves as a root. Thus every binary tree is a rooted tree. Two properties of binary trees follow directly from the definition:

1. The number of vertices n in a binary tree is always odd. This is because there is exactly one vertex of even degree, and the remaining $n - 1$ vertices are of odd degrees. Since from Theorem 1-1 the number of vertices of odd degrees is even, $n - 1$ is even. Hence n is odd.

2. Let p be the number of pendant vertices in a binary tree T. Then $n - p - 1$ is the number of vertices of degree three. Therefore, the number of edges in T equals

$$\frac{1}{2}[p + 3(n - p - 1) + 2] = n - 1;$$

hence

$$p = \frac{n + 1}{2}. \tag{3-1}$$

A nonpendant vertex in a tree is called an *internal vertex*. It follows from Eq. (3-1) that the number of internal vertices in a binary tree is one less than the number of pendant vertices. In a binary tree a vertex v_i is said to be at *level* l_i if v_i is at a distance of l_i from the root. Thus the root is at level 0. A 13-vertex, four-level binary tree is shown in Fig. 3-12. The number of vertices at levels 1, 2, 3, and 4 are 2, 2, 4, and 4, respectively.

One of the most straightforward applications of binary trees is in search procedures. Each vertex of a binary tree represents a test with two possible

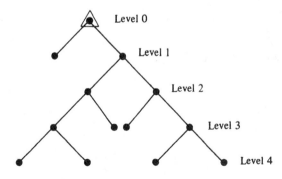

Fig. 3-12 A 13-vertex, 4-level binary tree.

outcomes. We start at the root, and the outcome of the test at the root sends us to one of the two vertices at the next level, where further tests are made, and so on. Reaching a specified pendant vertex (the goal of the search) terminates the search. For such a search procedure it is often important to construct a binary tree in which, for a given number of vertices n, the vertex farthest from the root is as close to the root as possible. Clearly, there can be only one vertex (the root) at level 0, at most two vertices at level 1, at most four vertices at level 2, and so on. Therefore, the maximum number of vertices possible in a k-level binary tree is

$$2^0 + 2^1 + 2^2 + \cdots + 2^k \geq n.$$

The maximum level, l_{max}, of any vertex in a binary tree is called the *height* of the tree. It is easy to see that the minimum possible height of an n-vertex binary tree is

$$\min l_{max} = \lceil \log_2 (n + 1) - 1 \rceil, \qquad (3\text{-}2)$$

where $\lceil n \rceil$ denotes the smallest integer greater than or equal to n.

On the other hand, to construct a binary tree for a given n such that the farthest vertex is as far as possible from the root, we must have exactly two vertices at each level, except at the 0 level. Therefore,

$$\max l_{max} = \frac{n-1}{2}. \qquad (3\text{-}3)$$

For $n = 11$, binary trees realizing both these extremes are shown in Fig. 3-13.

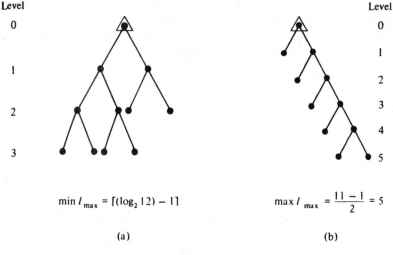

Fig. 3-13 Two 11-vertex binary trees.

In analysis of algorithms we are generally interested in computing the sum of the levels of all pendant vertices. This quantity, known as *the path length* (or external path length) *of a tree*, can be defined as the sum of the path lengths from the root to all pendant vertices. The path length of the binary tree in Fig. 3-12, for example, is

$$1 + 3 + 3 + 4 + 4 + 4 + 4 = 23.$$

The path lengths of trees in Figs. 3-13(a) and (b) are 16 and 20, respectively. The importance of the path length of a tree lies in the fact that this quantity is often directly related to the execution time of an algorithm.

It can be shown that the type of binary tree in Fig. 3-13(a) (i.e., a tree with $2^{l_{max}-1}$ vertices at level $l_{max} - 1$) yields the minimum path length for a given n.

Weighted Path Length: In some applications, every pendant vertex v_j of a binary tree has associated with it a positive real number w_j. Given w_1, $w_2, \ldots, w_m$ the problem is to construct a binary tree (with m pendant vertices) that minimizes

$$\sum w_j l_j,$$

where l_j is the level of pendant vertex v_j, and the sum is taken over all pendant vertices. Let us illustrate the significance of this problem with a simple example.

A Coke machine is to identify, by a sequence of tests, the coin that is put into the machine. Only pennies, nickels, dimes, and quarters can go through the slot. Let us assume that the probabilities of a coin being a penny, a nickel, a dime, and a quarter are .05, .15, .5, and .30, respectively. Each test has the effect of partitioning the four types of coins into two complementary sets and asserting the unknown coin to be in one of the two sets. Thus for four coins we have $2^3 - 1$ such tests. If the time taken for each test is the same, what sequence of tests will minimize the expected time taken by the Coke machine to identify the coin?

The solution requires the construction of a binary tree with four pendant vertices (and therefore three internal vertices) $v_1, v_2, v_3,$ and v_4 and corresponding weights $w_1 = .05$, $w_2 = .15$, $w_3 = .5$, and $w_4 = .3$, such that the quantity $\sum l_i w_i$ is minimized. The solution is given in Fig. 3-14(a), for which the expected time is $1.7t$, where t is the time taken for each test. Contrast this with Fig. 3-14(b), for which the expected time is $2t$. An algorithm for constructing a binary tree with minimum weighted path length can be found in [3-6].

In this problem of a Coke machine, many interesting variations are possible. For example, not all possible tests may be available, or they may not all consume the same time.

Binary trees with minimum weighted path length have also been used in

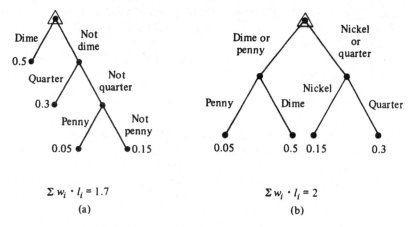

Fig. 3-14 Two binary trees with weighted pendant vertices.

constructing variable-length binary codes, where the letters of the alphabet (A, B, C, ..., Z) are represented by binary digits. Since different letters have different frequencies of occurrence (frequencies are interpreted as weights $w_1, w_2, \ldots, w_{26}$), a binary tree with minimum weighted path length corresponds to a binary code of minimum cost; see [3-6]. For more on minimum-path binary trees and their applications the reader is referred to [3-5] and [3-7].

3-6. ON COUNTING TREES

In 1857, Arthur Cayley discovered trees while he was trying to count the number of structural isomers of the saturated hydrocarbons (or paraffin series) C_kH_{2k+2}. He used a connected graph to represent the C_kH_{2k+2} molecule. Corresponding to their chemical valencies, a carbon atom was represented by a vertex of degree four and a hydrogen atom by a vertex of degree one (pendant vertices). The total number of vertices in such a graph is

$$n = 3k + 2,$$

and the total number of edges is

$$e = \frac{1}{2}(\text{sum of degrees}) = \frac{1}{2}(4k + 2k + 2)$$
$$= 3k + 1.$$

Since the graph is connected and the number of edges is one less than the number of vertices, it is a tree. Thus the problem of counting structural

SEC. 3-6   ON COUNTING TREES

isomers of a given hydrocarbon becomes the problem of counting trees (with certain qualifying properties, to be sure).

The first question Cayley asked was: what is the number of different trees that one can construct with n distinct (or labeled) vertices? If $n = 4$, for instance, we have 16 trees, as shown in Fig. 3-15. The reader can satisfy himself that there are no more trees of four vertices. (Of course, some of these trees are isomorphic—to be discussed later.)

A graph in which each vertex is assigned a unique name or label (i.e., no two vertices have the same label), as in Fig. 3-15, is called a *labeled graph*. The distinction between a labeled and an unlabeled graph is very important when we are counting the number of different graphs. For instance, the four graphs in the first row in Fig. 3-15 are counted as four different trees (even though they are isomorphic) only because the vertices are labeled. If there

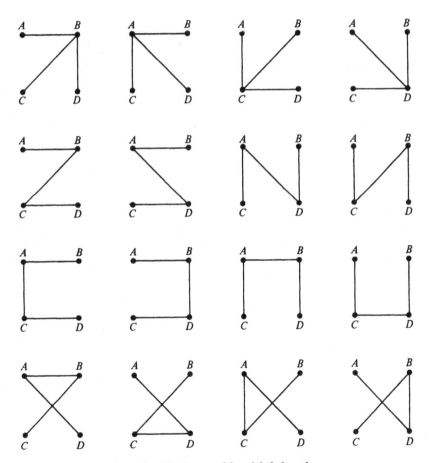

Fig. 3-15 All 16 trees of four labeled vertices.

were no distinction made between A, B, C, or D, these four trees would be counted as one. A careful inspection of the graphs in Fig. 3-15 will reveal that the number of unlabeled trees with four vertices (no distinction made between A, B, C, and D) is only two. But first we shall continue with counting labeled trees.

The following well-known theorem for counting trees was first stated and proved by Cayley, and is therefore called Cayley's theorem.

THEOREM 3-10

The number of labeled trees with n vertices ($n \geq 2$) is n^{n-2}.

Proof: The result was first stated and proved by Cayley. Many different proofs with various approaches (all somewhat involved) have been published since. An excellent summary of 10 such proofs is given by Moon [3-9]. We will give one proof in Chapter 10.

Unlabeled Trees: In the actual counting of isomers of $C_k H_{2k+2}$, Theorem 3-10 is not enough. In addition to the constraints on the degree of the vertices, two observations should be made:

1. Since the vertices representing hydrogen are pendant, they go with carbon atoms only one way, and hence make no contribution to isomerism. Therefore, we need not show any hydrogen vertices.

2. Thus the tree representing $C_k H_{2k+2}$ reduces to one with k vertices, each representing a carbon atom. In this tree no distinction can be made between vertices, and therefore it is unlabeled.

Thus for butane, $C_4 H_{10}$, there are only two distinct trees (Fig. 3-16). As every organic chemist knows, there are indeed exactly two different types of butanes: *n*-butane and isobutane. It may be noted in passing that the four trees in the first row of Fig. 3-15 are isomorphic to the one in Fig. 3-16(a); and the other 12 are isomorphic to Fig. 3-16(b).

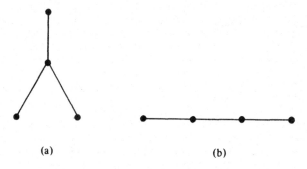

Fig. 3-16 All trees of four unlabeled vertices.

The problem of counting trees of different types will be taken up again and discussed more thoroughly in Chapter 10.

3-7. SPANNING TREES

So far we have discussed the tree and its properties when it occurs as a graph by itself. Now we shall study the tree as a subgraph of another graph. A given graph has numerous subgraphs—from e edges, 2^e distinct combinations are possible. Obviously, some of these subgraphs will be trees. Out of these trees we are particularly interested in certain types of trees, called *spanning trees*—as defined next.

A tree T is said to be a *spanning tree* of a connected graph G if T is a subgraph of G and T contains all vertices of G. For instance, the subgraph in heavy lines in Fig. 3-17 is a spanning tree of the graph shown.

Since the vertices of G are barely hanging together in a spanning tree, it is a sort of skeleton of the original graph G. This is why a spanning tree is sometimes referred to as a *skeleton* or *scaffolding* of G. Since spanning trees are the largest (with maximum number of edges) trees among all trees in G, it is also quite appropriate to call a spanning tree a *maximal tree subgraph* or *maximal tree* of G.

It is to be noted that a spanning tree is defined only for a connected graph, because a tree is always connected, and in a disconnected graph of n vertices we cannot find a connected subgraph with n vertices. Each component (which by definition is connected) of a disconnected graph, however, does have a spanning tree. Thus a disconnected graph with k components has a *spanning forest* consisting of k spanning trees. (A collection of trees is called a *forest*.)

Finding a spanning tree of a connected graph G is simple. If G has no circuit, it is its own spanning tree. If G has a circuit, delete an edge from the circuit. This will still leave the graph connected (Problem 2-10). If there are more circuits, repeat the operation till an edge from the last circuit is deleted—leaving a connected, circuit-free graph that contains all the vertices of G. Thus we have

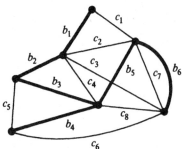

Fig. 3-17 Spanning tree.

THEOREM 3-11

Every connected graph has at least one spanning tree.

An edge in a spanning tree T is called a *branch* of T. An edge of G that is not in a given spanning tree T is called a *chord*. In electrical engineering a chord is sometimes referred to as a *tie* or a *link*. For instance, edges b_1, b_2, b_3, b_4, b_5, and b_6 are branches of the spanning tree shown in Fig. 3-17, while edges $c_1, c_2, c_3, c_4, c_5, c_6, c_7$, and c_8 are chords. It must be kept in mind that branches and chords are defined only with respect to a given spanning tree. An edge that is a branch of one spanning tree T_1 (in a graph G) may be a chord with respect to another spanning tree T_2.

It is sometimes convenient to consider a connected graph G as a union of two subgraphs, T and $\bar{T}$; that is,

$$T \cup \bar{T} = G,$$

where T is a spanning tree, and $\bar{T}$ is the complement of T in G. Since the subgraph $\bar{T}$ is the collection of chords, it is quite appropriately referred to as the *chord set* (or *tie set* or *cotree*) of T. From the definition, and from Theorem 3-3, the following theorem is evident.

THEOREM 3-12

With respect to any of its spanning trees, a connected graph of n vertices and e edges has $n - 1$ tree branches and $e - n + 1$ chords.

For example, the graph in Fig. 3-17 (with $n = 7$, $e = 14$), has six tree branches and eight chords with respect to the spanning tree $\{b_1, b_2, b_3, b_4, b_5, b_6\}$. Any other spanning tree will yield the same numbers.

If we have an electric network with e elements (edges) and n nodes (vertices), what is the minimum number of elements we must remove to eliminate all circuits in the network? The answer is $e - n + 1$. Similarly, if we have a farm consisting of six walled plots of land, as shown in Fig. 3-18, and these plots are full of water, how many walls will have to be broken so that all the water can be drained out? Here $n = 10$ and $e = 15$. We shall have to select

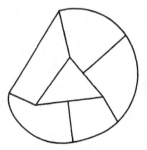

Fig. 3-18 Farm with walled plots of land.

a set of six $(15 - 10 + 1 = 6)$ walls such that the remaining nine constitute a spanning tree. Breaking these six walls will drain the water out.

Rank and Nullity: When someone specifies a graph G, the first thing he is most likely to mention is n, the number of vertices in G. Immediately following comes e, the number of edges in G. Then k, the number of components G has. If $k = 1$, G is connected. How are these three numbers of a graph related? Since every component of a graph must have at least one vertex, $n \geq k$. Moreover, the number of edges in a component can be no less than the number of vertices in that component minus one. Therefore, $e \geq n - k$. Apart from the constraints $n - k \geq 0$ and $e - n + k \geq 0$, these three numbers n, e, and k are independent, and they are fundamental numbers in graphs. (Needless to mention, these numbers alone are not enough to specify a graph, except for trivial cases.)

From these three numbers are derived two other important numbers called *rank* and *nullity*, defined as

$$\text{rank} \quad r = n - k,$$
$$\text{nullity} \quad \mu = e - n + k.$$

The rank of a connected graph is $n - 1$, and the nullity, $e - n + 1$. Although the real significance of these numbers will be clear in Chapter 7, it may be observed here that

$$\text{rank of } G = \text{number of branches in any spanning tree (or forest) of } G,$$
$$\text{nullity of } G = \text{number of chords in } G,$$
$$\text{rank} + \text{nullity} = \text{number of edges in } G.$$

The nullity of a graph is also referred to as its *cyclomatic number*, or *first Betti number*.

3-8. FUNDAMENTAL CIRCUITS

You may have noticed that if we add an edge between any two vertices of a tree (say, in Fig. 3-1) a circuit is created. This is because there already exists one path between any two vertices of a tree; adding an edge between them creates an additional path, and hence a circuit. Along this line of reasoning, it is not difficult to prove

THEOREM 3-13

A connected graph G is a tree if and only if adding an edge between any two vertices in G creates exactly one circuit.

Let us now consider a spanning tree T in a connected graph G. Adding any one chord to T will create exactly one circuit. Such a circuit, formed by adding a chord to a spanning tree, is called a *fundamental circuit*.

How many fundamental circuits does a graph have? Exactly as many as the number of chords, μ ($= e - n + k$). How many circuits does a graph have in all? We know that one circuit is created by adding any one chord to a tree. Suppose that we add one more chord. Will it create exactly one more circuit? What happens if we add all the chords simultaneously to the tree?

Let us look at the tree $\{b_1, b_2, b_3, b_4, b_5, b_6\}$ in Fig. 3-17. Adding c_1 to it, we get a subgraph $\{b_1, b_2, b_3, b_4, b_5, b_6, c_1\}$, which has one circuit (fundamental circuit), $\{b_1, b_2, b_3, b_5, c_1\}$. Had we added the chord c_2 (instead of c_1) to the tree, we would have obtained a different fundamental circuit, $\{b_2, b_3, b_5, c_2\}$. Now suppose that we add both chords c_1 and c_2 to the tree. The subgraph $\{b_1, b_2, b_3, b_4, b_5, b_6, c_1, c_2\}$ has not only the fundamental circuits we just mentioned, but it has also a third circuit, $\{b_1, c_1, c_2\}$, which is not a fundamental circuit. Although there are 75 circuits in Fig. 3-17 (enumerated by computer), only eight are fundamental circuits, each formed by one chord (together with the tree branches).

Two comments may be appropriate here. First, a circuit is a fundamental circuit only with respect to a given spanning tree. A given circuit may be fundamental with respect to one spanning tree, but not with respect to a different spanning tree of the same graph. Although the number of fundamental circuits (as well as the total number of circuits) in a graph is fixed, the circuits that become fundamental change with the spanning trees.

Second, in most applications we are not interested in all the circuits of a graph, but only in a set of fundamental circuits, which fortuitously are a lot easier to track. The concept of a fundamental circuit, introduced by Kirchhoff, is of enormous significance in electrical network analysis. What Kirchhoff showed, which now every sophomore in electrical engineering knows, is that no matter how many circuits a network contains we need consider only fundamental circuits with respect to any spanning tree. The rest of the circuits (as we shall prove rigorously in Chapter 7) are combinations of some fundamental circuits.

3-9. FINDING ALL SPANNING TREES OF A GRAPH

Usually, in a given connected graph there are a large number of spanning trees. In many applications we require all spanning trees. One reasonable way to generate spanning trees of a graph is to start with a given spanning tree, say tree T_1 (*a b c d* in Fig. 3-19). Add a chord, say *h*, to the tree T_1. This forms a fundamental circuit (*b c h d* in Fig. 3-19). Removal of any branch, say *c*, from the fundamental circuit *b c h d* just formed will create a new

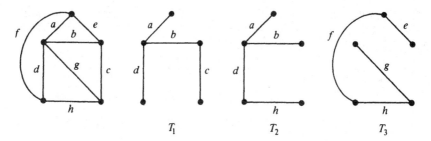

Fig. 3-19 Graph and three of its spanning trees.

spanning tree T_2. This generation of one spanning tree from another, through addition of a chord and deletion of an appropriate branch, is called a *cyclic interchange* or *elementary tree transformation*. (Such a transformation is a standard operation in the iteration sequence for solving certain transportation problems.)

In the above procedure, instead of deleting branch c, we could have deleted d or b and thus would have obtained two additional spanning trees $a\ b\ c\ h$ and $a\ c\ h\ d$. Moreover, after generating these three trees, each with chord h in it, we can restart with T_1 and add a different chord ($e, f,$ or g) and repeat the process of obtaining a different spanning tree each time a branch is deleted from the fundamental circuit formed. Thus we have a procedure for generating spanning trees for any given graph.

As we shall see in Chapter 13, the topological analysis of a linear electrical network essentially reduces to the generation of trees in the corresponding graph. Therefore, finding an efficient procedure for generating all trees of a graph is a very important practical problem.

The procedure outlined above raises many questions. Can we start from any spanning tree and get a desired spanning tree by a number of *cyclic exchanges*? Can we get all spanning trees of a given graph in this fashion? How long will we have to continue exchanging edges? Out of all possible spanning trees that we can start with, is there a preferred one for starting? Let us try to answer some of these questions; others will have to wait until Chapters 7, 10, and 11.

The distance between two spanning trees T_i and T_j of a graph G is defined as the number of edges of G present in one tree but not in the other. This distance may be written as $d(T_i, T_j)$. For instance, in Fig. 3-19 $d(T_2, T_3) = 3$.

Let $T_i \oplus T_j$ be the ring sum of two spanning trees T_i and T_j of G (as defined in Chapter 2, $T_i \oplus T_j$ is the subgraph of G containing all edges of G that are either in T_i or in T_j but not in both). Let $N(g)$ denote the number of edges in a graph g. Then, from definition,

$$d(T_i, T_j) = \frac{1}{2}N(T_i \oplus T_j).$$

It is not difficult to see that the number $d(T_i, T_j)$ is the minimum number of cyclic interchanges involved in going from T_i to T_j. The reader is encouraged to prove the following two theorems.

THEOREM 3-14

The distance between the spanning trees of a graph is a *metric*. That is, it satisfies

$d(T_i, T_j) \geq 0$ and $d(T_i, T_j) = 0$ if and only if $T_i = T_j$,
$d(T_i, T_j) = d(T_j, T_i)$,
$d(T_i, T_j) \leq d(T_i, T_k) + d(T_k, T_j)$.

THEOREM 3-15

Starting from any spanning tree of a graph G, we can obtain every spanning tree of G by successive cyclic exchanges.

Since in a connected graph G of rank r (i.e., of $r + 1$ vertices) a spanning tree has r edges, we have the following results:

The maximum distance between any two spanning trees in G is

$$\max d(T_i, T_j) = \frac{1}{2} \max N(T_i \oplus T_j)$$

$$\leq r, \text{ the rank of } G.$$

Also, if μ is the nullity of G, we know that no more than μ edges of a spanning tree T_i can be replaced to get another tree T_j.

Hence $\max d(T_i, T_j) \leq \mu$;

combining the two,

$$\max d(T_i, T_j) \leq \min(\mu, r),$$

where $\min(\mu, r)$ is the smaller of the two numbers μ and r of the graph G.

Central Tree: For a spanning tree T_0 of a graph G, let $\max_i d(T_0, T_i)$ denote the maximal distance between T_0 and any other spanning tree of G. Then T_0 is called a *central tree* of G if

$$\max_i d(T_0, T_i) \leq \max_j d(T, T_j) \quad \text{for every tree } T \text{ of } G.$$

The concept of a central tree is useful in enumerating all trees of a given graph. A central tree in a graph is, in general, not unique. For more on central trees the reader should see [3-1] and [3-4].

Tree Graph: The *tree graph* of a given graph G is defined as a graph in which each vertex corresponds to a spanning tree of G, and each edge corresponds to a *cyclic interchange* between the spanning trees of G represented by the two end vertices of the edge. From Theorem 3-15 we know that starting from any spanning tree we can obtain all other spanning trees through cyclic interchanges (or elementary tree transformations). Therefore, the tree graph of any given (finite, connected) graph is connected. For additional properties of tree graphs, the reader should see [3-3].

3-10. SPANNING TREES IN A WEIGHTED GRAPH

As discussed earlier in this chapter, a spanning tree in a graph G is a minimal subgraph connecting all the vertices of G. If graph G is a weighted graph (i.e., if there is a real number associated with each edge of G), then the *weight of a spanning tree* T of G is defined as the sum of the weights of all the branches in T. In general, different spanning trees of G will have different weights. Among all the spanning trees of G, one with the *smallest* weight is of practical significance. (There may be several spanning trees with the smallest weight; for instance, in a graph of n vertices in which every edge has unit weight, all spanning trees have a weight of $n - 1$ units.) A spanning tree with the smallest weight in a weighted graph is called a *shortest spanning tree* or *shortest-distance spanning tree* or *minimal spanning tree*.

One possible application of the shortest spanning tree is as follows: Suppose that we are to connect n cities $v_1, v_2, \ldots, v_n$ through a network of roads. The cost c_{ij} of building a direct road between v_i and v_j is given for all pairs of cities where roads can be built. (There may be pairs of cities between which no direct road can be built.) The problem is then to find the least expensive network that connects all n cities together. It is immediately evident that this connected network must be a tree: otherwise, we can always remove some edges and get a connected graph with smaller weight. Thus the problem of connecting n cities with a least expensive network is the problem of finding a shortest spanning tree in a connected weighted graph of n vertices. A necessary and sufficient condition for a spanning tree to be shortest is given by

THEOREM 3-16

A spanning tree T (of a given weighted connected graph G) is a shortest spanning tree (of G) if and only if there exists no other spanning tree (of G) at a distance of one from T whose weight is smaller than that of T.

Proof: The necessary or the "only if" condition is obvious; otherwise, we shall get another tree shorter than T by a cyclic interchange. The fact that this condition is also sufficient is remarkable and is not obvious. It can be proved as follows:

Let T_1 be a spanning tree in G satisfying the hypothesis of the theorem (i.e., there is no spanning tree at a distance of one from T_1 which is shorter than T_1).

The proof will be completed by showing that if T_2 is a shortest spanning tree (different from T_1) in G, the weight of T_1 will also be equal to that of T_2. Let T_2 be a shortest spanning tree in G. Clearly, T_2 must also satisfy the hypothesis of the theorem (otherwise there will be a spanning tree shorter than T_2 at a distance of one from T_2, violating the assumption that T_2 is shortest).

Consider an edge e in T_2 which is not in T_1. Adding e to T_1 forms a fundamental circuit with branches in T_1. Some, but not all, of the branches in T_1 that form the fundamental circuit with e may also be in T_2; each of these branches in T_1 has a weight smaller than or equal to that of e, because of the assumption on T_1. Amongst all those edges in this circuit which are not in T_2 at least one, say b_j, must form some fundamental circuit (with respect to T_2) containing e. Because of the minimality assumption on T_2 weight of b_j cannot be less than that of e. Therefore b_j must have the same weight as e. Hence the spanning tree $T'_1 = (T_1 \cup e - b_j)$, obtained from T_1 through one cycle exchange, has the same weight as T_1. But T_1 has one edge more in common with T_2, and it satisfies the condition of Theorem 3-16. This argument can be repeated, producing a series of trees of equal weight, $T_1, T_1, T_1, \ldots$, each a unit distance closer to T_2, until we get T_2 itself.

This proves that if none of the spanning trees at a unit distance from T is shorter than T, no spanning tree shorter than T exists in the graph. ∎

Algorithm for Shortest Spanning Tree: There are several methods available for actually finding a shortest spanning tree in a given graph, both by hand and by computer. One algorithm due to Kruskal [3-8] is as follows: List all edges of the graph G in order of nondecreasing weight. Next, select a smallest edge of G. Then for each successive step select (from all remaining edges of G) another smallest edge that makes no circuit with the previously selected edges. Continue until $n - 1$ edges have been selected, and these edges will constitute the desired shortest spanning tree. The validity of the method follows from Theorem 3-16.

Another algorithm, which does not require listing all edges in order of nondecreasing weight or checking at each step if a newly selected edge forms a circuit, is due to Prim [3-10]. For Prim's algorithm, draw n isolated vertices and label them $v_1, v_2, \ldots, v_n$. Tabulate the given weights of the edges of G in an n by n table. (Note that the entries in the table are symmetric with respect to the diagonal, and the diagonal is empty.) Set the weights of nonexistent edges (corresponding to those pairs of cities between which no direct road can be built) as very large.

Start from vertex v_1 and connect it to its nearest neighbor (i.e., to the vertex which has the smallest entry in row 1 of the table), say v_k. Now consider v_1 and v_k as one subgraph, and connect this subgraph to its closest neighbor (i.e., to a vertex other than v_1 and v_k that has the smallest entry among all entries in rows 1 and k). Let this new vertex be v_l. Next regard the tree with vertices $v_1, v_k,$ and v_l as one subgraph, and continue the process until all n vertices have been connected by $n - 1$ edges. Let us now illustrate this method of finding a shortest spanning tree.

SEC. 3-10   SPANNING TREES IN A WEIGHTED GRAPH

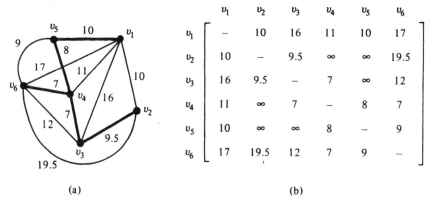

Fig. 3-20   Shortest spanning tree in a weighted graph.

A connected weighted graph with 6 vertices and 12 edges is shown in Fig. 3-20(a). The weight of its edges is tabulated in Fig. 3-20(b). We start with v_1 and pick the smallest entry in row 1, which is either (v_1, v_2) or (v_1, v_5). Let us pick (v_1, v_5). [Had we picked (v_1, v_2) we would have obtained a different shortest tree with the same weight.] The closest neighbor of subgraph (v_1, v_5) is v_4, as can be seen by examining all the entries in rows 1 and 5. The three remaining edges selected following the above procedure turn out to be (v_4, v_6), (v_4, v_3), and (v_3, v_2) in that sequence. The resulting tree—a shortest spanning tree—is shown in Fig. 3-20(a) in heavy lines. The weight of this tree is 41.5 units.

Degree-Constrained Shortest Spanning Tree: In a shortest spanning tree resulting from the preceding construction, a vertex v_i can end up with any degree; that is, $1 \leq d(v_i) \leq n - 1$. In some practical cases an upper limit on the degree of every vertex (of the resulting spanning tree) has to be imposed. For instance, in an electrical wiring problem, one may be required to wire together n pins (using as little wire as possible) with no more than three wires wrapped around any individual pin. Thus, in this particular case,

$$d(v_i) \leq 3 \quad \text{for every } v_i.$$

Such a spanning tree is called a *degree-constrained shortest spanning tree*.

In general, the problem may be stated as follows: Given a weighted connected graph G, find a shortest spanning tree T in G such that

$$d(v_i) \leq k \quad \text{for every vertex } v_i \text{ in } T.$$

If $k = 2$, this problem, in fact, reduces to the problem of finding the shortest Hamiltonian path, as well as the traveling-salesman problem (without the salesman returning to his home base), discussed at the end of Chapter 2. So far, no efficient method of finding an arbitrarily degree-constrained shortest spanning tree has been found.

SUMMARY

This chapter dealt with a particular type of connected graph called a tree. Because of their wide applications, trees form the most important topic in graph theory. Different types of trees, such as labeled and unlabeled, rooted and unrooted, were discussed, together with their properties and applications.

Of special interest are those trees that are subgraphs of a given connected graph G containing all vertices of G. Such trees are called spanning trees of G. Finding all spanning trees of a given graph is of great practical importance, and so is the problem of finding a shortest spanning tree in a given weighted graph.

Other related concepts, such as centers, radius, and diameter of a tree, rank and nullity of a graph, fundamental circuits, branches and chords, cyclic interchange, distance between spanning trees, and tree graphs, were also introduced and studied. Trees, spanning trees, and fundamental circuits will continue to appear from time to time in most of the succeeding chapters.

REFERENCES

Every textbook on graph theory has a chapter or two on trees. Especially recommended are

1. Berge [1-1], Chapters 12, 13, and 16.
2. Busacker and Saaty [1-2], Sections 1-8, 2-6, 3-6, 3-7, 6-19, 6-22, and 6-31.
3. Harary [1-5], Chapters 4, 15, and Appendix 3.
4. Ore [1-9], Chapter 4, Sections 2.4 and 6.5.
5. Ore [1-10], Chapter 3.

The importance of trees in information storage and processing can be seen in Knuth's book [3-7], pages 305–422, and in [3-2]. For counting trees of various types, see Chapter 6 of Riordan's book [3-11], and also see references given in Chapter 10 of this book. An elegant algorithm for finding a binary tree with minimum weighted path length has been given by Huffman in [3-6]. On algorithms for generating spanning trees, see Chapter 11 in this book and the references cited there. Moon [3-9] gives 10 different proofs of Cayley's formula for counting trees. Appendix 3 in Harary's book [1-5] has diagrams of all trees with $n \leq 10$. For treatment of distance between trees and central trees one should refer to Deo [3-4] and Amoia and Cottafava [3-1]. Tree graphs were first introduced and studied by Cummins [3-3]. Kruskal's [3-8] and Prim's [3-10] papers are sources for the study of shortest spanning trees.

3-1. AMOIA, V., and G. COTTAFAVA, "Invariance Properties of Central Trees," *IEEE Trans. Circuit Theory*, Vol. CT-18, No. 4, July 1971, 465–467.

3-2. COLLINS, N. L., and D. MICHIE (eds.) *Machine Intelligence*, Vol. 1, American Elsevier Publishing Company, Inc., New York, 1967.

3-3. CUMMINS, R. L., "Hamiltonian Circuits in Tree Graphs," *IEEE Trans. Circuit Theory*, Vol. CT-13, No. 1, March 1966, 82–90.

3-4. DEO, N., "A Central Tree," *IEEE Trans. Circuit Theory*, Vol. CT-13, No. 4, Dec. 1966, 439–440.

3-5. HU, T. C., and A. C. TUCKER, "Optimal Binary Search Tree," MRC Report No. 1049, University of Wisconsin, Madison, March 1970; also in *SIAM J. Appl. Math.*, Vol. 21, No. 4, Dec. 1971, 514–532.

3-6. HUFFMAN, D. A., "A Method for the Construction of Minimum-Redundancy Codes," *Proc. I.R.E.*, Vol. 40, Sept. 1952, 1098–1101.

3-7. KNUTH, D. E., *The Art of Computer Programming*, Vol. 1, Addison-Wesley Publishing Company, Inc., Reading, Mass., 1968.

3-8. KRUSKAL, J. B., Jr., "On the Shortest Spanning Subtree of Graph and the Traveling Salesman Problem," *Proc. Am. Math. Soc.*, Vol. 7, 1956, 48–50.

3-9. MOON, J. W., "Various Proofs of Cayley's Formula for Counting Trees," Chapter II in *A Seminar on Graph Theory*, (F. Harary, ed.), Holt, Rinehart and Winston, Inc., New York, 1967, 70–78.

3-10. PRIM, R. C., "Shortest Connection Networks and Some Generalizations," *Bell System Tech. J.*, Vol. 36, Nov. 1957, 1389–1401.

3-11. RIORDAN, J., *An Introduction to Combinatorial Analysis*, John Wiley & Sons, Inc., New York, 1958.

PROBLEMS

3-1. Draw all trees of n labeled vertices for $n = 1, 2, 3, 4$, and 5.

3-2. Draw all trees of n unlabeled vertices for $n = 1, 2, 3, 4$, and 5.

3-3. Draw all unlabeled rooted trees of n vertices for $n = 1, 2, 3, 4$, and 5.

3-4. It can be shown that there are only six different (nonisomorphic) trees of six vertices. Two such trees are given in Fig. 2-4. Draw the other four.

3-5. Prove Theorem 3-4.

3-6. Show a tree in which its diameter is not equal to twice the radius. Under what condition does this inequality hold? Elaborate.

3-7. Cite three different situations (games, activities, or problems) that can be represented by trees. Explain.

3-8. How many isomers does pentane C_5H_{12} have? Hexane, C_6H_{14}?

3-9. Suppose you are given eight coins and are told that seven of them are of equal weight, and one coin is either heavier or lighter than the rest. You are provided with an equal-arm balance, which you may use only three times, for comparing coins. Sketch a strategy in the form of a decision tree for identifying the nonconforming coin, as well as for finding out whether it is heavier or lighter than the rest.

3-10. Sketch all (unlabeled) binary trees with six pendant vertices. Find the path length of each. [*Hint:* Distribute the 11 vertices (because $n = 6 + 5$) among different levels. Observe that level 0 has exactly one vertex, level 1 has exactly two vertices; level 2 can have either two or four vertices; and so on. There are six such trees, and two of them are shown in Fig. 3-13.]

3-11. Sketch all spanning trees of the graph in Fig. 2-1.

3-12. Show that a path is its own spanning tree.

3-13. Prove that a pendant edge (an edge whose one end vertex is of degree one) in a connected graph G is contained in every spanning tree of G.

3-14. Prove that any subgraph g of a connected graph G is contained in some spanning tree of G if and only if g contains no circuit.

3-15. What is the nullity of a complete graph of n vertices?

3-16. Show that a Hamiltonian path is a spanning tree.

3-17. Prove that any circuit in a graph G must have at least one edge in common with a chord set.

3-18. Prove Theorem 3-13.

3-19. Find a spanning tree at a distance of four from spanning tree $\{b_1, b_2, b_3, b_4, b_5, b_6\}$ in Fig. 3-17. List all fundamental circuits with respect to this new spanning tree.

3-20. Show that the distance between two spanning trees as defined in this chapter is a metric.

3-21. Can you construct a graph if you are given all its spanning trees? How?

3-22. Prove that the nullity of a graph does not change when you either insert a vertex in the middle of an edge, or remove a vertex of degree two by merging two edges incident on it.

3-23. Prove that any given edge of a connected graph G is a branch of some spanning tree of G. Is it also true that any arbitrary edge of G is a chord for some spanning tree of G?

3-24. Suggest a method for determining the total number of spanning trees of a connected graph without actually listing them.

3-25. Prove that two colors are necessary and sufficient to paint all n vertices ($n \geq 2$) of a tree, such that no edge in the tree has both of its end vertices of the same color. (This fact is expressed by the statement that the chromatic number of a tree is two.)

3-26. Suppose that you are given a set of n positive integers. State some necessary conditions of this set so that the set can be the degrees of all the n vertices of a tree. Are these conditions sufficient also?

3-27. Let v be a vertex in a connected graph G. Prove that there exists a spanning tree T in G such that the distance of every vertex from v is the same both in G and in T.

3-28. Let T_1 and T_2 be two spanning trees of a connected graph G. If edge e is in T_1 but not in T_2, prove that there exists another edge f in T_2 but not in T_1 such that subgraphs $(T_1 - e) \cup f$ and $(T_2 - f) \cup e$ are also spanning trees of G.

3-29. Construct a tree graph (with 16 vertices, each corresponding to a tree in Fig. 3-15) of a labeled complete graph of four vertices.

3-30. In the tree graph obtained in Problem 3-29, observe the following properties (discovered by R. L. Cummins). A tree graph has at least one Hamiltonian circuit, and an arbitrary edge of a tree graph can be included in a Hamiltonian circuit.

3-31. In a given connected weighted graph G, suppose there exists an edge e_s whose weight is smaller than that of any other in G. Prove that every shortest spanning tree in G must contain e_s.

3-32. Let G be a connected weighted graph in which every edge belongs to some circuit. If e_l is the edge with weight greater than that of any other edge in G, show that no shortest spanning tree in G will contain e_l.

3-33. Show by constructing counterexamples that in Problems 3-31 and 3-32 the same cannot be said of the second smallest and the second largest edges, respectively.

3-34. Use the algorithm of Kruskal, as outlined in this chapter, to find a shortest spanning tree in the graph of Fig. 3-20(a).

3-35. Pick 15 large cities in the United States and obtain the 105 intercity distances from an atlas. Find the shortest spanning tree connecting these cities by using (a) Kruskal's method, and (b) Prim's method. Compare their relative efficiencies.

4 CUT-SETS AND CUT-VERTICES

In Chapter 3 we studied the spanning tree—a special type of subgraph of a connected graph *G*—which kept all the vertices of *G* together. In this chapter we shall study the *cut-set*—another type of subgraph of a connected graph *G* whose removal from *G* separates some vertices from others in *G*. Properties of cut-sets and their applications will be covered. Other related topics, such as connectivity, separability, and vulnerability of graphs, will also be discussed.

4-1. CUT-SETS

In a connected graph *G*, a *cut-set* is a set of edges[†] whose removal from *G* leaves *G* disconnected, provided removal of no proper subset of these edges disconnects *G*. For instance, in Fig. 4-1 the set of edges $\{a, c, d, f\}$ is a cut-set. There are many other cut-sets, such as $\{a, b, g\}, \{a, b, e, f\}$, and $\{d, h, f\}$. Edge $\{k\}$ alone is also a cut-set. The set of edges $\{a, c, h, d\}$, on the other hand, is *not* a cut-set, because one of its proper subsets, $\{a, c, h\}$, is a cut-set.

To emphasize the fact that no proper subset of a cut-set can be a cut-set, some authors refer to a cut-set as a *minimal cut-set*, a *proper cut-set*, or a *simple cut-set*. Sometimes a cut-set is also called a *cocycle*. We shall just use the term *cut-set*.

A cut-set always "cuts" a graph into two. Therefore, a cut-set can also be defined as a minimal set of edges in a connected graph whose removal reduces the rank of the graph by one. The rank of the graph in Fig. 4.1(b), for in-

---

[†]Since a set of edges (together with their end vertices) constitutes a subgraph, a cut-set in *G* is a subgraph of *G*.

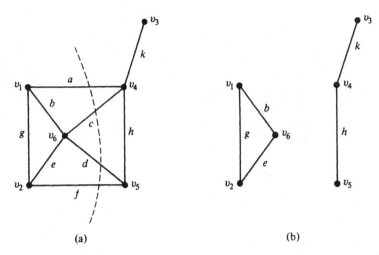

Fig. 4-1 Removal of a cut-set $\{a, c, d, f\}$ from a graph "cuts" it into two.

stance, is four, one less than that of the graph in Fig. 4.1(a). Another way of looking at a cut-set is this: if we partition all the vertices of a connected graph G into two mutually exclusive subsets, a cut-set is a minimal number of edges whose removal from G destroys all paths between these two sets of vertices. For example, in Fig. 4-1(a) cut-set $\{a, c, d, f\}$ connects vertex set $\{v_1, v_2, v_6\}$ with $\{v_3, v_4, v_5\}$. (Note that one or both of these two subsets of vertices may consist of just one vertex.) Since removal of any edge from a tree breaks the tree into two parts, *every edge of a tree is a cut-set.*

Cut-sets are of great importance in studying properties of communication and transportation networks. Suppose, for example, that the six vertices in Fig. 4-1(a) represent six cities connected by telephone lines (edges). We wish to find out if there are any weak spots in the network that need strengthening by means of additional telephone lines. We look at all cut-sets of the graph, and the one with the smallest number of edges is the most vulnerable. In Fig. 4-1(a), the city represented by vertex v_3 can be severed from the rest of the network by the destruction of just one edge. After some additional study of the properties of cut-sets, we shall return to their applications.

4-2. SOME PROPERTIES OF A CUT-SET

Consider a spanning tree T in a connected graph G and an arbitrary cut-set S in G. Is it possible for S not to have any edge in common with T? The answer is *no*. Otherwise, removal of the cut-set S from G would not disconnect the graph. Therefore,

THEOREM 4-1

Every cut-set in a connected graph G must contain at least one branch of every spanning tree of G.

Will the converse also be true? In other words, will any minimal set of edges containing at least one branch of every spanning tree be a cut-set? The answer is *yes*, by the following reasoning:

In a given connected graph G, let Q be a minimal set of edges containing at least one branch of every spanning tree of G. Consider $G - Q$, the subgraph that remains after removing the edges in Q from G. Since the subgraph $G - Q$ contains no spanning tree of G, $G - Q$ is disconnected (one component of which may just consist of an isolated vertex). Also, since Q is a minimal set of edges with this property, any edge e from Q returned to $G - Q$ will create at least one spanning tree. Thus the subgraph $G - Q + e$ will be a connected graph. Therefore, Q is a minimal set of edges whose removal from G disconnects G. This, by definition, is a cut-set. Hence

THEOREM 4-2

In a connected graph G, any minimal set of edges containing at least one branch of every spanning tree of G is a cut-set.

THEOREM 4-3

Every circuit has an even number of edges in common with any cut-set.

Proof: Consider a cut-set S in graph G (Fig. 4-2). Let the removal of S partition the vertices of G into two (mutually exclusive or disjoint) subsets V_1 and V_2. Consider a circuit Γ in G. If all the vertices in Γ are entirely within vertex set V_1 (or V_2), the number of edges common to S and Γ is zero; that is, $N(S \cap \Gamma) = 0$, an even number.†

If, on the other hand, some vertices in Γ are in V_1 and some in V_2, we traverse

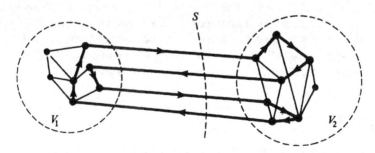

Circuit Γ shown in heavy lines, and is traversed along the direction of the arrows

Fig. 4-2 Circuit and a cut-set in G.

†As in Chapter 3, $N(g)$ stands for the number of edges in subgraph g.

back and forth between the sets V_1 and V_2 as we traverse the circuit (see Fig. 4-2). Because of the closed nature of a circuit, the number of edges we traverse between V_1 and V_2 must be even. And since very edge in S has one end in V_1 and the other in V_2, and no other edge in G has this property (of separating sets V_1 and V_2), the number of edges common to S and Γ is even. ∎

4-3. ALL CUT-SETS IN A GRAPH

In Section 4-1 it was shown how cut-sets are used to identify weak spots in a communication net. For this purpose we list all cut-sets of the corresponding graph, and find which ones have the smallest number of edges. It must also have become apparent to you that even in a simple example, such as in Fig. 4-1, there is a large number of cut-sets, and we must have a systematic method of generating all relevant cut-sets.

In the case of circuits, we solved a similar problem by the simple technique of finding a set of *fundamental circuits* and then realizing that other circuits in a graph are just *combinations* of two or more fundamental circuits. We shall follow a similar strategy here. Just as a spanning tree is essential for defining a set of fundamental circuits, so is a spanning tree essential for a set of *fundamental cut-sets*. It will be beneficial for the reader to look for the parallelism between circuits and cut-sets.

Fundamental Cut-Sets: Consider a spanning tree T of a connected graph G. Take any branch b in T. Since $\{b\}$ is a cut-set in T, $\{b\}$ partitions all vertices of T into two disjoint sets—one at each end of b. Consider the same partition of vertices in G, and the cut set S in G that corresponds to this partition. Cut-set S will contain only one branch b of T, and the rest (if any) of the edges in S are chords with respect to T. Such a cut-set S containing exactly one branch of a tree T is called a *fundamental cut-set* with respect to T. Sometimes a fundamental cut-set is also called a *basic cut-set*. In Fig. 4-3, a spanning tree

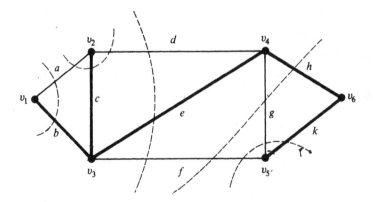

Fig. 4-3 Fundamental cut-sets of a graph.

T (in heavy lines) and all five of the fundamental cut-sets with respect to T are shown (broken lines "cutting" through each cut-set).

Just as every chord of a spanning tree defines a *unique* fundamental circuit, every branch of a spanning tree defines a *unique* fundamental cut-set. It must also be kept in mind that the term fundamental cut-set (like the term fundamental circuit) has meaning only with respect to a *given* spanning tree.

Now we shall show how other cut-sets of a graph can be obtained from a given set of cut-sets.

THEOREM 4-4

The ring sum of any two cut-sets in a graph is either a third cut-set or an edge-disjoint union of cut-sets.

Outline of Proof: Let S_1 and S_2 be two cut-sets in a given connected graph G. Let V_1 and V_2 be the (unique and disjoint) partitioning of the vertex set V of G corresponding to S_1. Let V_3 and V_4 be the partitioning corresponding to S_2. Clearly [see Figs. 4-4(a) and (b)],

$$V_1 \cup V_2 = V \quad \text{and} \quad V_1 \cap V_2 = \emptyset,$$
$$V_3 \cup V_4 = V \quad \text{and} \quad V_3 \cap V_4 = \emptyset.$$

Now let the subset $(V_1 \cap V_4) \cup (V_2 \cap V_3)$ be called V_5, and this by definition is the same as the ring sum $V_1 \oplus V_3$. Similarly, let the subset $(V_1 \cap V_3) \cup (V_2 \cap V_4)$ be called V_6, which is the same as $V_2 \oplus V_3$. See Fig. 4-4(c).

The ring sum of the two cut-sets $S_1 \oplus S_2$ can be seen to consist only of edges that join vertices in V_5 to those in V_6. Also, there are no edges outside $S_1 \oplus S_2$ that join vertices in V_5 to those in V_6.

Thus the set of edges $S_1 \oplus S_2$ produces a partitioning of V into V_5 and V_6 such that

$$V_5 \cup V_6 = V \quad \text{and} \quad V_5 \cap V_6 = \emptyset.$$

Hence $S_1 \oplus S_2$ is a cut-set if the subgraphs containing V_5 and V_6 each remain connected after $S_1 \oplus S_2$ is removed from G. Otherwise, $S_1 \oplus S_2$ is an edge-disjoint union of cut-sets.

Example: In Fig. 4-3 let us consider ring sums of the following three pairs of cut-sets.

$$\{d, e, f\} \oplus \{f, g, h\} = \{d, e, g, h\}, \quad \text{another cut-set,}$$
$$\{a, b\} \oplus \{b, c, e, f\} = \{a, c, e, f\}, \quad \text{another cut-set,}$$
$$\{d, e, g, h\} \oplus \{f, g, k\} = \{d, e, f, h, k\}$$
$$= \{d, e, f\} \cup \{h, k\}, \text{ an edge-disjoint}$$
$$\text{union of cut-sets.} \quad \blacksquare$$

SEC. 4-4   FUNDAMENTAL CIRCUITS AND CUT-SETS   73

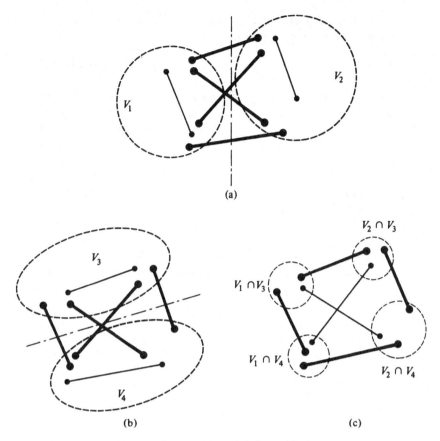

Fig. 4-4 Two cut-sets and their partitionings.

So we have a method of generating additional cut-sets from a number of given cut-sets. Obviously, we cannot start with any two cut-sets in a given graph and hope to obtain all its cut-sets by this method. What then is a minimal set of cut-sets from which we can obtain every cut-set of G by taking ring sums? The answer (to be proved in Chapter 6) is the set of all fundamental cut-sets with respect to a given spanning tree.

4-4. FUNDAMENTAL CIRCUITS AND CUT-SETS

Consider a spanning tree T in a given connected graph G. Let c_i be a chord with respect to T, and let the fundamental circuit made by c_i be called Γ, consisting of k branches $b_1, b_2, \ldots, b_k$ in addition to the chord c_i; that is,

$\Gamma = \{c_i, b_1, b_2, \ldots, b_k\}$ is a fundamental circuit with respect to T.

Every branch of any spanning tree has a fundamental cut-set associated with it. Let S_1 be the fundamental cut-set associated with b_1, consisting of q chords in addition to the branch b_1; that is,

$$S_1 = \{b_1, c_1, c_2, \ldots, c_q\} \quad \text{is a fundamental cut-set with respect to } T.$$

Because of Theorem 4-3, there must be an even number of edges common to Γ and S_1. Edge b_1 is in both Γ and S_1, and there is only one other edge in Γ (which is c_i) that can possibly also be in S_1. Therefore, we must have two edges b_1 and c_i common to S_1 and Γ. Thus the chord c_i is one of the chords $c_1, c_2, \ldots, c_q$.

Exactly the same argument holds for fundamental cut-sets associated with $b_2, b_3, \ldots,$ and b_k. Therefore, the chord c_i is contained in every fundamental cut-set associated with branches in Γ.

Is it possible for the chord c_i to be in any other fundamental cut-set S' (with respect to T, of course) besides those associated with $b_1, b_2, \ldots$ and b_k? The answer is *no*. Otherwise (since none of the branches in Γ are in S'), there would be only one edge c_i common to S' and Γ, a contradiction to Theorem 4-3. Thus we have an important result.

THEOREM 4-5

With respect to a given spanning tree T, a chord c_i that determines a fundamental circuit Γ occurs in every fundamental cut-set associated with the branches in Γ and in no other.

As an example, consider the spanning tree $\{b, c, e, h, k\}$, shown in heavy lines, in Fig. 4-3. The fundamental circuit made by chord f is

$$\{f, e, h, k\}.$$

The three fundamental cut-sets determined by the three branches e, h, and k are

determined by branch e: $\{d, e, f\}$,
determined by branch h: $\{f, g, h\}$,
determined by branch k: $\{f, g, k\}$.

Chord f occurs in each of these three fundamental cut-sets, and there is no other fundamental cut-set that contains f. The converse of Theorem 4-5 is also true.

THEOREM 4-6

With respect to a given spanning tree T, a branch b_i that determines a fundamental cut-set S is contained in every fundamental circuit associated with the chords in S, and in no others.

Proof: The proof consists of arguments similar to those that led to Theorem 4-5. Let the fundamental cut-set S determined by a branch b_i be

$$S = \{b_i, c_1, c_2, \ldots, c_p\},$$

and let Γ_1 be the fundamental circuit determined by chord c_1:

$$\Gamma_1 = \{c_1, b_1, b_2, \ldots, b_q\}.$$

Since the number of edges common to S and Γ_1 must be even, b_i must be in Γ_1. The same is true for the fundamental circuits made by chords $c_2, c_3, \ldots, c_p$.

On the other hand, suppose that b_i occurs in a fundamental circuit Γ_{p+1} made by a chord other than $c_1, c_2, \ldots, c_p$. Since none of the chords $c_1, c_2, \ldots, c_p$ is in Γ_{p+1}, there is only one edge b_i common to a circuit Γ_{p+1} and the cut-set S, which is not possible. Hence the theorem. ∎

Turning again for illustration to the graph in Fig. 4-3, consider branch e of spanning tree $\{b, c, e, h, k\}$. The fundamental cut-set determined by e is

$$\{e, d, f\}.$$

The two fundamental circuits determined by chords d and f are

determined by chord d: $\{d, c, e\}$,
determined by chord f: $\{f, e, h, k\}$.

Branch e is contained in both these fundamental circuits, and none of the remaining three fundamental circuits contains branch e.

4-5. CONNECTIVITY AND SEPARABILITY

Edge Connectivity: Each cut-set of a connected graph G consists of a certain number of edges. The number of edges in the smallest cut-set (i.e., cut-set with fewest number of edges) is defined as the *edge connectivity* of G. Equivalently, the edge connectivity of a connected graph† can be defined as the minimum number of edges whose removal (i.e., deletion) reduces the rank of the graph by one. The edge connectivity of a tree, for instance, is one. The edge connectivities of the graphs in Figs. 4-1(a), 4-3, 4-5 are one, two, and three, respectively.

Vertex Connectivity: On examining the graph in Fig. 4-5, we find that although removal of no single edge (or even a pair of edges) disconnects the

---

†Although we shall talk of edge connectivity and vertex connectivity only for a connected graph, some authors define both the edge connectivity and vertex connectivity of a disconnected graph as *zero*.

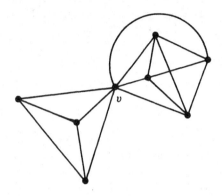

Fig. 4-5 Separable graph.

graph, the removal of the single vertex v does.† Therefore, we define another analogous term called *vertex connectivity*. The *vertex connectivity* (or simply *connectivity*) of a connected graph G is defined as the minimum number of vertices whose removal from G leaves the remaining graph disconnected.‡ Again, the vertex connectivity of a tree is one. The vertex connectivities of the graphs in Figs. 4-1(a), 4-3, and 4-5 are one, two, and one, respectively. Note that from the way we have defined it vertex connectivity is meaningful only for graphs that have three or more vertices and are not complete.

Separable Graph: A connected graph is said to be *separable* if its vertex connectivity is one. All other connected graphs are called *nonseparable*. An equivalent definition is that a connected graph G is said to be separable if there exists a subgraph g in G such that $\bar{g}$ (the complement of g in G) and g have only one vertex in common. That these two definitions are equivalent can be easily seen (Problem 4-7). In a separable graph a vertex whose removal disconnects the graph is called a *cut-vertex*, a *cut-node*, or an *articulation point*. For example, in Fig. 4-5 the vertex v is a cut-vertex, and in Fig. 4-1(a) vertex v_4 is a cut-vertex. It can be shown (Problem 4-18) that in a tree every vertex with degree greater than one is a cut-vertex. Moreover:

THEOREM 4-7

A vertex v in a connected graph G is a cut-vertex if and only if there exist two vertices x and y in G such that every path between x and y passes through v.

The proof of the theorem is quite easy and is left as an exercise (Problem 4-17). The implication of the theorem is very significant. It states that v is a crucial vertex in the sense that any communication between x and y (if G represented a communication network) must "pass through" v.

---

†Recall that removal of a vertex implies the removal of all the edges incident on that vertex, because without both the end vertices an edge does not exist. On the other hand, when we delete or remove an edge from a graph, the end vertices of the edge are still left in the graph.

‡See the footnote on p. 75.

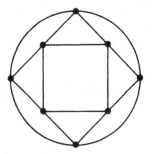

Fig. 4-6 Graph with 8 vertices and 16 edges.

An Application: Suppose we are given n stations that are to be connected by means of e lines (telephone lines, bridges, railroads, tunnels, or highways) where $e \geq n - 1$. What is the best way of connecting? By "best" we mean that the network should be as invulnerable to destruction of individual stations and individual lines as possible. In other words, construct a graph with n vertices and e edges that has the maximum possible edge connectivity and vertex connectivity.

For example, the graph in Fig. 4-5 has $n = 8$, $e = 16$, and has vertex connectivity of one and edge connectivity of three. Another graph with the same number of vertices and edges (8 and 16, respectively) can be drawn as shown in Fig. 4-6.

It can easily be seen that the edge connectivity as well as the vertex connectivity of this graph is four. Consequently, even after any three stations are bombed, or any three lines destroyed, the remaining stations can still continue to "communicate" with each other. Thus the network of Fig. 4-6 is better connected than that of Fig. 4-5 (although both consist of the same number of lines—16).

The next question is what is the highest vertex and edge connectivity we can achieve for a given n and e? The following theorems constitute the answer.

THEOREM 4-8

The edge connectivity of a graph G cannot exceed the degree of the vertex with the smallest degree in G.

Proof: Let vertex v_i be the vertex with the smallest degree in G. Let $d(v_i)$ be the degree of v_i. Vertex v_i can be separated from G by removing the $d(v_i)$ edges incident on vertex v_i. Hence the theorem. ■

THEOREM 4-9

The vertex connectivity of any graph G can never exceed the edge connectivity of G.

Proof: Let α denote the edge connectivity of G. Therefore, there exists a cut-set S in G with α edges. Let S partition the vertices of G into subsets V_1 and V_2. By removing at most α vertices from V_1 (or V_2) on which the edges in S are incident, we can effect the removal of S (together with all other edges incident on these vertices) from G. Hence the theorem. ∎

COROLLARY

Every cut-set in a nonseparable graph with more than two vertices contains at least two edges.

THEOREM 4-10

The maximum vertex connectivity one can achieve with a graph G of n vertices and e edges ($e \geq n - 1$) is the integral part of the number $2e/n$; that is, $\lfloor 2e/n \rfloor$.

Proof: Every edge in G contributes two degrees. The total ($2e$ degrees) is divided among n vertices. Therefore, there must be at least one vertex in G whose degree is equal to or less than the number $2e/n$. The vertex connectivity of G cannot exceed this number, in light of Theorems 4-8 and 4-9.

To show that this value can actually be achieved, one can first construct an n-vertex regular graph of degree $\lfloor 2e/n \rfloor$ and then add the remaining $e - (n/2)\cdot\lfloor 2e/n \rfloor$ edges arbitrarily. The completion of the proof is left as an exercise. ∎

The results of Theorems 4-8, 4-9, and 4-10 can be summarized as follows:

$$\text{vertex connectivity} \leq \text{edge connectivity} \leq \frac{2e}{n},$$

and

$$\text{maximum vertex connectivity possible} = \left\lfloor \frac{2e}{n} \right\rfloor.$$

Thus, for a graph with 8 vertices and 16 edges (Figs. 4-5 and 4-6), for example, we can achieve a vertex connectivity (and therefore edge connectivity) as high as four ($= 2\cdot 16/8$).

A graph G is said to be *k-connected* if the vertex connectivity of G is k; therefore, a *1-connected* graph is the same as a separable graph.

THEOREM 4-11

A connected graph G is k-connected if and only if every pair of vertices in G is joined by k or more paths that do not intersect,† and at least one pair of vertices is joined by exactly k nonintersecting paths.

THEOREM 4-12

The edge connectivity of a graph G is k if and only if every pair of vertices in G is joined by k or more edge-disjoint paths (i.e., paths that may intersect, but have no edges in common), and at least one pair of vertices is joined by exactly k edge-disjoint paths.

†Paths with no common vertices, except the two terminal vertices, are called nonintersecting paths or vertex-disjoint paths.

The reader is referred to Chapter 5 of [1-5] for the proofs of Theorems 4-11 and 4-12. Note that our definition of k-connectedness is slightly different from the one given in [1-5]. A special result of Theorem 4-11 is that a graph G is nonseparable if and only if any pair of vertices in G can be placed in a circuit (Problem 4-13).

The reader is encouraged to verify these theorems by enumerating all edge-disjoint and vertex-disjoint paths between each of the 15 pairs of vertices in Fig. 4-3.

4-6. NETWORK FLOWS

In a network of telephone lines, highways, railroads, pipelines of oil (or gas or water), and so on, it is important to know the maximum rate of flow that is possible from one station to another in the network. This type of network is represented by a weighted connected graph in which the vertices are the stations and the edges are lines through which the given commodity (oil, gas, water, number of messages, number of cars, etc.) flows. The weight, a real positive number, associated with each edge represents the capacity of the line, that is, the maximum amount of flow possible per unit of time. The graph in Fig. 4-7, for example, represents a flow network consisting of 12 stations and 31 lines. The capacity of each of these lines is also indicated in the figure.

It is assumed that at each intermediate vertex the total rate of commodity entering is equal to the rate leaving. In other words, there is no accumulation or generation of the commodity at any vertex along the way. Furthermore, the flow through a vertex is limited only by the capacities of the edges incident on it. In other words, the vertex itself can handle as much flow as allowed through the edges. Finally, the lines are lossless.

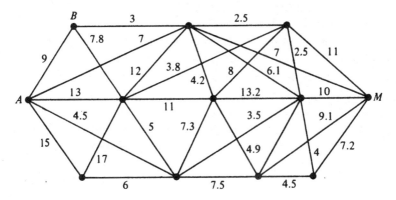

Fig. 4-7 Graph of a flow network.

In such a flow problem the questions to be answered are

1. What is the maximum flow possible through the network between a specified pair of vertices—say, from B to M in Fig. 4-7?

2. How do we achieve this flow (i.e., determine the actual flow through each edge when the maximum flow exists)?

Theorem 4-13, perhaps the most important result in the theory of transport networks, answers the first question. The second question is answered implicitly by a constructive proof of the theorem. To facilitate the statement and proof of the theorem, let us define a few terms.

A *cut-set with respect to a pair of vertices* a and b in a connected graph G puts a and b into two different components (i.e., separates vertices a and b). For instance, in Fig. 4-3 cut-set $\{d, e, f\}$ is a cut-set with respect to v_1 and v_6. The set $\{f, g, h\}$ is also a cut-set with respect to v_1 and v_6. But the cut-set $\{f, g, h\}$ is *not* a cut-set with respect to v_1 and v_6. The *capacity of cut-set* S in a weighted connected graph G (in which the weight of each edge represents its flow capacity) is defined as the sum of the weights of all the edges in S.

THEOREM 4-13

The maximum flow possible between two vertices a and b in a network is equal to the minimum of the capacities of all cut-sets with respect to a and b.

Proof: Consider any cut-set S with respect to vertices a and b in G. In the subgraph $G - S$ (the subgraph left after removing S from G) there is no path between a and b. Therefore, every path in G between a and b must contain at least one edge of S. Thus every flow from a to b (or from b to a) must pass through one or more edges of S. Hence the total flow rate between these two vertices cannot exceed the capacity of S. Since this holds for all cut-sets with respect to a and b, the flow rate cannot exceed the minimum of their capacities. ∎

To show that this flow can actually be achieved is somewhat involved. It requires some concepts that are to be introduced later. The complete proof will therefore be deferred till Chapter 14, where flow problems will be treated in much greater detail.

4-7. 1-ISOMORPHISM

A separable graph consists of two or more nonseparable subgraphs. Each of the largest nonseparable subgraphs is called a *block*. (Some authors use the term *component*, but to avoid confusion with components of a disconnected graph, we shall use the term block.) The graph in Fig. 4-5 has two blocks. The graph in Fig. 4-8 has five blocks (and three cut-vertices a, b, and c); each block

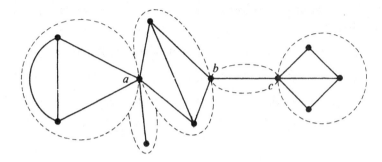

Fig. 4-8 Separable graph with three cut-vertices and five blocks.

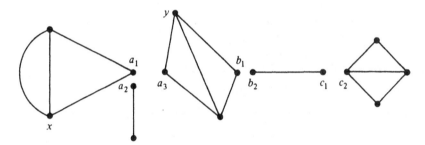

Fig. 4-9 Disconnected graph 1-isomorphic to Fig. 4-8.

is shown enclosed by a broken line. Note that a nonseparable connected graph consists of just one block.

Visually compare the disconnected graph in Fig. 4-9 with the one in Fig. 4-8. These two graphs are certainly not isomorphic (they do not have the same number of vertices), but they are related by the fact that the blocks of the graph in Fig. 4-8 are isomorphic to the components of the graph in Fig. 4-9. Such graphs are said to be *1-isomorphic*. More formally:

Two graphs G_1 and G_2 are said to be *1-isomorphic* if they become isomorphic to each other under repeated application of the following operation.

Operation 1: "Split" a cut-vertex into two vertices to produce two disjoint subgraphs.

From this definition it is apparent that two nonseparable graphs are 1-isomorphic if and only if they are isomorphic.

THEOREM 4-14

If G_1 and G_2 are two 1-isomorphic graphs, the rank of G_1 equals the rank of G_2 and the nullity of G_1 equals the nullity of G_2.

Proof: Under operation 1, whenever a cut-vertex in a graph G is "split" into two vertices, the number of components in G increases by one. Therefore, the rank of G which is

number of vertices in G − number of components in G

remains invariant under operation 1.

Also, since no edges are destroyed or new edges created by operation 1, two 1-isomorphic graphs have the same number of edges. Two graphs with equal rank and with equal numbers of edges must have the same nullity, because

nullity = number of edges − rank. ∎

What if we join two components of Fig. 4-9 by "gluing" together two vertices (say vertex x to y)? We obtain the graph shown in Fig. 4-10.

Clearly, the graph in Fig. 4-10 is 1-isomorphic to the graph in Fig. 4-9. Since the blocks of the graph in Fig. 4-10 are isomorphic to the blocks of the graph in Fig. 4-8, these two graphs are also 1-isomorphic. Thus the three graphs in Figs. 4-8, 4-9, and 4-10 are 1-isomorphic to one another.

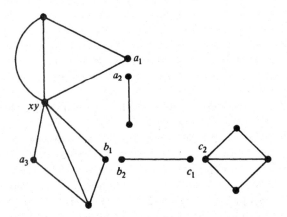

Fig. 4-10 Graph 1-isomorphic to Figs. 4-8 and 4-9.

4-8. 2-ISOMORPHISM

In Section 4-7 we generalized the concept of isomorphism by introducing 1-isomorphism. A graph G_1 was 1-isomorphic to graph G_2 if the blocks of G_1 were isomorphic to the blocks of G_2. Since a nonseparable graph is just one block, 1-isomorphism for nonseparable graphs is the same as isomorphism. However, for separable graphs (i.e., graphs with vertex connectivity of one), 1-isomorphism is different from isomorphism. Graphs that are isomorphic are also 1-isomorphic, but 1-isomorphic graphs may not be isomorphic. This generalized isomorphism is very useful in the study of separable graphs.

We can generalize this concept further to broaden its scope for 2-connected graphs (i.e., graphs with vertex connectivity of two), as follows:

In a 2-connected graph G let vertices x and y be a pair of vertices whose removal from G will leave the remaining graph disconnected. In other words, G consists of a subgraph g_1 and its complement $\bar{g}_1$ such that g_1 and $\bar{g}_1$ have exactly two vertices, x and y, in common. Suppose that we perform the following *operation 2* on G (after which, of course, G no longer remains the original graph).

Operation 2: "Split" the vertex x into x_1 and x_2 and the vertex y into y_1 and y_2 such that G is split into g_1 and $\bar{g}_1$. Let vertices x_1 and y_1 go with g_1 and x_2 and y_2 with $\bar{g}_1$. Now rejoin the graphs g_1 and $\bar{g}_1$ by merging x_1 with y_2 and x_2 with y_1. (Clearly, edges whose end vertices were x and y in G could have gone with g_1 or $\bar{g}_1$, without affecting the final graph.)

Two graphs are said to be *2-isomorphic* if they become isomorphic after undergoing operation 1 (in Section 4-7) or operation 2, or both operations any number of times. For example, Fig. 4-11 shows how the two graphs in Figs. 4-11(a) and (d) are 2-isomorphic. Note that in (a) the degree of vertex x is four, but in (d) no vertex is of degree four.

From the definition it follows immediately that isomorphic graphs are always 1-isomorphic, and 1-isomorphic graphs are always 2-isomorphic. But 2-isomorphic graphs are not necessarily 1-isomorphic, and 1-isomorphic

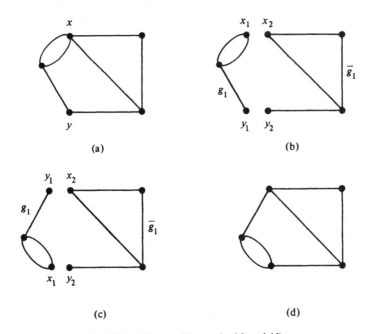

Fig. 4-11 2-isomorphic graphs (a) and (d).

graphs are not necessarily isomorphic. However, for graphs with connectivity three or more, isomorphism, 1-isomorphism, and 2-isomorphism are synonymous.

It is clear that no edges or vertices are created or destroyed under operation 2. Therefore, the rank and nullity of a graph remain unchanged under operation 2. And as shown in Section 4-7, the rank or nullity of a graph does not change under operation 1. Therefore, 2-isomorphic graphs are equal in rank and equal in nullity. The fact that the rank r and nullity μ are not enough to specify a graph within 2-isomorphism can easily be shown by constructing a counterexample (Problem 4-23).

Circuit Correspondence: Two graphs G_1 and G_2 are said to have a *circuit correspondence* if they meet the following condition: There is a one-to-one correspondence between the edges of G_1 and G_2 and a one-to-one correspondence between the circuits of G_1 and G_2, such that a circuit in G_1 formed by certain edges of G_1 has a corresponding circuit in G_2 formed by the corresponding edges of G_2, and vice versa. Isomorphic graphs, obviously, have circuit correspondence.

Since in a separable graph G every circuit is confined to a particular block (Problem 4-15), every circuit in G retains its edges as G undergoes *operation 1* (in Section 4-7). Hence 1-isomorphic graphs have circuit correpondence.

Similarly, let us consider what happens to a circuit in a graph G when it undergoes operation 2, as defined in this section. A circuit Γ in G will fall in one of three categories:

1. Γ is made of edges all in g_1, or
2. Γ is made of edges all in $\bar{g}_1$, or
3. Γ is made of edges from both g_1 and $\bar{g}_1$, and in that case Γ must include both vertices x and y.

In cases 1 and 2, Γ is unaffected by operation 2. In case 3, Γ still has the original edges, except that the path between vertices x and y in g_1, which constituted a part of Γ, is "flipped around." Thus every circuit in a graph undergoing operation 2 retains its original edges. Therefore, 2-isomorphic graphs also have circuit correspondence.

Theorem 4-15, which is considered the most important re ult for 2-isomorphic graphs, is due to H. Whitney.

THEOREM 4-15

Two graphs are 2-isomorphic if and only if they have circuit correspondence.

Proof: The "only if" part has already been shown in the argument preceding the theorem. The "if" part is more involved, and the reader is referred to Whitney's original paper [4-7].

As we shall observe in subsequent chapters, the ideas of 2-isomorphism and circuit correspondence play important roles in the theory of contact networks, electrical networks, and in duality of graphs.

SUMMARY

Our main concern in this chapter was with answering the following question about a connected graph: Which part of a connected graph, when removed, breaks the graph apart? Clearly, the answer to this question does specify a graph in many aspects and tells a great deal about it. Some of these properties are of considerable significance both in theory and applications of graphs.

In pursuit of the answer to the above question, we came across the concepts of cut-sets, cut-vertices, connectivity, and so on. Many of the theorems showed relationships between these characteristics of a graph.

In contrast to a spanning tree (which keeps the vertices together), a cut-set separates the vertices. Consequently, there was bound to be a close relationship between a spanning tree and a cut-set. Some of the theorems (and the problems at the end of this chapter) describe this relationship between spanning trees and cut-sets.

In terms of the minimum number of vertices whose removal disconnects a graph, all graphs can be classified according to Fig. 4-12.

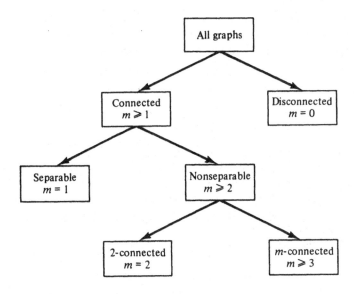

(m is the vertex-connectivity)

Fig. 4-12 Classification of graphs according to their connectivity.

A very important and practical result of this chapter was the *max-flow min-cut theorem* (Theorem 4-13).

REFERENCES

Most of the material in this chapter is based on the classic work of Hassler Whitney conducted in the early 1930s, [4-6, 4-7]. Menger [4-5], in 1927, showed that the vertex connectivity of a graph was related to the number of vertex-disjoint paths between two vertices in a graph. Many variations of Menger's theorem have appeared. Harary [1-5], Chapter 5, gives an excellent survey of Mengerian results and shows how there are 18 different theorems possible (of which we have given the two most important—Theorems 4-11 and 4-12). The max-flow min-cut theorem—also a variation of Mengerian results —was discovered independently by Ford and Fulkerson [4-2] and Elias, Feinstein, and Shannon [4-1]. The best reference for network flow problems is the authoritative book by Ford and Fulkerson [4-3]. Other references recommended for this chapter are

1. Berge [1-1], Chapter 20.
2. Busacker and Saaty [1-2], Chapter 7.
3. Kim and Chien [4-4], Part V, Chapters 2 and 3.

4-1. ELIAS, P., A. FEINSTEIN, and C. E. SHANNON, "A Note on the Maximal Flow Through a Network," *IRE Trans. Inform. Theory*, Vol. IT-2, Dec. 1956, 117–119.
4-2. FORD, L. R., and D. R. FULKERSON, "Maximal Flow Through a Network," *Can. J. Math.*, Vol. 8, 1956, 399–404.
4-3. FORD, L. R., and D. R. FULKERSON, *Flows in Networks*, Princeton University Press, Princeton, N.J., 1962.
4-4. KIM, W. H., and R. T. CHIEN, *Topological Analysis and Synthesis of Communication Networks*, Columbia University Press, New York, 1962.
4-5. MENGER, K., "Zur allgemeinen Kurventheorie," *Fund. Math.*, Vol. 10, 1927, 96–115.
4-6. WHITNEY, H., "Congruent Graphs and the Connectivity of Graphs," *Am. J. Math.*, Vol. 54, 1932, 150–168.
4-7. WHITNEY, H., "2-Isomorphic Graphs," *Am. J. Math.*, Vol. 55, 1933, 245–254.

PROBLEMS

4-1. Pick an arbitrary spanning tree in the graph given in Fig. 4-6. List all seven (because $n - 1 = 7$) fundamental cut-sets with respect to this tree.

4-2. By taking the ring sum of the seven fundamental cut-sets obtained in Problem 4-1, list all other cut-sets of the graph.

4-3. List all cut-sets with respect to the vertex pair v_2, v_3 in the graph in Fig. 4-1(a).

4-4. Show that the edge connectivity and vertex connectivity of the graphs in Fig. 2-2 are each equal to three.

4-5. What is the edge connectivity of the complete graph of n vertices?

4-6. Prove that in a connected graph G the complement of a cut-set in G does not contain a spanning tree and the complement of a spanning tree (i.e., chord set) does not contain a cut-set.

4-7. Show that the two definitions of separability in Section 4-5 are equivalent.

4-8. Prove that in a nonseparable graph G the set of edges incident on each vertex of G is a cut-set.

4-9. Why is the result of Problem 4-8 not applicable to separable graphs also? Explain.

4-10. Prove that in a connected graph G a vertex v is a cut-vertex if and only if there exist two (or more) edges x and y incident on v such that no circuit in G includes both x and y.

4-11. Prove that every connected graph with three or more vertices has at least two vertices which are not cut-vertices.

4-12. Prove that a nonseparable graph has a nullity $\mu = 1$ if and only if the graph is a circuit.

4-13. Show that a graph G is nonseparable if and only if every vertex pair in G can be placed in some circuit in G.

4-14. Show that a simple graph is nonseparable if and only if for any two given arbitrary edges a circuit can always be found that will include these two edges.

4-15. How can you utilize the result of Problem 4-13 to obtain an algorithm for identifying every block of a large separable graph?

4-16. What is a necessary and sufficient condition that any $n - 1$ cut-sets in Problem 4-8 constitute a set of fundamental cut-sets in G?

4-17. Prove Theorem 4-7.

4-18. Prove that in a tree every vertex of degree greater than one is a cut-vertex.

4-19. Show that a graph with n vertices and with vertex connectivity k must have at least $kn/2$ edges. (A special case of this result is that the degree of every vertex in a nonseparable graph is at least two.)

4-20. Is every regular graph of degree d ($d \geq 3$) nonseparable? If not, give a simple regular graph of degree three that is separable.

4-21. Complete the proof of Theorem 4-10.

4-22. In a connected graph G, let Q be a set of edges with the following properties:
(a) Q has an even number (zero included) of edges in common with every cut-set of G.
(b) There is no proper subset of Q that satisfies property (a).
Prove that Q is a circuit.

4-23. Construct a graph G with the following properties: Edge connectivity of $G = 4$, vertex connectivity of $G = 3$, and degree of every vertex of $G \geq 5$.

4-24. Show (by drawing them) that two graphs with the same rank and the same nullity need not be 2-isomorphic.

4-25. In Fig. 4-7, between vertices A and M, pick out a complete set of
(a) Edge-disjoint paths.
(b) Vertex-disjoint paths.
From this, verify Theorems 4-11 and 4-12.

4-26. Suppose that a singles tennis tournament is to be arranged among n players and the number of matches planned is a fixed number e (where $n - 1 < e < n(n - 1)/2$). For the sake of fairness, how will you make sure that some players do not group together and isolate an individual (or a small group of players)?

4-27. Let us define a new term called *edge isomorphism* as follows: Two graphs G_1 and G_2 are *edge isomorphic* if there is a one-to-one correspondence between the edges of G_1 and G_2 such that two edges are incident (at a common vertex) in G_1 if and only if the corresponding edges are also incident in G_2. Discuss the properties of *edge isomorphism*. Construct an example to prove that edge-isomorphic graphs may not be isomorphic.

4-28. Prove that an Euler graph cannot have a cut-set with an odd number of edges. (*Hint:* Use Theorem 1-1.)

5 PLANAR AND DUAL GRAPHS

In Chapters 2, 3, and 4 we studied properties of subgraphs, such as paths, circuits, spanning trees, and cut-sets, in a given connected graph G. In this chapter we shall subject the entire graph G to the following important question: Is it possible to draw G in a plane without its edges crossing over?

This question of planarity is of great significance from a theoretical point of view. In addition, planarity and other related concepts are useful in many practical situations. For instance, in the design of a printed-circuit board, the electrical engineer must know if he can make the required connections without an extra layer of insulation. The solution to the puzzle of three utilities, posed in Chapter 1, requires the knowledge of whether or not the corresponding graph can be drawn in a plane.

But before we attempt to draw a graph in a plane, let us examine the meaning of "drawing" a graph.

5-1. COMBINATORIAL VERSUS GEOMETRIC GRAPHS

As mentioned in Chapter 1, a graph exists as an abstract object, devoid of any geometric connotation of its ability of being drawn in a three-dimensional Euclidean space. For example, an abstract graph G_1 can be defined as

$$G_1 = (V, E, \Psi),$$

where the set V consists of the five objects named a, b, c, d, and e, that is,

$$V = \{a, b, c, d, e\},$$

and the set E consists of seven objects (none of which is in set V) named 1, 2, 3, 4, 5, 6, and 7, that is,

$$E = \{1, 2, 3, 4, 5, 6, 7\},$$

and the relationship between the two sets is defined by the mapping Ψ, which consists of

$$\Psi = \begin{bmatrix} 1 \longrightarrow (a, c) \\ 2 \longrightarrow (c, d) \\ 3 \longrightarrow (a, d) \\ 4 \longrightarrow (a, b) \\ 5 \longrightarrow (b, d) \\ 6 \longrightarrow (d, e) \\ 7 \longrightarrow (b, e) \end{bmatrix}.$$

Here, the symbol $1 \longrightarrow (a, c)$ says that object 1 from set E is mapped onto the (unordered) pair (a, c) of objects from set V.

Now it so happens that this combinatorial abstract object G_1 can also be *represented* by means of a geometric figure. In fact, the sketch in Fig. 2-13 is one such geometric representation of this graph. Moreover, it is also true that any graph can be represented by means of such a configuration in three-dimensional Euclidean space.

It is important to realize that what is sketched in Fig. 2-13 is merely one (out of infinitely many) representation of the graph G_1 and *not the graph G_1 itself*. We could have, for instance, twisted some of the edges or could have placed e within the triangle a, d, b and thereby obtained a different figure representing G_1. However, when there is no chance of confusion, a pictorial representation of the graph has been and will be regarded as the graph itself.

This convenient slurring over is done deliberately for the sake of simplicity and clarity. Learning graph theory for the first time without any diagrams would be extremely difficult and little fun.†

Unlike in the last four chapters, in this chapter it will often be necessary to make a distinction between the abstract (or combinatorial) graph and a geometric representation of a graph.

---

†At this point I cannot resist quoting the following comment by Hadamard: "Descartes distrusts that intervention of imagination, and wishes to eliminate it completely from science.... More recently, another rigorous treatment of... geometry... freed from any appeal to intuition, has been developed... by the celebrated mathematician Hilbert. Logically, every intervention of geometrical sense is eliminated. But is it the same from the psychological point of view? Certainly not.... Diagrams appear at practically every page (of Hilbert's book)."

5-2. PLANAR GRAPHS

A graph G is said to be *planar* if there exists some geometric representation of G which can be drawn on a plane such that no two of its edges intersect.† A graph that cannot be drawn on a plane without a crossover between its edges is called *nonplanar*.

A drawing of a geometric representation of a graph on any surface such that no edges intersect is called *embedding*. Thus, to declare that a graph G is nonplanar, we have to show that of all possible geometric representations of G none can be embedded in a plane. Equivalently, a geometric graph G is planar if there exists a graph isomorphic to G that is embedded in a plane. Otherwise, G is nonplanar. An embedding of a planar graph G on a plane is called a *plane representation* of G.

For instance, consider the graph represented by Fig. 1-3. The geometric representation shown in Fig. 1-3 clearly is not embedded in a plane, because the edges e and f are intersecting. But if we redraw edge f outside the quadrilateral, leaving the other edges unchanged, we have embedded the new geometric graph in the plane, thus showing that the graph which is being represented by Fig. 1-3 is planar. As another example, the two isomorphic diagrams in Fig. 2-2 are different geometric representations of one and the same graph. One of the diagrams is a plane representation; the other one is not. The graph, of course, is planar. On the other hand, you will not be able to draw any of the three configurations in Fig. 2-3 on a plane without edges intersecting. The reason is that the graph which these three different diagrams in Fig. 2-3 represent is nonplanar.

A natural question now is: How can we tell if a graph G [which may be given by an abstract notation $G = (V, E, \Psi)$ or by one of its geometric representations] is planar or nonplanar? To answer this question, let us first discuss two specific nonplanar graphs which are of fundamental importance. These are called Kuratowski's graphs, after the Polish mathematician Kasimir Kuratowski, who discovered their unique property.

5-3. KURATOWSKI'S TWO GRAPHS

THEOREM 5-1

The complete graph of five vertices is nonplanar.

Proof: Let the five vertices in the complete graph be named v_1, v_2, v_3, v_4, and v_5. A complete graph, as you may recall, is a simple graph in which every vertex is joined to every other vertex by means of an edge. This being the case, we must

†Note that the "meeting" of edges at a vertex is not considered an intersection.

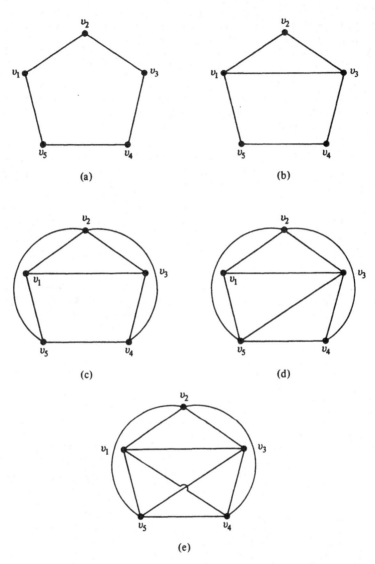

Fig. 5-1 Building up of the five-vertex complete graph.

have a circuit going from v_1 to v_2 to v_3 to v_4 to v_5 to v_1—that is, a pentagon. See Fig. 5-1(a). This pentagon must divide the plane of the paper into two regions, one *inside* and the other *outside* (Jordan curve theorem).

Since vertex v_1 is to be connected to v_3 by means of an edge, this edge may be drawn inside or outside the pentagon (without intersecting the five edges drawn previously). Suppose that we choose to draw a line from v_1 to v_3 inside the pentagon. See Fig. 5-1(b). (If we choose outside, we end up with the same argument.)

Now we have to draw an edge from v_2 to v_4 and another one from v_2 to v_5. Since neither of these edges can be drawn inside the pentagon without crossing over the edge already drawn, we draw both these edges outside the pentagon. See Fig. 5-1(c). The edge connecting v_3 and v_5 cannot be drawn outside the pentagon without crossing the edge between v_2 and v_4. Therefore, v_3 and v_5 have to be connected with an edge inside the pentagon. See Fig. 5-1(d).

Now we have yet to draw an edge between v_1 and v_4. This edge cannot be placed inside or outside the pentagon without a crossover. Thus the graph cannot be embedded in a plane. See Fig. 5-1(e). ∎

Some readers may find this proof somewhat unsatisfactory because it depends so heavily on visual intuition. Do not despair; we shall provide you with an algebraic nonvisual proof in the next section.

A complete graph with five vertices is the first of the two graphs of Kuratowski. The second graph of Kuratowski is a regular† connected graph with six vertices and nine edges, shown in its two common geometric representations in Figs. 5-2(a) and (b), where it is fairly easy to see that the graphs are isomorphic.

Employing visual geometric arguments similar to those used in proving Theorem 5-1, it can be shown that the second graph of Kuratowski is also nonplanar. The proof of Theorem 5-2 is, therefore, left as an exercise (Problem 5-1).

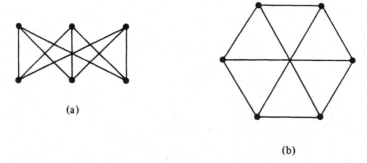

(a)

(b)

Fig. 5-2 Kuratowski's second graph.

THEOREM 5-2

Kuratowski's second graph is also nonplanar.

You may have noticed several properties common to the two graphs of Kuratowski. These are

1. Both are regular graphs.

†Recall that a graph in which all vertices are of equal degree is called a *regular graph*.

2. Both are nonplanar.
3. Removal of one edge or a vertex makes each a planar graph.
4. Kuratowski's first graph is the nonplanar graph with the smallest number of vertices, and Kuratowski's second graph is the nonplanar graph with the smallest number of edges. Thus both are the simplest nonplanar graphs.

In the literature, Kuratowski's first graph is usually denoted by K_5 and the second graph by $K_{3,3}$—letter K being for Kuratowski.

5-4. DIFFERENT REPRESENTATIONS OF A PLANAR GRAPH

In following the proof of Theorem 5-1, it may have appeared that one's ability to draw a planar graph in a plane depended on his ability to draw many crooked lines through devious routes. This is not the case. The following important and somewhat surprising result, due to Fary, tells us that there is no need to bend edges in drawing a planar graph to avoid edge intersections.

THEOREM 5-3

Any simple planar graph can be embedded in a plane such that every edge is drawn as a straight line segment.

Proof: The proof is involved and does not contribute much to the understanding of planarity. The interested reader is, therefore, referred to pages 74–77 in [1-2] or to the original paper of Fary [5-4]. As an illustration, the graph in Fig. 5-1(d) can be redrawn using straight line segments to look like Fig. 5-3. In this theorem, it is necessary for the graph to be simple because a self-loop or one of two parallel edges cannot be drawn by a straight line segment. ∎

Region: A plane representation of a graph divides the plane into *regions* (also called *windows*, *faces*, or *meshes*), as shown in Fig. 5-4. A region is

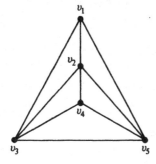

Fig. 5-3 Straight-line representation of the graph in Fig. 5-1(d).

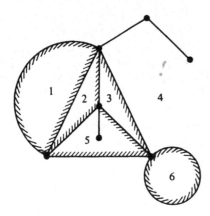

Fig. 5-4 Plane representation (the numbers stand for regions).

characterized by the set of edges (or the set of vertices) forming its *boundary*. Note that a region is not defined in a nonplanar graph or even in a planar graph not embedded in a plane. For example, the geometric graph in Fig. 1-3 does not have regions. Thus a region is a property of the specific plane representation of a graph and not of an abstract graph per se.

Infinite Region: The portion of the plane lying outside a graph embedded in a plane, such as region 4 in Fig. 5-4, is infinite in its extent. Such a region is called the *infinite, unbounded, outer,* or *exterior* region for that particular plane representation. Like other regions, the infinite region is also characterized by a set of edges (or vertices). Clearly, by changing the embedding of a given planar graph, we can change the infinite region. For instance, Figs. 5-1(d) and 5-3 are two different embeddings of the same graph. The finite region $v_1 \, v_3 \, v_5$ in Fig. 5-1(d) becomes the infinite region in Fig. 5-3. In fact, we shall shortly show that any region can be made the infinite region by proper embedding.

Embedding on a Sphere: To eliminate the distinction between finite and infinite regions, a planar graph is often embedded in the surface of a sphere. It is accomplished by stereographic projection of a sphere on a plane. Put the sphere on the plane and call the point of contact SP (south pole). At point SP, draw a straight line perpendicular to the plane, and let the point where this line intersects the surface of the sphere be called NP (north pole). See Fig. 5-5.

Now, corresponding to any point p on the plane, there exists a unique point p' on the sphere and vice versa, where p' is the point at which the straight line from point p to point NP intersects the surface of the sphere. Thus there is a one-to-one correspondence between the points of the sphere and the finite points on the plane, and points at infinity in the plane correspond to the point NP on the sphere.

From this construction, it is clear that any graph that can be embedded in

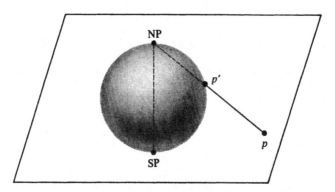

Fig. 5-5 Stereographic projection.

a plane (i.e., drawn on a plane such that its edges do not intersect) can also be embedded in the surface of the sphere, and vice versa. Hence

THEOREM 5-4

A graph can be embedded in the surface of a sphere if and only if it can be embedded in a plane.

A planar graph embedded in the surface of a sphere divides the surface into different regions. Each region on the sphere is finite, the infinite region on the plane having been mapped onto the region containing the point NP. Now it is clear that by suitably rotating the sphere we can make any specified region map onto the infinite region on the plane. From this we obtain

THEOREM 5-5

A planar graph may be embedded in a plane such that any specified region (i.e., specified by the edges forming it) can be made the infinite region.

Thinking in terms of the regions on the sphere, we see that there is no real difference between the infinite region and the finite regions on the plane. Therefore, when we talk of the regions in a plane regresentation of a graph, we include the infinite region. Also, since there is no essential difference between an embedding of a planar graph on a plane or on a sphere (a plane may be regarded as the surface of a sphere of infinitely large radius), the term "plane representation" of a graph is often used to include spherical as well as planar embedding.

Euler's Formula: Since a planar graph may have different plane representations, we may ask if the number of regions resulting from each embedding is the same. The answer is *yes*. Theorem 5-6, known as Euler's formula, gives the number of regions in any planar graph.

Theorem 5-6

A connected planar graph with n vertices and e edges has $e - n + 2$ regions.

Proof: It will suffice to prove the theorem for a simple graph, because adding a self-loop or a parallel edge simply adds one region to the graph and simultaneously increases the value of e by one. We can also disregard (i.e., remove) all edges that do not form boundaries of any region. Three such edges are shown in Fig. 5-4. Addition (or removal) of any such edge increases (or decreases) e by one and increases (or decreases) n by one, keeping the quantity $e - n$ unaltered.

Since any simple planar graph can have a plane representation such that each edge is a straight line (Theorem 5-3), any planar graph can be drawn such that each region is a polygon (a polygonal net). Let the polygonal net representing the given graph consist of f regions or faces, and let k_p be the number of p-sided regions. Since each edge is on the boundary of exactly two regions,

$$3 \cdot k_3 + 4 \cdot k_4 + 5 \cdot k_5 + \cdots + r \cdot k_r = 2 \cdot e, \tag{5-1}$$

where k_r is the number of polygons, with maximum edges.
Also,

$$k_3 + k_4 + k_5 + \cdots + k_r = f. \tag{5-2}$$

The sum of all angles subtended at each vertex in the polygonal net is

$$2\pi n. \tag{5-3}$$

Recalling that the sum of all interior angles of a p-sided polygon is $\pi(p - 2)$, and the sum of the exterior angles is $\pi(p + 2)$, let us compute the expression in (5-3) as the grand sum of all interior angles of $f - 1$ finite regions plus the sum of the exterior angles of the polygon defining the infinite region. This sum is

$$\pi(3 - 2) \cdot k_3 + \pi(4 - 2) \cdot k_4 + \cdots + \pi(r - 2) \cdot k_r + 4\pi$$
$$= \pi(2e - 2f) + 4\pi. \tag{5-4}$$

Equating (5-4) to (5-3), we get

$$2\pi(e - f) + 4\pi = 2\pi n,$$
or
$$e - f + 2 = n.$$

Therefore, the number of regions is

$$f = e - n + 2. \quad \blacksquare$$

Corollary

In any simple, connected planar graph with f regions, n vertices, and e edges ($e > 2$), the following inequalities must hold:

$$e \geq \frac{3}{2} f, \tag{5-5}$$

$$e \leq 3n - 6. \tag{5-6}$$

Proof: Since each region is bounded by at least three edges and each edge belongs to exactly two regions,

$$2e \geq 3f$$

or

$$e \geq \frac{3}{2}f.$$

Substituting for f from Euler's formula in inequality (5-5),

$$e \geq \frac{3}{2}(e - n + 2)$$

or

$$e \leq 3n - 6. \quad \blacksquare$$

Inequality (5-6) is often useful in finding out if a graph is nonplanar. For example, in the case of K_5, the complete graph of five vertices [Fig. 5-1(e)],

$$n = 5, \quad e = 10, \quad 3n - 6 = 9 < e.$$

Thus the graph violates inequality (5-6), and hence it is not planar.

Incidentally, this is an alternative and independent proof of the nonplanarity of Kuratowski's first graph, as promised in Section 5-3.

The reader must be warned that inequality (5-6) is only a necessary, but *not* a sufficient, condition for the planarity of a graph. In other words, although every simple planar graph must satisfy (5-6), the mere satisfaction of this inequality does not guarantee the planarity of a graph. For example, Kuratowski's second graph, $K_{3,3}$, satisfies (5-6), because

$$e = 9,$$
$$3n - 6 = 3 \cdot 6 - 6 = 12.$$

Yet the graph is nonplanar.

To prove the nonplanarity of Kuratowski's second graph, we make use of the additional fact that no region in this graph can be bounded with fewer than four edges. Hence, if this graph were planar, we would have

$$2e \geq 4f,$$

and, substituting for f from Euler's formula,

$$2e \geq 4(e - n + 2),$$

or

$$2 \cdot 9 \geq 4(9 - 6 + 2),$$

or

$$18 \geq 20, \quad \text{a contradiction.}$$

Hence the graph cannot be planar.

Plane Representation and Connectivity: In a disconnected graph the embedding of each component can be considered independently. Therefore, it is clear that a disconnected graph is planar if and only if each of its components is planar. Similarly, in a separable (or 1-connected) graph the embedding of each block (i.e., maximal nonseparable subgraph) can be considered independently. Hence a separable graph is planar if and only if each of its blocks is planar.

Therefore, in questions of embedding or planarity, one need consider only nonseparable graphs.

Does a nonseparable planar graph G have a unique embedding on a sphere? Before answering this question, we must define the meaning of *unique embedding*. Two embeddings of a planar graph on spheres are *not* distinct if the embeddings can be made to coincide by suitably rotating one sphere with respect to the other and possibly distorting regions (without letting a vertex cross an edge). If of all possible embeddings on a sphere no two are distinct, the graph is said to have a *unique embedding* on a sphere (or a unique plane representation).

For example, consider two embeddings of the same graph in Fig. 5-6. The embedding (b) has a region bounded with five edges, but embedding (a) has no region with five edges. Thus, rotating the two spheres on which (a) and (b) are embedded will not make them coincide. Hence the two embeddings are distinct, and the graph has no unique plane representation.

On the other hand, the embeddings in Figs. 5-1(d) and 5-3, when considered on a sphere, can be made to coincide. (Remember that edges can be bent, and in a spherical embedding there is no infinite region.) Theorem 5-7, due to Whitney, tells us exactly when a graph is uniquely embeddable in a sphere. For a proof of the theorem, the reader is referred to [5-9].

THEOREM 5-7

The spherical embedding of every planar 3-connected graph is unique.

This theorem plays a very important role in determining if a graph is

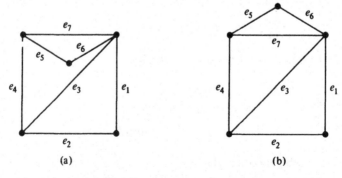

Fig. 5-6 Two distinct plane representations of the same graph.

5-5. DETECTION OF PLANARITY

How to tell if a given graph G is planar or nonplanar is an important problem, and "find out by drawing it" is obviously not a good answer. We must have some simple and efficient criterion. Toward that goal, we take the following simplifying steps:

Elementary Reduction

Step 1: Since a disconnected graph is planar if and only if each of its components is planar, we need consider only one component at a time. Also, a separable graph is planar if and only if each of its blocks is planar. Therefore, for the given arbitrary graph G, determine the set

$$G = \{G_1, G_2, \ldots, G_k\},$$

where each G_i is a nonseparable block of G. Then we have to test each G_i for planarity.

Step 2: Since addition or removal of self-loops does not affect planarity, remove all self-loops.

Step 3: Since parallel edges also do not affect planarity, eliminate edges in parallel by removing all but one edge between every pair of vertices.

Step 4: Elimination of a vertex of degree two by merging two edges in series† does not affect planarity. Therefore, eliminate all edges in series.

Repeated application of steps 3 and 4 will usually reduce a graph drastically. For example, Fig. 5-7 illustrates the series-parallel reduction of the graph of Fig. 5-6(b).

Let the nonseparable connected graph G_i be reduced to a new graph H_i after the repeated application of steps 3 and 4. What will graph H_i look like? Theorem 5-8 has the answer.

THEOREM 5-8

Graph H_i is

1. A single edge, or
2. A complete graph of four vertices, or
3. A nonseparable, simple graph with $n \geq 5$ and $e \geq 7$.

†In a graph, two edges are said to be in series if they have exactly one vertex in common and if this vertex is of degree two. Edges e_5 and e_6 (and also e_1 and e_2) are in series in Fig. 5-6.

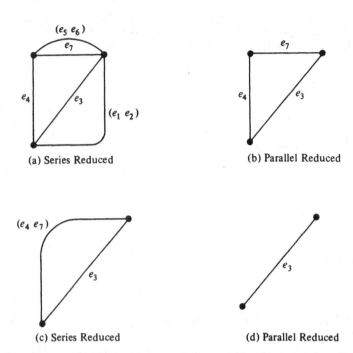

Fig. 5-7 Series-parallel reduction of the graph in Fig. 5-6(b).

Proof: The theorem can be proved by considering all connected nonseparable graphs of six edges or less. The proof is left as an exercise (Problem 5-9).

In Theorem 5-8, all H_i falling in categories 1 or 2 are planar and need not be checked further.

From now on, therefore, we need to investigate only *simple, connected, nonseparable graphs of at least five vertices and with every vertex of degree three or more*. Next, we can check to see if $e \leq 3n - 6$. If this inequality is not satisfied, the graph H_i is nonplanar. If the inequality is satisfied, we have to test the graph further and, with this, we come to Kuratowski's theorem (Theorem 5-9), perhaps the most important result of this chapter.

Homeomorphic Graphs: Two graphs are said to be *homeomorphic* if one graph can be obtained from the other by the creation of edges in series (i.e., by insertion of vertices of degree two) or by the merger of edges in series. The three graphs in Fig. 5-8 are homeomorphic to each other, for instance. A graph G is planar if and only if every graph that is homeomorphic to G is planar. (This is a restatement of series reduction, step 4 in this section.)

THEOREM 5-9

A necessary and sufficient condition for a graph G to be planar is that G does not contain either of Kuratowski's two graphs or any graph homeomorphic to either of them.

Fig. 5-8 Three graphs homeomorphic to each other.

Proof: The necessary condition is clear, because a graph G cannot be embedded in a plane if G has a subgraph that cannot be embedded. That this condition is also sufficient is surprising, and its proof is involved. Several different proofs of the theorem have appeared since Kuratowski stated and proved it in 1930. For a complete proof of the theorem, the reader is referred to Harary [1-5], pages 108–112, Berge [1-1], pages 211–213, or Busacker and Saaty [1-2], pages 70–73.

Note that it is *not* necessary for a nonplanar graph to have either of the Kuratowski graphs as a subgraph, as this theorem is sometimes misstated. The nonplanar graph may have a subgraph homeomorphic to a Kuratowski graph. For example, the graph in Fig. 5-9(a) is nonplanar, and yet it does not have either of the Kuratowski graphs as a subgraph. However, if we remove

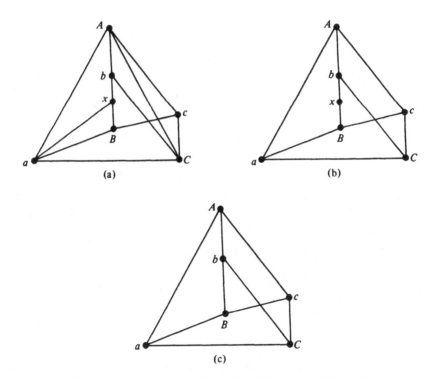

Fig. 5-9 Nonplanar graph with a subgraph homeomorphic to $K_{3,3}$.

edges (a, x) and (A, C) from this graph, we get a subgraph, as shown in Fig. 5-9(b). This subgraph is homeomorphic (merge two series edges at vertex x) to the one shown in Fig. 5-9(c). The graph of Fig. 5-9(c) clearly is isomorphic to $K_{3,3}$, Kuratowski's second graph, and this demonstrates the nonplanarity of the graph in Fig. 5-9(a).

The example just discussed also points out that although Theorem 5-9 (Kuratowski's theorem) gives an elegant and simple-looking criterion for planarity of a graph, the theorem is difficult to apply in the actual testing of a large graph (say, a simple, nonseparable graph of 25 vertices, each of degree three or more). There have been several alternative characterizations of a planar graph. One of these characterizations, the existence of a dual graph, is the subject of the next two sections.

5-6. GEOMETRIC DUAL

Consider the plane representation of a graph in Fig. 5-10(a), with six regions or faces F_1, F_2, F_3, F_4, F_5, and F_6. Let us place six points $p_1, p_2, \ldots, p_6$, one in each of the regions, as shown in Fig. 5-10(b). Next let us join these six points according to the following procedure:

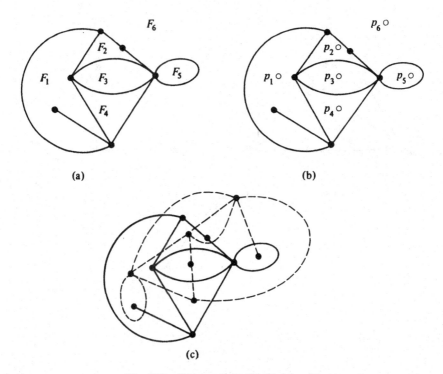

Fig. 5-10 Construction of a dual graph.

If two regions F_i and F_j are adjacent (i.e., have a common edge), draw a line joining points p_i and p_j that intersects the common edge between F_i and F_j exactly once. If there is more than one edge common between F_i and F_j, draw one line between points p_i and p_j for each of the common edges. For an edge e lying entirely in one region, say F_k, draw a self-loop at point p_k intersecting e exactly once.

By this procedure we obtain a new graph G^* [in broken lines in Fig. 5-10(c)] consisting of six vertices, $p_1, p_2, \ldots, p_6$ and of edges joining these vertices. Such a graph G^* is called a *dual* (or strictly speaking, a *geometric dual*) of G.

Clearly, there is a one-to-one correspondence between the edges of graph G and its dual G^*—one edge of G^* intersecting one edge of G. Some simple observations that can be made about the relationship between a planar graph G and its dual G^* are

1. An edge forming a self-loop in G yields a pendant edge† in G^*.
2. A pendant edge in G yields a self-loop in G^*.
3. Edges that are in series in G produce parallel edges in G^*.
4. Parallel edges in G produce edges in series in G^*.
5. Remarks 1–4 are the result of the general observation that the number of edges constituting the boundary of a region F_i in G is equal to the degree of the corresponding vertex p_i in G^*, and vice versa.
6. Graph G^* is also embedded in the plane and is therefore planar.
7. Considering the process of drawing a dual G^* from G, it is evident that G is a dual of G^* [see Fig. 5-10(c)]. Therefore, instead of calling G^* a dual of G, we usually say that G and G^* are dual graphs.
8. If n, e, f, r, and μ denote as usual the numbers of vertices, edges, regions, rank, and nullity of a connected planar graph G, and if n^*, e^*, f^*, r^*, and μ^* are the corresponding numbers in dual graph G^*, then

$$n^* = f,$$
$$e^* = e,$$
$$f^* = n.$$

Using the above relationship, one can immediately get

$$r^* = \mu,$$
$$\mu^* = r.$$

Uniqueness of Dual Graphs: Is a (geometric) dual of a graph unique? In

†An edge incident on a pendant vertex is called a *pendant edge*.

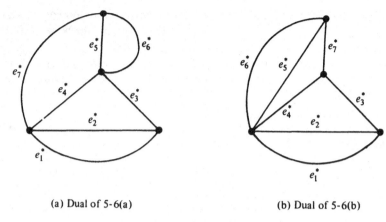

(a) Dual of 5-6(a) (b) Dual of 5-6(b)

Fig. 5-11 Duals of graphs in Fig. 5-6.

other words, are all duals of a given graph isomorphic? From the method of constructing a dual, it is reasonable to expect that a planar graph G will have a unique dual if and only if it has a unique plane representation or unique embedding on a sphere.

For instance, in Fig. 5-6 the same graph (isomorphic) had two distinct embeddings, (a) and (b). Consequently, the duals of these isomorphic graphs are nonisomorphic, as shown in Fig. 5-11.

The graphs in Fig. 5-11, however, are 2-isomorphic. Theorem 5-10, stated without proof, is a generalization of this example.

THEOREM 5-10

All duals of a planar graph G are 2-isomorphic; and every graph 2-isomorphic to a dual of G is also a dual of G.

With this qualification in mind, it is quite appropriate to refer to *a dual* as *the dual* of a planar graph.

Since a 3-connected planar graph has a unique embedding on a sphere, its dual must also be unique. In other words, all duals of a 3-connected graph are isomorphic.

5-7. COMBINATORIAL DUAL

So far we have defined and discussed duality of planar graphs in a purely geometric sense. The following provides us with an equivalent definition of duality independent of geometric notions.

THEOREM 5-11

A necessary and sufficient condition for two planar graphs G_1 and G_2 to be duals of each other is as follows: There is a one-to-one correspondence between the edges in G_1 and the edges in G_2 such that a set of edges in G_1 forms a circuit if and only if the corresponding set in G_2 forms a cut-set.

Proof: Let us consider a plane representation of a planar graph G. Let us also draw (geometrically) a dual G^* of G. Then consider an arbitrary circuit Γ in G. Clearly, Γ will form some closed simple curve in the plane representation of G—dividing the plane into two areas. (Jordan Curve Theorem). Thus the vertices of G^* are partitioned into two nonempty, mutually exclusive subsets—one inside Γ and the other outside. In other words, the set of edges Γ^* in G^* corresponding to the set Γ in G is a cut-set in G^*. (No proper subset of Γ^* will be a cut-set in G^*; why?). Likewise it is apparent that corresponding to a cut-set S^* in G^* there is a unique circuit consisting of the corresponding edge-set S in G such that S is a circuit. This proves the necessity portion of Theorem 5-11.

To prove the sufficiency, let G be a planar graph and let G' be a graph for which there is a one-to-one correspondence between the cut-sets of G and circuits of G', and vice versa. Let G^* be a dual graph of G. There is a one-to-one correspondence between the circuits of G' and cut-sets of G, and also between the cut-sets of G and circuits of G^*. Therefore there is a one-to-one correspondence between the circuits of G' and G^*, implying that G' and G^* are 2-isomorphic (Theorem 4-15). According to Theorem 5-10, G' must be a dual of G. ∎

Dual of a Subgraph: Let G be a planar graph and G^* be its dual. Let a be an edge in G, and the corresponding edge in G^* be a^*. Suppose that we delete edge a from G and then try to find the dual of $G - a$. If edge a was on the boundary of two regions, removal of a would merge these two regions into one. Thus the dual $(G - a)^*$ can be obtained from G^* by deleting the corresponding edge a^* and then fusing the two end vertices of a^* in $G^* - a^*$. On the other hand, if edge a is not on the boundary, a^* forms a self-loop. In that case $G^* - a^*$ is the same as $(G - a)^*$. Thus if a graph G has a dual G^*, the dual of any subgraph of G can be obtained by successive application of this procedure.

Dual of a Homeomorphic Graph: Let G be a planar graph and G^* be its dual. Let a be an edge in G, and the corresponding edge in G^* be a^*. Suppose that we create an additional vertex in G by introducing a vertex of degree two in edge a (i.e., a now becomes two edges in series). How will this addition affect the dual? It will simply add an edge parallel to a^* in G^*. Likewise, the reverse process of merging two edges in series (step 4 in Section 5-5) will simply eliminate one of the corresponding parallel edges in G^*. Thus if a graph G has a dual G^*, the dual of any graph homeomorphic to G can be obtained from G^* by the above procedure.

So far we have been studying duality for planar graphs only. This was forced upon us because the very definition of duality depended on the graph being embedded in a plane. However, now that Theorem 5-11 provides us with an equivalent abstract definition of duality (namely, the correspondence between circuits and cut-sets), which does not depend on a plane representation of a graph, we will see if the concept of duality can be extended to nonplanar graphs also. In other words, given a nonplanar graph G, can we find another graph G' with one-to-one correspondence between their edges such that every circuit in G corresponds to a unique cut-set in G', and vice versa? The answer to this question is *no*, as shown in the following important theorem, due to Whitney.

THEOREM 5-12

A graph has a dual if and only if it is planar.

Proof: We need prove just the "only if" part. That is, we have only to prove that a nonplanar graph does not have a dual. Let G be a nonplanar graph. Then according to Kuratowski's theorem, G contains K_5 or $K_{3,3}$ or a graph homeomorphic to either of these. We have already seen that a graph G can have a dual only if every subgraph g of G and every graph homeomorphic to g has a dual. Thus if we can show that neither K_5 nor $K_{3,3}$ has a dual, we have proved the theorem. This we shall prove by contradiction as follows:

(a) Suppose that $K_{3,3}$ has a dual D. Observe that the cut-sets in $K_{3,3}$ correspond to circuits in D and vice versa (Theorem 5-10). Since $K_{3,3}$ has no cut-set consisting of two edges, D has no circuit consisting of two edges. That is, D contains no pair of parallel edges. Since every circuit in $K_{3,3}$ is of length four or six, D has no cut-set with less than four edges. Therefore, the degree of every vertex in D is at least four. As D has no parallel edges and the degree of every vertex is at least four, D must have at least five vertices each of degree four or more. That is, D must have at least $(5 \times 4)/2 = 10$ edges. This is a contradiction, because $K_{3,3}$ has nine edges and so must its dual. Thus $K_{3,3}$ cannot have a dual. Likewise,

(b) Suppose that the graph K_5 has a dual H. Note that K_5 has (1) 10 edges, (2) no pair of parallel edges, (3) no cut-set with two edges, and (4) cut-sets with only four or six edges. Consequently, graph H must have (1) 10 edges, (2) no vertex with degree less than three, (3) no pair of parallel edges, and (4) circuits of length four and six only. Now graph H contains a hexagon (a circuit of length six), and no more than three edges can be added to a hexagon without creating a circuit of length three or a pair of parallel edges [see Fig. 5-2(b)]. Since both of these are forbidden in H and H has 10 edges, there must be at least seven vertices in H. The degree of each of these vertices is at least three. This leads to H having at least 11 edges. A contradiction. ∎

This proof of theorem 5-12 is not the one originally given by Whitney. Whitney's proof, though more rigorous, is much more involved. Our proof is based on one given by Parson [5-7].

There is yet another equivalent combinatorial definition of duality, also given by Whitney and proved equivalent to the earlier two definitions [5-10].

Two planar graphs G and G^* are said to be duals (or *combinatorial duals*) of each other if there is a one-to-one correspondence between the edges of G and G^* such that if g is any subgraph of G and h is the corresponding subgraph of G^*, then

$$\text{rank of } (G^* - h) = \text{rank of } G^* - \text{nullity of } g. \tag{5-7}$$

This relationship is shown diagrammatically in Fig. 5-12.

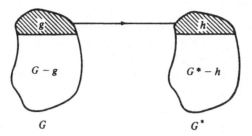

Rank of $(G^* - h)$ = Rank of G^* − Nullity of g

Fig. 5-12 Combinatorial duals.

As an example, consider the graph in Fig. 5-6(a) and its dual in Fig. 5-11(a). Take the subgraph $\{e_4, e_5, e_6, e_7\}$ in Fig. 5-6(a) and the corresponding subgraph $\{e_4^*, e_5^*, e_6^*, e_7^*\}$ in Fig. 5-11(a).

$$\text{rank of } (G^* - \{e_4^*, e_5^*, e_6^*, e_7^*\}) = \text{rank of } \{e_1^*, e_2^*, e_3^*\} = 2,$$
$$\text{rank of } G^* = 3,$$
$$\text{nullity of } \{e_4, e_5, e_6, e_7\} = 1,$$

and

$$2 = 3 - 1.$$

Clearly, this definition is also independent of the geometric connotation. It is therefore often preferred for proving results in purely algebraic fashion. However, in deciding whether or not two given graphs are dual the combinatorial definitions are difficult to use.

The proof of equivalence of combinatorial and geometric duals is quite involved. The interested reader is referred to the original papers of Whitney [5-10, 5-12] or to Seshu and Reed [1-13], pages 45–50. Since the geometric and combinatorial duals are one and the same, we simply refer to them as the dual, rather than the geometric or combinatorial dual.

Self-Dual Graphs: If a planar graph G is isomorphic to its own dual, it is called a *self-dual* graph. It can be easily shown that the four-vertex complete graph is a self-dual graph (Problem 5-20). Self-dual graphs have interesting properties and pose some unsolved problems.

5-8. MORE ON CRITERIA OF PLANARITY

Theorems 5-9 (Kuratowski's theorem) and 5-12 (Whitney's theorem) provided us with two different and alternative ways of characterizing a planar graph. The third classic planarity criterion, due to MacLane [5-6], is given next.

Set of Basic Circuits: A set C of circuits in a graph is said to be *a complete set of basic circuits* if (i) every circuit in the graph can be expressed as a ring sum of some or all circuits in C, and (ii) no circuit in C can be expressed as a ring sum of others in C. The significance of complete sets of basic circuits will be clearer in Chapter 6, in relation to the vector space of a graph. It may, however, be mentioned here that whereas a set of fundamental circuits (as defined in Chapter 3 with respect to a spanning tree) always constitutes a complete set of basic circuits, the converse does not hold for all graphs (Problem 5-15).

In a planar graph a complete set of basic circuits has an additional property, which we will observe next.

In a plane representation of a planar, connected graph G the set of circuits forming the interior regions constitutes a complete set of basic circuits. For any circuit Γ in G can be expressed as the ring sum of the circuits defining the regions contained in Γ. Observe that every edge appears in at most two of these basic circuits. Thus for every planar graph G we can find a complete set of basic circuits such that no edge appears in more than two of these basic circuits. This result and its converse (proof of which can be found in [5-6]) lead to another well-known characterization of planar graphs.

THEOREM 5-13

A graph G is planar if and only if there exists a complete set of basic circuits (i.e., all μ of them, μ being the nullity of G) such that no edge appears in more than two of these circuits.

All three of these classic characterizations suffer from two shortcomings. First, they are extremely difficult to implement for a large graph. Second, in case the graph is planar they do not give a plane representation of the graph.

These drawbacks have prompted recent discoveries of several *map-construction* methods, where the testing of planarity itself is based on an attempt to produce a plane representation of the graph. One such method is given by Tutte [5-9]. Several other construction methods, some of them quite similar, have been implemented on digital computers [5-2, 5-8]. In most of these methods, the given graph is first reduced to one or more *simple, nonseparable graphs with every vertex of degree three or more* and with $e \leq 3n - 6$. Then the construction algorithm is applied such that either one succeeds in obtaining a planar realization of the graph or the graph is nonplanar. More will be said on such algorithms in Chapter 11.

Some algorithms are better than others, but all are laborious and time-consuming. The search for a simple, elegant, and practical characterization of a planar graph is far from over.

5-9. THICKNESS AND CROSSINGS

Having found that a given graph G is nonplanar, it is natural to ask, what is the minimum number of planes necessary for embedding G? The least number of planar subgraphs whose union is the given graph G is called the *thickness* of G. In a printed-circuit board, for instance, the number of insulation layers necessary is the thickness of the corresponding graph.

By definition, then, the thickness of a planar graph is one. The thickness of each of Kuratowski's graphs is clearly two. The reader can show, by sketching them, that the thickness of the complete graph of eight vertices is two, while the thickness of the complete graph of nine vertices is three (Problem 5-19). Although there are several results available on the thickness of special types of graphs [1-5, pages 120–121], the thickness of an arbitrary graph is in general, difficult to determine.

Another question one might ask about a nonplanar graph is: What is the fewest number of crossings (or intersections) necessary in order to "draw" the graph in a plane?

The crossing number of a planar graph is, by definition, zero, and of either of Kuratowski's graphs, it is one. The crossing numbers of only a few graphs have been determined. No formula exists to give the crossing number of an arbitrary graph.

SUMMARY

Can a given graph be placed in a plane without its edges crossing over? This is clearly a geometric question about the graph—an object that exists in two different worlds, purely combinatorial and purely geometric. To quote Harary [1-5], page 106, "one of the most fascinating areas of study... is the interplay between considering a graph as a combinatorial object and as a geometric figure."

On probing a bit further, we discovered that we needed to investigate only simple, nonseparable graphs which have no vertex of degree less than three. Moreover, we found that any graph with the number of edges $e > 3n - 6$ need not be investigated any further, because such a graph is nonplanar.

Three equivalent, but very different, planarity characterizations, those of Kuratowski, Whitney, and MacLane, were presented and their significance and drawbacks discussed. For graphs that are nonplanar, additional relevant properties, such as thickness and number of crossings, were defined and discussed. There are many unsolved problems in this field of study. Because of the current interest in such areas as automatic wiring of complex systems, technology of printed circuits, and design of large-scale integrated circuits, these geometrical properties of graphs are of practical importance.

The existence of a dual graph, in addition to being a condition equivalent to that of planarity, is important in its own right. The underlying structural relationship between dual graphs becomes very clear in terms of the vector space of the graph, a subject for the next chapter.

REFERENCES

Starting from Kuratowski's celebrated paper in 1930, a large number of papers on planarity of graphs have appeared. An excellent survey of this work, especially on characterization of planar graphs and practical methods of embedding, is given in [5-8], which also contains a bibliography of about 120 papers on the subject. Recommended readings from textbooks are Harary [1-5], Chapter 11, Seshu and Reed [1-13], Chapter 3, Ore [1-10], Chapter 8, Berge [1-1], Chapter 21, and Busacker and Saaty [1-2], Chapter 4. Bruno, Steiglitz, and Weinberg [5-2] give an efficient computer algorithm for testing of planarity, as does Shirey [5-8]. Works of Whitney [5-10, 5-11, 5-12], MacLane [5-6], Fary [5-4], and Tutte [5-9] have already been referred to earlier. More on self-dual graphs can be found in [5-1] and [5-3]. A thorough survey on thickness of graphs with relevant references is to be found in [5-5]. For more on computer algorithms for planarity testing, see Chapter 11.

5-1. BENEDICT, C. P., "On Self-Dualism in Graphs and Networks," Ph.D. Dissertation, University of Waterloo, Waterloo, Canada 1969.

5-2. BRUNO, J., K. STEIGLITZ, and L. WEINBERG, "A New Planarity Test Based on 3-Connectivity," *IEEE Trans. Circuit Theory*, Vol. CT-17, No. 2, May 1970, 197–206.

5-3. DEO, N., "Self-Dual Graphs and Digraphs," *Proc. Sixth Annual Allerton Conf. on Circuit and System Theory*, Oct. 1968, 832–840.

5-4. FARY, I., "On Straight Line Representation of Planar Graphs," *Acta Sci. Math. Szeged*, Vol. 11, 1948, 229–233.

5-5. HOBBS, A. M., "A Survey of Thickness," in *Recent Progress in Combinatorics* (W. T. Tuttle, ed.), Academic Press, Inc., New York, 1969, 255–264.

5-6. MACLANE, S., "A Combinatorial Condition for Planar Graphs," *Fund. Math.*, Vol. 28, 1937, 22–32.

5-7. PARSON, T. D., "On Planar Graphs," *Am. Math. Monthly*, Vol. 78, No. 2, 1971, 176–178.

5-8. SHIREY, R. W., "Implementation and Analysis of Efficient Graph Planarity Testing Algorithms," Ph.D. Dissertation, Computer Sciences, University of Wisconsin, Madison, Wisc., 1969.

5-9. TUTTE, W. T., "How to Draw a Graph," *Proc. London Math. Soc.*, Ser. 3, Vol. 13, 1963, 743–768.

5-10. WHITNEY, H., "Non-separable and Planar Graphs," *Trans. Am. Math. Soc.*, Vol. 34, 1932, 339–362.

5-11. WHITNEY, H., "A Set of Topological Invariants for Graphs," *Am. J. Math.*, Vol. 55, 1933, 231–235.

5-12. WHITNEY, H., "Planar Graphs," *Fund. Math.*, Vol. 21, 1933, 73–84.

PROBLEMS

5-1. Using geometric arguments similar to those used in proving Theorem 5-1, prove that Kuratowski's second graph is also nonplanar.

5-2. If every region of a simple planar graph (with n vertices and e edges) embedded in a plane is bounded by k edges, show that

$$e = \frac{k(n-2)}{k-2}.$$

5-3. A simple planar graph to which no edge can be added without destroying its planarity (while keeping the graph simple, of course) is called a *maximal planar graph*. Prove that every region in a maximal planar graph is a triangle.

5-4. Prove that a planar graph of n vertices ($n \geq 4$) has at least four vertices with degree five or less. This will also prove that there are no 6-connected planar graphs. (*Hint:* Use the result of Problem 5-3.)

5-5. A planar graph G is said to be *completely regular* if the degrees of all vertices of G are equal and every region is bounded by the same number of edges. The graphs in Figs. 2-20(a) and 2-21(b) are completely regular, for example. Show that there are only five possible simple completely regular planar graphs, excluding the trivial graphs with degree ≤ 2. Sketch them. (*Hint:* Use Euler's formula.)

5-6. Prove that an infinite pattern formed of a regular polygon repeating itself, such as those found in mosaics and tiled floors (see infinite graphs in Fig. 1-10), can consist of only three types of polygons—square, triangular, and hexagonal.

5-7. Redraw the graph in Fig. 5-4 such that region 2 becomes the infinite region.

5-8. Using Kuratowski's theorem, show that the graphs in Fig. 2-3 (known as Petersen's graph) are nonplanar.

5-9. By sketching all (don't panic, their number is small) simple, nonseparable graphs with $n \leq 4$ and $e \leq 6$, prove Theorem 5-8.

5-10. Draw the geometric dual of the graph in Fig. 5-4.

5-11. Show by actual construction that the geometric dual of the two (2-isomorphic) graphs in Figs. 4-11(a) and (d) are isomorphic.

5-12. Construct an example to demonstrate that G^{**}, the dual of a dual of a graph G, may not be isomorphic to G, but is 2-isomorphic to it.

5-13. Prove that the geometric dual of a self-loop-free nonseparable planar graph is also nonseparable.

5-14. Prove that a self-loop-free planar graph is 2-connected if and only if its dual is also 2-connected.

5-15. Give an example of a graph which has at least one complete set of basic circuits not constituting a set of fundamental circuits (with respect to any spanning tree).

5-16. Show that the edges forming a spanning tree in a planar graph G correspond to the edges forming a set of chords in the dual G^*.

5-17. Show that a set of fundamental circuits in a planar graph G corresponds to a set of fundamental cut-sets in its dual G^*.

5-18. Determine the number of crossings and the thickness of the graph in Fig. 2-3.

5-19. Show, by sketching, that the thickness of the eight-vertex complete graph is two, whereas that of the nine-vertex complete graph is three.

5-20. Show that the complete graph of four vertices is self-dual. Give another example of a self-dual graph.

6 VECTOR SPACES OF A GRAPH

Modern abstract algebra is a powerful tool in the theory as well as in the applications of graphs. It is essential for a thorough understanding of graphs and a must for those wishing to do research in the field. Moreover, since digital computers do not (at least internally) work on pictorial graphs, it is necessary to represent a graph algebraically and to manipulate it algebraically, if one wishes to enlist the aid of a computer in solving graph-theory problems.

6-1. SETS WITH ONE OPERATION

Set: A *set* is a collection of objects (called the *elements* of the set).

Note that there is no specification on the nature of the elements or the number of elements. Nor do the elements have anything to do with each other, except belong to the same set. Braces are used to enclose the elements of a set. For instance, a set S consisting of five objects a, b, c, x, and y may be written as $S = \{a, b, c, x, y\}$. Since the order in which these elements appear is of no significance, we could have written the same set as $S = \{x, b, a, y, c\}$, for instance. The symbol $a \in S$ is used to indicate that element a is in set S.

A *subset* S' of a set S is a collection of some of the elements of S. If S has at least one element that is not in S', then S' is called a *proper subset* of S. The empty set or null set, written $\emptyset$, has no element in it and is considered a subset of every set. The two most common combinations of sets are the union $\cup$ and intersection $\cap$, defined as

$S_1 \cup S_2 = S_3$, a set containing all the elements of S_1 and S_2,

$S_1 \cap S_2 = S_4$, a set containing only those elements that are both in S_1 and in S_2.

In this chapter we shall be concerned with the combination of two elements within a set rather than the combination of two different sets.

Operation: Let us introduce a rule of combination called *binary operation* (also called *binary composition, law of composition*, or *internal law of composition*) between two elements of a set. Addition, multiplication, subtraction, and division are some of the familiar binary operations between two elements in a set of numbers. To keep the binary operation general enough, we shall use the symbol * (rather than using $+$, $-$, $\times$, $\div$, etc.) to denote the binary operation. A set with operations defined on it is called an *algebraic system* or just *algebra*.

Special Types of Algebras: Now we have a set, say $S = \{a, b, c, \ldots\}$, and a binary operation * (written as $a * b$) between the elements of S. Depending on the nature of the binary operation *, set S can be classified as one of several special types of algebras. For instance, if * satisfies postulates 1 and 2 below, set S is called a *semigroup*:

1. *Closure:* If a and b are in S, then $a * b$ is also in S.

2. *Associative:* If the elements a, b, and c are in S, then $(a * b) * c = a * (b * c)$.

Semigroups have many interesting properties and have been studied in great detail. In fact, there are several thick books written on the theory of semigroups. But since semigroups as such are not applicable to the business at hand, we shall move on to more specialized semigroups.

A semigroup that satisfies postulate 3, below, is called a *monoid*.

3. *Identity element:* There exists a unique element e in S such that for any element x in S, $x * e = e * x = x$.

A monoid that satisfies postulate 4, below, is called a *group*.

4. *Inverse:* For every element x in S there exists a unique element x' in S such that $x * x' = x' * x = e$. Element x' is called the inverse of x, with respect to operation *.

A semigroup that satisfies postulate 5, below, is called an *abelian semigroup* or *commutative semigroup*.

5. *Commutative:* If a and b are in S, then $a * b = b * a$.

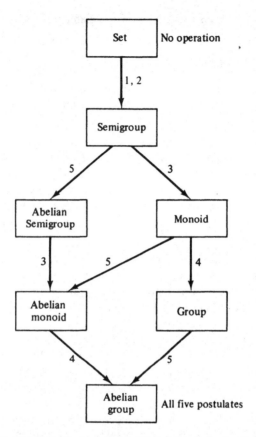

Fig. 6-1 Algebraic systems with one internal operation.

If an abelian semigroup also has an identity element, it is called an *abelian monoid* (or an *abelian semigroup with identity element*).

A set S with an operation * that satisfies all these five postulates is called an *abelian group* (or a *commutative group*).

Figure 6-1 summarizes the definitions of these "algebraic systems" and shows the relationships among them. The arrows point toward the direction of increasing restriction on the set S. The number next to a line indicates the particular postulate that converts one algebraic system into another.

It ought to be mentioned here that an algebraic system in which the binary operation does not satisfy even the closure and associative rules is of little mathematical interest. Another observation that may be made is that postulate 4 cannot be satisfied before 3. In view of these two remarks, Fig. 6-1 does show all possible combinations of the five postulates.

Examples: Some examples are in order now.

Consider the set of all positive integers, $S_1 = \{1, 2, 3, \ldots\}$. Set S_1 satisfies closure and associative rules if the binary operation ∗ is the ordinary addition operation +. Moreover, it also satisfies the *commutative* requirement. Hence S_1 under addition is a *commutative semigroup*. Note that in S_1 there is no identity element (an element when added to any other element results in the latter element).

Consider the same set $S_1 = \{1, 2, 3, \ldots\}$ under the ordinary division operation ÷. Since S_1 contains no fractions, clearly S_1 does not satisfy the *closure* rule, and hence is *not* a semigroup.

Again, the same set S_1 under multiplication · is an *abelian monoid*, because it has an identity element, 1. The set S_1, however, is not a group under the multiplication operation because S_1 does not have the inverse of every element (because S_1 has no fractions).

The set of all integers $S_2 = \{\ldots, -3, -2, -1, 0, 1, 2, 3, \ldots\}$ is an abelian group under the addition operation (hence an *additive abelian group*).

The reader can verify (Problem 6-2) that the set consisting of the four fourth roots of unity, which is $\{1, -1, i, -i\}$ (where $i = \sqrt{-1}$), is an abelian group under the multiplication operation (therefore, a *multiplicative abelian group*).

Groups of Subgraphs: Now we shall show that sets of certain subgraphs of any given graph G satisfy the preceding postulates and thus form their groups. These are very fundamental and important results in graph theory.

THEOREM 6-1

The ring sum of two circuits in a graph G is either a circuit or an edge-disjoint union of circuits.

Proof: Let Γ_1 and Γ_2 be any two circuits in a graph G. If the two circuits have no edges or vertices in common, their ring sum $\Gamma_1 \oplus \Gamma_2$ is a disconnected subgraph of G, and is obviously an edge-disjoint union of circuits. If, on the other hand, Γ_1 and Γ_2 do have edges and/or vertices in common, we have the following possible situations:

Since the degree of every vertex in a graph that is a circuit is two, every vertex v in subgraph $\Gamma_1 \oplus \Gamma_2$ has degree $d(v)$, where

$d(v) = 2$ if v is in Γ_1 only, or in Γ_2 only; or if one of the edges formerly incident on v was in both Γ_1 and Γ_2; or

$d(v) = 4$ if Γ_1 and Γ_2 just intersect at v (without a common edge).

There is no other type of vertex in $\Gamma_1 \oplus \Gamma_2$. Thus $\Gamma_1 \oplus \Gamma_2$ is an Euler graph, and therefore consists of either a circuit or an edge-disjoint union of circuits (Theorem 2-6). ∎

It is immediate from Theorem 6-1 that the ring sum of any two edge-disjoint unions of circuits is also a circuit or another edge-disjoint union of circuits.

Theorem 6-2

The set consisting of all the circuits and the edge-disjoint unions of circuits (including the null set $\emptyset$) in a graph G is an abelian group under the ring-sum operation $\oplus$.

Proof: It is required to prove that this set under the operation $\oplus$ satisfies postulates 1–5 in this section. That the closure postulate is satisfied has just been proved in Theorem 6-1. Associative and commutative postulates are also clearly satisfied. The null graph serves as the identity element $\emptyset$, because $\emptyset \oplus g = g$, for any subgraph g of G. What about the inverse?

A circuit or an edge-disjoint union of circuits Γ is its own inverse, because

$$\Gamma \oplus \Gamma = \emptyset.$$

Hence the theorem. ∎

Theorem 6-3

The set consisting of all the cut-sets and the edge-disjoint unions of cut-sets (including the null set $\emptyset$) in a graph G is an abelian group under the ring sum operation.

Proof: It is follows from Theorem 4-4 that this set satisfies the closure axiom. Associativity and commutativity are also immediately apparent. And so is the existence of the identity element $\emptyset$. Just as in the case of circuits, a cut-set or an edge-disjoint union of cut-sets is its own inverse. Thus the theorem. ∎

6-2. SETS WITH TWO OPERATIONS

Now suppose that on the elements of an abelian group we impose another binary operation $\odot$, in addition to the operation $*$ imposed in Section 6-1. The five postulates on $\odot$ can be written as follows (note that these are the same postulates as in Section 6-1, but they are for a different binary operation $\odot$):

6. *Closure:* If a and b are in S, then $a \odot b$ is also in S.

7. *Associative:* If a, b, and c are in S, then $(a \odot b) \odot c = a \odot (b \odot c)$.

8. *Identity element:* There exists a unique element i in S such that for any element x in S, $x \odot i = i \odot x = x$. This element i is called the identity element (or unity) with respect to operation $\odot$.

9. *Inverse:* For every element (except for the identity element e of postulate 3 in Section 6-1) x in S, there exists a unique element x^{-1} in S such that $x \odot x^{-1} = x^{-1} \odot x = i$. Element x^{-1} is called the inverse of x, with respect to operation $\odot$.

10. *Commutative:* If a and b are in S, then $a \odot b = b \odot a$.

And to relate these two different binary operations, postulate 11 is introduced.

11. *Distributive:* The operation $\odot$ is distributive with respect to the operation $*$; that is, for elements a, b, and c in S

$$a \odot (b * c) = a \odot b * a \odot c,$$
and $$(b * c) \odot a = b \odot a * c \odot a.$$

Just as in Section 6-1, the different combinations of these postulates, in addition to postulates 1–5, will render different types of algebraic systems. These are

Ring: An abelian group with respect to $*$ that satisfies postulates 6, 7, and 11 is called a *ring*.

Ring with Unity: A ring that has a unity or identity element i with respect to the second operation $\odot$.

Commutative Ring: A ring that satisfies the commutative postualate (10) with respect to $\odot$.

Commutative Ring with Unity: A commutative ring that has an identity element (8) with respect to $\odot$.

Division Ring (or Skew Field or S-Field): A ring with unity that also satisfies the inverse postulate (9) with respect to $\odot$.

Field (sometimes called Commutative Field): A division ring that satisfies the commutative postulate (10) with respect to $\odot$. Thus a field satisfies all eleven postulates, and therefore may be regarded as the "strongest" algebraic system considered here.

The relationship among these algebraic systems is summarized in Fig. 6-2.

Examples: As mentioned in Section 6-1, the set of all integers

$$S_2 = \{\ldots, -3, -2, -1, 0, 1, 2, 3, \ldots\}$$

is an abelian group under $+$, the usual addition operation. Moreover, ordinary multiplication between elements of S_2 also satisfies the closure, associative, distributive, and commutative postulates, and there is a unity element, 1, in S_2. Thus S_2 is a commutative ring with unity. However, since S_2 does not contain fractions, it does not satisfy postulate 9, and hence S_2 is not a field.

The set of all *rational numbers* does satisfy postulate 9, in addition to the other ten satisfied by S_2. Therefore, the set of *all rational numbers* is a field under addition and multiplication. The set of *all real numbers* also forms a field under addition and multiplication. *All complex numbers* also form a field under the usual addition and multiplication.

In this book we shall mainly be concerned with groups and fields. The rest of the algebraic systems are defined simply for your general interest.

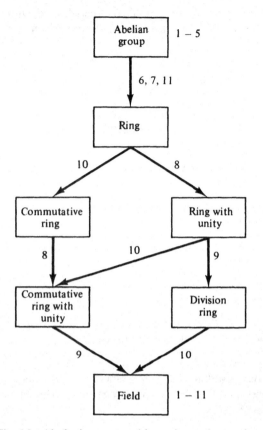

Fig. 6-2 Algebraic systems with two internal operations.

6-3. MODULAR ARITHMETIC AND GALOIS FIELDS

Consider a system of numbers that has only three numbers in it, ordinary 0, 1, and 2. And let the rules for addition and multiplication in this system be the same as ordinary addition and multiplication with the following exception: If a number q (resulting from addition or multiplication operations) equals or exceeds 3, it is to be divided by 3, the quotient is discarded, and the remainder is used in place of q. The addition and multiplication tables for such a number system are given in Fig. 6-3, and are called *addition modulo* 3 and *multiplication modulo* 3. Together they are called *modulo* 3 *arithmetic*. For example, in modulo 3 arithmetic,

$$1 + 1 + 2 \cdot 2 + 1 + 2 + 1 = 1 \quad (\text{mod } 3).$$

Similarly, we can define any modulo m arithmetic system consisting of m

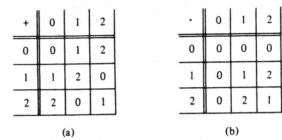

Fig. 6-3 Addition and multiplication tables for arithmetic modulo 3.

elements $0, 1, 2, \ldots, m-1$ and the relationship for any $q > m - 1$:

$$q = m \cdot p + r = r \quad (\bmod\ m) \quad \text{and} \quad r < m.$$

It is suggested that the reader write down arithmetic tables for $m = 4, 5, 6,$ and 7 (Problem 6-7).

Finite Fields: From the tables in Fig. 6-3, it can be verified that the set $\{0, 1, 2\}$ with addition and multiplication modulo 3 is a *field*. There is an identity 0 with respect to modulo 3 addition, and an identity 1 with respect to modulo 3 multiplication. Every element has a unique additive inverse, and every element other than 0 has a multiplicative inverse.

By means of actual tables, like those in Fig. 6-3, it can be easily verfied that modulo 2, 5, and 7 systems are also fields. On the other hand, the set $\{0, 1, 2, 3\}$ with modulo 4 addition and multiplication is not a field, because no inverse of 2 exists with respect to modulo 4 multiplication (Problem 6-8).

In fact, it turns out that every finite set

$$Z_m = \{0, 1, 2, \ldots, m - 1\}$$

with modulo m addition and multiplication is a field if and only if m is a *prime number*. Such a field is called a *Galois field modulo m, or GF(m)*.

As we shall see shortly, in representing graphs we are concerned only with GF(2), Galois field modulo 2. It consists of $\{0, 1\}$ and the addition modulo 2 and multiplication modulo 2 operations. The two arithmetic tables are given in Fig. 6-4. (Those familiar with computer logic will readily recognize that in

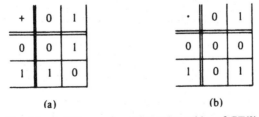

Fig. 6-4 Addition and multiplication tables of GF(2).

Fig. 6-4, + is the same as "EXCLUSIVE OR" and · is the same as "AND" of Boolean logic.)

6-4. VECTORS AND VECTOR SPACES

In an ordinary two-dimensional (Euclidean) plane, a point is represented by an ordered pair of numbers $X = (x_1, x_2)$. Point X can also be regarded as a vector emanating from the origin $0 = (0, 0)$ to the point (x_1, x_2). Similarly, in three-dimensional Euclidean space the triplet $(7, 2.1, -3)$ represents a vector. Sometimes, instead of row notation a column notation is used, for example,

$$\begin{pmatrix} 7 \\ 2.1 \\ -3 \end{pmatrix}$$

The three components 7, 2.1, and -3 in the example above are from the *field of real numbers*. Every point (of the infinitely many points) in E_3, the three-dimensional Euclidean space, corresponds to a unique ordered triplet (of the infinitely many triplets) consisting of three real numbers.

Now suppose that we are working with GF(2), the field of integers modulo 2. Then every number in a triplet can only be either 0 or 1. Thus there are only eight ($2^3 = 8$) vectors possible (instead of infinitely many as in the real number system) in a three-dimensional space if our numbers are restricted to GF(2). These are

(0, 0, 0), (1, 0, 0), (0, 1, 0), (0, 0, 1), (1, 1, 0), (1, 0, 1), (0, 1, 1), (1, 1, 1).

This concept of representing vectors can be extended to representation of a vector in *k-dimensional* space by means of an ordered *k-tuple*. For instance, the 7-tuple (0, 1, 1, 0, 1, 0, 1,) represents a vector in a *seven-dimensional vector space over the field GF(2)*.

The numbers in a field are sometimes called *scalars* (to distinguish them from vectors). The scalars in the field GF(2) are 0 and 1.

A *vector space*, in addition to being made up of *k*-tuples (from some specified field), must satisfy certain other conditions regarding combinations of two vectors, or operation of a vector with a scalar, and the like. These can be summarized in the following definition.

DEFINITION

A *k-dimensional vector space* (or a *linear vector space*) *over the field F*, is an object consisting of

1. A field F (with its set of elements S, and two operations $*$ and $\odot$).

2. A set W of k-tuples (all numbers taken from F).
3. A binary operation $\boxplus$ (called *vector sum*) between the elements of the set W, such that W is an abelian group under this operation $\boxplus$.
4. A binary operation $\boxdot$ (called *scalar multiplication*), which when applied between any scalar c in F and a vector $X = (x_1, x_2, \ldots, x_k)$ in W produces another vector P in W. P is called the scalar product of c and X, and is given by

$$P = c \boxdot X = (c \odot x_1, c \odot x_2, \ldots, c \odot x_k).$$

Furthermore, scalar multiplication satisfies the following:

$$c_1 \boxdot (c_2 \boxdot X) = (c_1 \odot c_2) \odot X, \quad \text{where } c_1, c_2 \in F,$$
$$c_1 \boxdot (X \boxplus Y) = c_1 \boxdot X \boxplus c_1 \boxdot Y,$$
$$(c_1 * c_2) \boxdot X = c_1 \boxdot X \boxplus c_2 \boxdot X,$$
$$1 \boxdot X = X, \quad \text{where 1 is the identity with respect to operation } \odot \text{ in } F.$$

Let us now leave the general vector space, and concern ourselves with the specific vector space associated with a graph G.

6-5. VECTOR SPACE ASSOCIATED WITH A GRAPH

Let us consider the graph G in Fig. 6-5 with four vertices and five edges e_1, e_2, e_3, e_4, e_5. Any subset of these five edges (i.e., any subgraph g) of G can be represented by a 5-tuple:

$$X = (x_1, x_2, x_3, x_4, x_5)$$

such that

$$x_i = 1 \quad \text{if } e_i \text{ is in } g \text{ and}$$
$$x_i = 0 \quad \text{if } e_i \text{ is } not \text{ in } g.$$

For instance, the subgraph g_1 in Fig. 6-5 will be represented by (1, 0, 1, 0, 1).

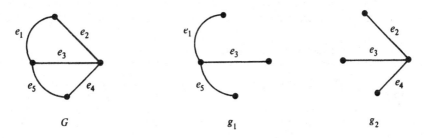

Fig. 6-5 Graph and two of its subgraphs.

Altogether there are 2^5 or 32 such 5-tuples possible, including the zero vector $0 = (0, 0, 0, 0, 0)$, which represents a null graph,† and $(1, 1, 1, 1, 1)$, which is G itself.

It is not difficult to see that the ring-sum operation between two subgraphs corresponds to the modulo 2 addition between the two 5-tuples representing the two subgraphs. For example, consider two subgraphs

$$g_1 = \{e_1, e_3, e_5\} \quad \text{represented by } (1, 0, 1, 0, 1), \text{ and}$$
$$g_2 = \{e_2, e_3, e_4\} \quad \text{represented by } (0, 1, 1, 1, 0).$$

The ring sum

$$g_1 \oplus g_2 = \{e_1, e_2, e_4, e_5\} \quad \text{represented by } (1, 1, 0, 1, 1),$$

which is clearly modulo 2 addition of the 5-tuples for g_1 and g_2.

Now, generalizing this example, we can make the most important observation of this chapter: There is a vector space W_G associated with every graph G, and this vector space consists of

1. Galois field modulo 2; that is, set $\{1, 0\}$ with operation *addition* modulo 2 written as $+$ such that $0 + 0 = 0$, $1 + 0 = 1 = 0 + 1$, $1 + 1 = 0$, and *multiplication* modulo 2 written as $\cdot$ such that $0 \cdot 0 = 0 = 1 \cdot 0 = 0 \cdot 1$, and $1 \cdot 1 = 1$.
2. 2^e vectors (e-tuples), where e is the number of edges in G.
3. An *addition operation* between two vectors X, Y in this space, defined as the vector sum‡

$$X \oplus Y = (x_1 + y_1, x_2 + y_2, \ldots, x_e + y_e),$$

$+$ being addition modulo 2.

4. And a *scalar multiplication* between a scalar c in Z_2 and a vector X, defined as $c \cdot X = (c \cdot x_1, \ldots, c \cdot x_e)$.

The reader can verify that the vector space W_G associated with a graph G, as defined above, does indeed satisfy all the requirements of a vector space

---

†In considering vector spaces of graphs, isolated vertices are of no consequence. Hence a null graph of four vertices is not distinguished from a null graph of 100 vertices.

‡The same symbol $\oplus$ has been used for the ring sum of two subgraphs, as well as for the vector sum between the two vectors representing the two subgraphs. This is done as much to eliminate an extra symbol as to remind the reader that a ring sum between two subgraphs amounts to the same thing as vector sum of the corresponding vectors. There will be no occasion for ambiguity.

(Problem 6-11). Note that the identity element (for the vector sum operation) in a vector space is 0, the zero vector.

6-6. BASIS VECTORS OF A GRAPH

Linear Dependence: A set of vectors $X_1, X_2, \ldots, X_r$ (over some field F) is said to be *linearly independent* if for scalars $c_1, c_2, \ldots, c_r$ in F the expression

$$c_1 X_1 + c_2 X_2 + \cdots + c_r X_r = 0$$

holds only if $c_1 = c_2 = \cdots = c_r = 0$. Otherwise, the set of vectors is said to be *linearly dependent*. For example, consider the set of three vectors, over the field of real numbers:

$$X_1 = \begin{pmatrix} 1 \\ 4 \\ 0 \end{pmatrix}, \quad X_2 = \begin{pmatrix} 0 \\ 1 \\ 2 \end{pmatrix}, \quad X_3 = \begin{pmatrix} 3 \\ 0 \\ 0 \end{pmatrix}.$$

An arbitrary *linear combination* of these three vectors set to zero gives

$$c_1 X_1 + c_2 X_2 + c_3 X_3 = \begin{pmatrix} c_1 \\ 4c_1 \\ 0 \end{pmatrix} + \begin{pmatrix} 0 \\ c_2 \\ 2c_2 \end{pmatrix} + \begin{pmatrix} 3c_3 \\ 0 \\ 0 \end{pmatrix} = \begin{pmatrix} c_1 + 3c_3 \\ 4c_1 + c_2 \\ 2c_2 \end{pmatrix} = \begin{pmatrix} 0 \\ 0 \\ 0 \end{pmatrix}.$$

That is, $2c_2 = 0$, $4c_1 + c_2 = 0$, and $c_1 + 3c_3 = 0$, which hold only if $c_1 = c_2 = c_3 = 0$. Thus the set of vectors $\{X_1, X_2, X_3\}$ is linearly independent.

On the other hand, consider another set of vectors (over the same field of real numbers):

$$X_4 = \begin{pmatrix} 0 \\ 2 \\ -2 \end{pmatrix}, \quad X_5 = \begin{pmatrix} 1 \\ 2 \\ 0 \end{pmatrix}, \quad X_6 = \begin{pmatrix} .5 \\ 0 \\ 1 \end{pmatrix}.$$

Setting an arbitrary linear combination of these vectors to zero,

$$c_4 X_4 + c_5 X_5 + c_6 X_6 = \begin{pmatrix} c_5 + .5c_6 \\ 2c_4 + 2c_5 \\ -2c_4 + c_6 \end{pmatrix} = \begin{pmatrix} 0 \\ 0 \\ 0 \end{pmatrix},$$

gives $c_4 = -c_5 = .5c_6 = \alpha$, where α can be any real number not necessarily zero. Therefore, the set $\{X_4, X_5, X_6\}$ is linearly dependent.

Basis Vectors: To the set of three linearly independent vectors $\{X_1, X_2, X_3\}$

in the first example, let us add another vector

$$Y = \begin{pmatrix} y_1 \\ y_2 \\ y_3 \end{pmatrix}.$$

Now you can show without much difficulty that the set $\{X_1, X_2, X_3, Y\}$ is linearly dependent regardless of what Y is. In other words, you can find a set of four real numbers a, b, c, and d (not all of which are zero) such that†

$$aX_1 + bX_2 + cX_3 + dY = 0. \tag{6-1}$$

Rewriting Eq. (6-1),

$$Y = -\frac{a}{d}X_1 - \frac{b}{d}X_2 - \frac{c}{d}X_3.$$

Thus a vector Y can be expressed as a linear combination of the vectors X_1, X_2, X_3. Such a set of k linearly independent vectors is called a *basis* (or the *coordinate system*) in the vector space. More formally:

If every vector in a vector space W can be expressed as a linear combination of a given set of vectors, this set is said to *span the vector space W*. The *dimension of the vector space W* is the minimal number of linearly independent vectors required to span W. Any set of k linearly independent vectors that spans W, a k-dimensional vector space, is called a *basis* for the vector space W.

For example, the following set of k unit vectors in a k-dimensional vector space is a basis. This is the most commonly used basis, and is often called the *natural* or *standard basis*.

$$\begin{pmatrix} 1 \\ 0 \\ 0 \\ \vdots \\ 0 \\ 0 \end{pmatrix}, \begin{pmatrix} 0 \\ 1 \\ 0 \\ \vdots \\ 0 \\ 0 \end{pmatrix}, \ldots, \begin{pmatrix} 0 \\ 0 \\ 0 \\ \vdots \\ 0 \\ 1 \end{pmatrix}.$$

It is clear that any vector in the k-dimensional vector space (over the field of real numbers) can be expressed as a linear combination of these k vectors.

Basis Vectors of a Graph: In Section 6-5 it was shown that there was a

†One possible solution (out of infinitely many) that satisfies Eq. (6-1) is

$$d = 1, \quad a = \left(\frac{y_3 - 2y_2}{8}\right), \quad b = -\left(\frac{y_3}{2}\right), \quad c = -\left(\frac{8y_1 - 2y_2 + y_3}{24}\right).$$

vector space W_G associated with every graph G. Corresponding to each subgraph of G there was a vector in W_G, represented by an e-tuple. The natural basis for this vector space W_G is a set of e linearly independent vectors, each representing a subgraph consisting of one edge of G. For instance, for the graph in Fig. 6-5, the set of the following five vectors serves as a basis for W_G.

$$(1, 0, 0, 0, 0),$$
$$(0, 1, 0, 0, 0),$$
$$(0, 0, 1, 0, 0),$$
$$(0, 0, 0, 1, 0),$$
$$(0, 0, 0, 0, 1).$$

Any of the possible 32 subgraphs (including G as well as the null graph) can be represented by a suitable (and unique) linear combination of these five basic vectors.

6-7. CIRCUIT AND CUT-SET SUBSPACES

A nonempty subset of vectors in a space is called a *subspace* if the subset satisfies the axioms of a vector space. To check whether a given subset of vectors is a subspace we have only to check for closure under scalar multiplication and vector addition. Since the scalar product of 0 and a vector X is the zero vector 0, the closure under scalar multiplication assures the presence of 0. Closure under scalar multiplication also assures the inverse of every vector [because the inverse of vector X is the vector $(-1) \cdot X$]. If the associative, commutative, and distributive axioms hold in the original space, they must also hold for every subset of vectors. Thus a subset of vectors closed under vector addition and multiplication by scalars is a subspace.

A vector space is trivially its own subspace. The null space, consisting of 0, is also a subspace. A Euclidean plane E_2 through the origin is a subspace of the three-dimensional Euclidean space E_3. A line E_1 through the origin is a subspace of both E_2 and E_3.

The *dimension of a subspace* is the number of linearly independent vectors required to span the subspace.

Subspaces in W_G

In the vector space W_G (over the Galois field modulo 2) associated with a graph G, let us consider the following two types of vectors: A *circuit vector* is a vector in W_G representing either a circuit or a union of edge-disjoint circuits in graph G. A *cut-set vector* is a vector in W_G representing either a cut-set or a union of edge-disjoint cut-sets in G.

We know that in the vector space W_G the linear combination of two vectors (which is simply modulo 2 addition of their components) corresponds to the ring sum of the corresponding subgraphs in G. From Theorem 6-2, the ring sum of two circuits (or unions of edge-disjoint circuits) is a circuit or a union of edge-disjoint circuits. Therefore, the linear combination of two circuit vectors is also a circuit vector. Hence

THEOREM 6-4

The set of all circuit vectors in W_G forms a subspace W_Γ.

Based on parallel arguments and on Theorem 6-3, we have an identical result for cut-set vectors.

THEOREM 6-5

The set of all cut-set vectors in W_G forms a subspace W_S.

Quite naturally, subspaces W_Γ and W_S are called the *circuit subspace* and *cut-set subspace*, respectively.

Bases of W_S and W_Γ

After having discovered that a particular set of vectors constitutes a subspace, the questions that one asks next are: What is the dimension of this subspace? How many vectors does the subspace contain? These questions about the subspaces W_Γ and W_S are answered by the following important results.

THEOREM 6-6

The set of circuit vectors corresponding to the set of fundamental circuits, with respect to any spanning tree, forms a basis for the circuit subspace W_Γ.

Proof: Consider a spanning tree, T, in a connected graph G, with $n - 1 = r$ tree branches and $e - n + 1 = \mu$ chords. Adding a chord c_1 to T produces a fundamental circuit, and the corresponding circuit vector can be included in the basis of W_Γ. Adding another chord c_2 to subgraph $T \cup c_1$ produces another fundamental circuit, with at least one edge that was not in the previous circuit. Therefore, the circuit vector representing the second fundamental circuit and the first circuit vector are linearly independent. Thus both these circuit vectors can be included in the basis. Adding a third chord to $T \cup c_1 \cup c_2$ will give another fundamental circuit with at least one edge not in either of the previous circuits. Therefore, this third circuit vector can also be included in the basis. Continuing with this argument, we see that all the μ vectors successively obtained this way are linearly independent, because each represents a circuit containing at least one edge not present in any of the previous ones. Therefore, these μ vectors, each corresponding to a fundamental circuit, are linearly independent.

SEC. 6-7   CIRCUIT AND CUT-SET SUBSPACES   127

Now we have to show that every circuit vector is a linear combination of these μ vectors.

Consider an arbitrary circuit Γ_1 in G, such that

$$\Gamma_1 = \{e_1, e_2, \ldots, e_i, e_{i+1}, \ldots, e_m\},$$

where edges $e_1, e_2, \ldots, e_i$ are chords with respect to T, and $e_{i+1}, e_{i+2}, \ldots, e_m$ are branches of T.

Let g be a subgraph obtained by taking the ring sum of the i fundamental circuits formed by the chords $e_1, e_2, \ldots,$ and e_i.

Because of Theorem 6-1, subgraph g must be a circuit or a union of edge-disjoint circuits. Assume $\Gamma_1 \neq g$. Then the subgraph $\Gamma_1 \oplus g$ must be either a circuit or a union of edge-disjoint circuits. But since both Γ_1 and g contain the chords $e_1, e_2, \ldots, e_i$ and no other chords, the subgraph $\Gamma_1 \oplus g$ will not contain any chord with respect to T. Hence $\Gamma_1 \oplus g$ has no circuit, a contradiction. So $\Gamma_1 = g$.

Thus we have shown that any circuit (and by extension a union of edge-disjoint circuits) in G can be expressed as a ring sum of some of the fundamental circuits with respect to T. The vectors corresponding to a set of fundamental circuits must therefore span W_Γ. ∎

As was brought out in Chapter 5, every set of fundamental circuits constitutes a basis in the circuit subspace W_Γ (i.e., forms a set of basic circuits), but every basis in the circuit subspace need not correspond to a set of fundamental circuits. (See Problems 5-15 and 6-18.)

COROLLARY

The dimension of the circuit subspace W_Γ is equal to the nullity μ of the graph, and the number of circuit vectors (including 0) in W_Γ is 2^μ.

Employing an argument parallel to that used in proving Theorem 6-6, it can be shown that the r cut-set vectors, each corresponding to a fundamental cut-set with respect to a spanning tree, are linearly independent.

Also, by a parallel argument it can be proved that any cut-set or a union of edge-disjoint cut-sets can be obtained by taking the ring sum of a subset of the r fundamental cut-sets with respect to a spanning tree. And thus we get a similar result for the cut-set subspace.

THEOREM 6-7

The set of cut-set vectors corresponding to the set of fundamental cut-sets, with respect to any spanning tree, forms a basis for the cut-set subspace W_S.

COROLLARY

The dimension of the cut-set subspace W_S is equal to the rank r of the graph, and the number of cut-set vectors (including 0) in W_S is 2^r.

Example: Let us now illustrate these results with an example.

For the graph G in Fig. 6-5

$$\text{number of edges, } e = 5,$$
$$\text{rank, } r = 3,$$
$$\text{nullity, } \mu = 2.$$

The number of vectors in the circuit subspace, therefore, is $2^2 = 4$, and these are

$$\begin{pmatrix}1\\1\\1\\0\\0\end{pmatrix}, \begin{pmatrix}0\\0\\1\\1\\1\end{pmatrix}, \begin{pmatrix}1\\1\\0\\1\\1\end{pmatrix}, \text{ and } \begin{pmatrix}0\\0\\0\\0\\0\end{pmatrix}.$$

$\underbrace{}_{\text{a basis of } W_r}$

The first two of these vectors correspond to the set of fundamental circuits with respect to either of the spanning trees in Fig. 6-5, and therefore they form a basis for W_r. (In fact, any two of the first three vectors form a basis of W_r.) The three subgraphs, each corresponding to a nonzero vector in W_r, are shown in Fig. 6-6.

The cut-set subspace W_s has a dimension of three, and therefore the number of vectors in W_s is $2^3 = 8$. These cut-set vectors are

$$\begin{pmatrix}1\\1\\0\\0\\0\end{pmatrix}, \begin{pmatrix}1\\0\\1\\0\\1\end{pmatrix}, \begin{pmatrix}0\\0\\0\\1\\1\end{pmatrix}, \begin{pmatrix}0\\1\\1\\1\\0\end{pmatrix}, \begin{pmatrix}1\\0\\1\\1\\0\end{pmatrix}, \begin{pmatrix}0\\1\\1\\0\\1\end{pmatrix}, \begin{pmatrix}1\\1\\0\\1\\1\end{pmatrix}, \begin{pmatrix}0\\0\\0\\0\\0\end{pmatrix}.$$

$\underbrace{}_{\text{a basis of } W_s}$

The first three vectors correspond to the three fundamental cut-sets with respect to the tree g_2 in Fig. 6-5. The rest of the vectors can easily be seen to be the vector sums of any two or three of these basis vectors. The seven

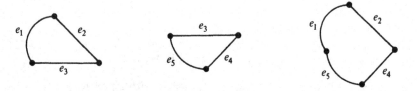

Fig. 6-6 Circuits in graph G of Fig. 6-5.

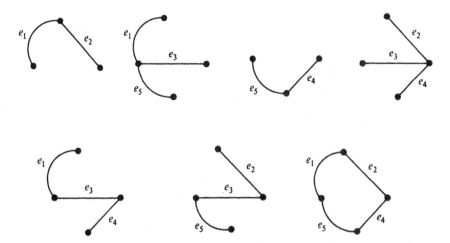

Fig. 6-7 Cut-sets and union of edge-disjoint cut-sets in graph G of Fig. 6-5.

subgraphs, each corresponding to a nonzero cut-set vector, are sketched in Fig. 6-7.

In this example you may have observed that the subgraph $\{e_1, e_2, e_4, e_5\}$ is both a circuit and a union of two edge-disjoint cut-sets. The vector $(1, 1, 0, 1, 1)$ corresponding to this subgraph, therefore, occurs in both subspaces W_Γ and W_S.

Another observation you may have also made is that there are at least $(2^e - 2^\mu - 2^r + 1)$ nonzero vectors which are neither in W_Γ nor in W_S. In this example we must have at least 21 ($= 2^5 - 2^3 - 2^2 + 1$) such vectors. Since there is one vector common to W_Γ and W_S, we have in fact 22 vectors in W_G that are neither circuit vectors nor cut-set vectors.

Having obtained some insight into the circuit subspace and cut-set subspace, let us now explore the relationship between these two subspaces.

6-8. ORTHOGONAL VECTORS AND SPACES

Consider two vectors $(4, 2)$ and $(-3, 6)$ in a plane (which is also called a two-dimensional Euclidean space E_2), as shown in Fig. 6-8. These vectors are orthogonal because their dot product $4 \cdot (-3) + 2 \cdot 6 = 0$. Generalizing this notion to a k-dimensional vector space, we have the following definitions:

Dot Product: The *dot product* of two vectors X and Y in a vector space W is a scalar quantity defined as

$$\begin{aligned} \mathbf{X} \cdot \mathbf{Y} &= (x_1, x_2, \ldots, x_k) \cdot (y_1, y_2, \ldots, y_k) \\ &= x_1 \cdot y_1 + x_2 \cdot y_2 + \cdots + x_k \cdot y_k. \end{aligned}$$

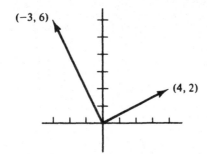

Fig. 6-8 Pair of orthogonal vectors in space E_2.

Orthogonal Vectors: Two vectors are called orthogonal if their dot product is zero; and two subspaces are said to be *orthogonal to* each other if every vector in one is orthogonal to every vector in the other.

Returning to the vector space associated with a graph G, the dot product of two vectors, each representing a subgraph of G, is the modulo 2 sum of the products of the corresponding entries in the two vectors. For example, the dot product of the vectors representing subgraphs g_1 and g_2 in Fig. 6-5 is

$$(1, 0, 1, 0, 1) \cdot (0, 1, 1, 1, 0) = 1 \cdot 0 + 0 \cdot 1 + 1 \cdot 1 + 0 \cdot 1$$
$$+ 1 \cdot 0 \quad \text{(mod 2 sum)}$$
$$= 0 + 0 + 1 + 0 + 0$$
$$= 1.$$

The number of nonzero entries in the sum of products above is the number of edges common to g_1 and g_2. Theorem 6-8 follows directly from the definition of the dot product of two vectors.

THEOREM 6-8

The dot product of two vectors, one representing a subgraph g and the other g', is zero if the number of edges common to g and g' is even; the dot product is 1 if the number of common edges is odd.

THEOREM 6-9

In the vector space of a graph, the circuit subspace and the cut-set subspace are orthogonal to each other.

Proof: According to Theorem 4-3, the number of edges common to a circuit and a cut-set is even. What about the number of edges common to a union of edge-disjoint circuits and a union of edge-disjoint cut-sets? That this is also even can be shown as follows:

Let g_1 be a union of three edge-disjoint circuits Γ_1, Γ_2, and Γ_3 in a graph G, and g_2 be a union of two edge-disjoint cut-sets S_1 and S_2 in G.

Let the number of edges common to

Γ_1 and S_1 be $2a$,
Γ_1 and S_2 be $2b$,
Γ_2 and S_1 be $2c$,
Γ_2 and S_2 be $2d$,
Γ_3 and S_1 be $2e$,
Γ_3 and S_2 be $2f$.

Since there is no edge common between S_1 and S_2, or between Γ_1 and Γ_2 and Γ_3, the six sets of common edges enumerated above are all distinct (some may be empty). Therefore, the number of edges common to g_1 and g_2 is

$$2a + 2b + 2c + 2d + 2e + 2f, \quad \text{an even number.}$$

This example can be extended to g_1 and g_2 to include the union of any finite numbers of edge-disjoint circuits and cut-sets, respectively. From Theorem 6-8, the dot product of a circuit vector and a cut-set vector is zero. Hence every vector in each of these subspaces is orthogonal to every vector in the other. Therefore, the theorem. ∎

For instance, the dot product of the cut-set vector $(0, 1, 1, 1, 0)$ and the circuit vector $(1, 1, 1, 0, 0)$ in the example in Section 6-7 (i.e., Fig. 6-5) is

$$(0, 1, 1, 1, 0) \cdot (1, 1, 1, 0, 0) = 0 \cdot 1 + 1 \cdot 1 + 1 \cdot 1 + 1 \cdot 0 + 0 \cdot 0$$
$$= 0 \quad (\text{mod } 2).$$

6-9. INTERSECTION AND JOIN OF W_Γ AND W_S

Given the two subspaces W_Γ and W_S of the vector space W_G, it is interesting to ask what is the largest set of vectors that belongs to both circuit subspace W_Γ and the cut-set subspace W_S; and what is the smallest set of vectors containing both W_Γ and W_S? Clearly, the null or zero vector 0 is in both W_Γ and W_S, but there may also be some nonzero vectors contained in the intersection $W_\Gamma \cap W_S$. For example, the vector

$$\begin{pmatrix} 1 \\ 1 \\ 0 \\ 1 \\ 1 \end{pmatrix}$$

for the graph in Fig. 6-5 is in both subspaces. It is not difficult to show that the set of vectors $W_\Gamma \cap W_S$ always forms a vector subspace in W_G.

On the other hand, the smallest subspace containing both W_Γ and W_S must contain the set union $W_\Gamma \cup W_S$, of course, but (because of the closure requirements for a subspace) it will usually contain some additional vectors not in $W_\Gamma \cup W_S$. For example, for the graph in Fig. 6-5 set $W_\Gamma \cup W_S$ contains 10 vectors (union of Figs. 6-6 and 6-7), while the smallest subspace containing set $W_\Gamma \cup W_S$, that is, the subspace spanned by the set of vectors in $W_\Gamma \cup W_S$, consists of 16 vectors. (What are the remaining six subgraphs not included in Figs. 6-6 and 6-7?) The subspace spanned by $W_\Gamma \cup W_S$ is called the *join* of W_Γ and W_S, and is written as $W_\Gamma \vee W_S$.

The following is a well-known result from linear algebra: If X and Y are two subspaces in a finite-dimensional vector space, then the dimension of their join, $\dim(X \vee Y)$, is given by

$$\dim(X \vee Y) = \dim X + \dim Y - \dim(X \cap Y).$$

Using this result, we get

$$\dim(W_\Gamma \vee W_S) = e - \dim(W_\Gamma \cap W_S).$$

Two subspaces of a vector space are said to be *orthogonal complements* if the subspaces are orthogonal to each other, and they together span the entire vector space. Thus we have the following interesting result.

THEOREM 6-10

Subspaces W_Γ and W_S are orthogonal complements if and only if

$$\dim(W_\Gamma \cap W_S) = 0, \quad \text{i.e.,} \quad W_\Gamma \cap W_S = 0.$$

In other words, a set of basis vectors of W_Γ together with a set of basis vectors of W_S form a basis for W_G if and only if $W_\Gamma \cap W_S = 0$. Consequently, any subgraph g of G can be uniquely expressed as a ring sum of two subgraphs, one a circuit or an edge-disjoint union of circuits and the other a cut-set or an edge-disjoint union of cut-sets, if and only if

$$W_\Gamma \cap W_S = 0.$$

These properties are illustrated in Fig. 6-9.

In the case

$$W_\Gamma \cap W_S \neq 0$$

we have nonzero vectors each orthogonal to itself. This seemingly peculiar situation arises from the finiteness of the field. In fact, the dot product of any

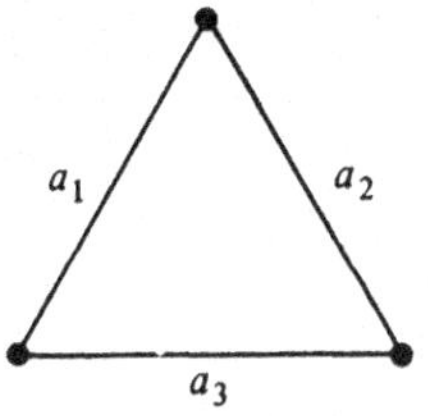

SEC. 6-9　　INTERSECTION AND JOIN OF W_Γ AND W_S

$\dim W_G = e = 3$
$\dim W_S = n - 1 = 2$
$\dim W_\Gamma = \mu = 1$

Graph G

W_Γ contains $\begin{pmatrix}1\\1\\1\end{pmatrix}$ and $\begin{pmatrix}0\\0\\0\end{pmatrix}$.

W_S contains $\begin{pmatrix}1\\1\\0\end{pmatrix}, \begin{pmatrix}0\\1\\1\end{pmatrix}, \begin{pmatrix}1\\0\\1\end{pmatrix}$ and $\begin{pmatrix}0\\0\\0\end{pmatrix}$

$W_\Gamma \cap W_S = \begin{pmatrix}0\\0\\0\end{pmatrix} = 0$

$W_\Gamma \vee W_S = W_G$

$W_\Gamma \cup W_S = \left\{\begin{pmatrix}1\\1\\0\end{pmatrix}, \begin{pmatrix}0\\1\\1\end{pmatrix}, \begin{pmatrix}1\\0\\1\end{pmatrix}, \begin{pmatrix}1\\1\\1\end{pmatrix}, \begin{pmatrix}0\\0\\0\end{pmatrix}\right\}$

The remaining three vectors are uniquely expressed as

$\begin{pmatrix}1\\0\\0\end{pmatrix} = \begin{pmatrix}1\\1\\1\end{pmatrix} \oplus \begin{pmatrix}0\\1\\1\end{pmatrix}$

$\begin{pmatrix}0\\1\\0\end{pmatrix} = \begin{pmatrix}1\\1\\1\end{pmatrix} \oplus \begin{pmatrix}1\\0\\1\end{pmatrix}$, and

$\begin{pmatrix}0\\0\\1\end{pmatrix} = \begin{pmatrix}1\\1\\1\end{pmatrix} \oplus \begin{pmatrix}1\\1\\0\end{pmatrix}$

Fig. 6-9　Graph and its different subspaces.

vector over GF(2) with itself is zero if and only if the vector contains an even number of 1's.

Now, since

$$\dim(W_\Gamma \cap W_S) \neq 0,$$

the two subspaces W_Γ and W_S are not orthogonal complements. Nor is it possible to express every vector in W_G as a sum of two vectors, one from W_Γ and the other from W_S. For example, in Fig. 6-5 no linear combination of vectors in W_Γ and W_S will yield the vector

$$\begin{pmatrix} 1 \\ 0 \\ 0 \\ 0 \\ 0 \end{pmatrix}.$$

In fact, for Fig. 6-5 there are 16 such vectors in W_G that are not in $W_\Gamma \vee W_S$, because

$$\dim(W_\Gamma \vee W_S) = e - \dim(W_\Gamma \cap W_S) = 5 - 1 = 4.$$

The reader is encouraged to sketch a figure like Fig. 6-9, using the graph in Fig. 6-5. Identify all 32 subgraphs, and place them in subspaces W_Γ, W_S, $(W_S \cap W_\Gamma)$, and $(W_S \vee W_\Gamma)$. For more on properties of these subspaces see [6-8] and [6-1].

SUMMARY

In this chapter various algebraic or number systems were introduced, and it was shown that to every graph G corresponds a vector space W_G over the field of integers modulo 2 [i.e., GF(2)]. For a graph G with e edges the dimension of W_G is e, and the number of vectors in W_G is 2^e, each corresponding to a subgraph of G.

Cut-sets and unions of edge-disjoint cut-sets formed an r-dimensional subspace W_S in W_G. The number of vectors in subspace W_S is naturally 2^r, each vector corresponding to a cut-set or a union of edge-disjoint cut-sets. Similarly, the circuits and union of edge-disjoint circuits correspond to a μ-dimensional vector space W_Γ, with 2^μ vectors. Out of many bases available, the set of μ vectors representing all fundamental circuits, with respect to any spanning tree, forms a convenient basis in the circuit subspace. Likewise, the set of r fundamental cut-sets, with respect to any spanning tree, provides a basis in the cut-set subspace.

The cut-set subspace and circuit subspace of a graph are orthogonal to each other. The intersection of these two subspaces is not necessarily $\{0\}$; that is, there may be nonzero vectors common to cut-set and circuit subspaces. Every one of these vectors in $W_S \cap W_\Gamma$ is orthogonal to itself, and they (including the origin 0) form another vector subspace. The set of vectors in the union $W_S \cup W_\Gamma$ does not necessarily form a vector space.

It was also shown that W_G has, in general, a large number of vectors ($2^e - 2^\mu - 2^r + 1$ vectors or more) which belong neither to the cut-set subspace nor to the circuit subspace.

On one hand, a graph provides an elegant and concrete example of "spaces" of more than three dimensions, which often appear frighteningly mysterious to many nonmathematicians. A graph also provides an example of a vector space over a field other than those of usual real or complex numbers. On the other hand, a study of the vector space of a graph and the nature of different subspaces shows us "what makes a graph tick." It gives us an additional mathematical footing in analysis and applications of graphs, such as in coding theory (to be covered in Chapter 12).

Vectors and matrices are closely related. In the next chapter we will explore various matrices associated with a graph, and tie the vector spaces and matrices of graphs together.

REFERENCES

A small but very important portion of abstract algebra has been presented in its barest essentials, and that too without much rigor. Since the classical book of van der Waerden (1931), many excellent texts have appeared on the subject. Of these, four are recommended for those wishing a more detailed and thorough coverage. For groups and fields see Chapters 2, 3, and 4 of Herstein [6-6] and Chapter 3 of Miller [6-7]. For vector spaces, consult Chapter 1 of Halmos [6-5], Chapter 7 of Dean [6-2], and Chapter 5 of Herstein [6-6].

For a study of vector spaces of graphs, Chapter 4 of Seshu and Reed [1-13], and papers by Gould [6-4] and Goldman and Rota [6-3] are suggested; in particular, for the material covered in Section 6-9, see Chen [6-1] and Williams and Maxwell [6-8]. (Note that in [6-8], *seg* is a cut-set or an edge-disjoint union of cut-sets; *circ* is a circuit or an edge-disjoint union of circuits.)

6-1. CHEN, W. K., "On Vector Spaces Associated with a Graph," *SIAM J. Appl. Math.*, Vol. 20, No. 3, May 1971, 526–529.
6-2. DEAN, R. A., *Elements of Abstract Algebra*, John Wiley & Sons, Inc., New York, 1966.
6-3. GOLDMAN, J., G. C. ROTA, "The Number of Subspaces of a Vector Space," in *Recent Progress in Combinatorics* (W. T. Tutte, ed.), Academic Press, Inc., New York, 1969.
6-4. GOULD, R., "Graphs and Vector Spaces," *J. Math. Phys.*, Vol. 37, 1958, 193–214.
6-5. HALMOS, P. R., "Finite-Dimensional Vector Spaces," Van Nostrand Reinhold Company, New York, 1958.
6-6. HERSTEIN, I. N., *Topics in Algebra*, Xerox College Publishing, Lexington, Mass., 1964.
6-7. MILLER, K. S., *Elements of Modern Abstract Algebra*, Harper & Row, Inc., New York, 1958.
6-8. WILLIAMS, T. W., and L. M. MAXWELL, "The Decomposition of a Graph and the Introduction of a New Class of Subgraphs," *SIAM J. Appl. Math.*, Vol. 20, No. 3, May 1971, 385–389.

PROBLEMS

6-1. Show that the usual operation of subtraction does not satisfy the associative axiom.
6-2. Show that the set of the four fourth roots of unity that is, $\{1, -1, i, -i\}$, satisfies all five criteria for being an abelian group under the ordinary multiplication operation.

6-3. Given a set $\{x, y, z\}$ of three elements, show that there is only one group possible with this set.

6-4. From the table in Fig. 6-3(a), show that each element in the set $\{0, 1, 2\}$ has a unique inverse under modulo 3 addition. What about under multiplication modulo 3? Use the table in Fig. 6-3(b).

6-5. Show that there are only two different groups possible with four elements, and that both these groups are abelian.

6-6. Given a set $\{a, b, c, d\}$ of four elements, construct two four by four tables for operations $*$ and $\odot$, such that the set is a field. Identify the letters playing the roles of identities with respect to $*$ and $\odot$ (i.e., 0 and 1).

6-7. Write down the addition and multiplication tables for each of modulo 4, 5, 6, and 7 arithmetics (similar to those in Figs. 6-3 and 6-4).

6-8. From the appropriate table in Problem 6-7, show that not every nonzero element (i.e., 1, 2, and 3) has a unique inverse under the modulo 4 multiplication operation.

6-9. Show that the modulo 6 system is an abelian ring with unity, but is not a field.

6-10. Prove that in any vector space the null vector 0 is orthogonal to every vector in the space.

6-11. Show that W_G, as defined in Section 6-5, satisfies all four conditions for being a vector space, as stated in Section 6-4.

6-12. In vector space W_G, do the vectors associated with the spanning trees of G form a vector space over GF(2)? Explain.

6-13. Let G be a graph consisting of a circuit of length four. Depict the four subspaces W_S, W_Γ, $W_\Gamma \cap W_S$, and $W_\Gamma \vee W_S$ as was done in Fig. 6-9. Draw the corresponding subgraphs. Have all 16 subgraphs of G been accounted for?

6-14. Repeat Problem 6-13 for a complete graph of four vertices. Find a basis for W_S and W_Γ.

6-15. If a graph G is a tree (or a forest), show that the cut-set subspace W_S fills the entire vector space W_G of graph G.

6-16. Characterize a graph for which the circuit space contains the vector $(1, 1, \ldots, 1)$.

6-17. Prove that the number of distinct bases possible in a cut-set subspace is

$$\frac{1}{r!}(2^r - 2^0)(2^r - 2^1)(2^r - 2^2) \ldots (2^r - 2^{r-1}),$$

where r is the rank of the graph.

6-18. Prove that the number of spanning trees in a connected labeled graph with nullity μ cannot exceed the number

$$\frac{1}{\mu!}(2^\mu - 2^0)(2^\mu - 2^1)(2^\mu - 2^2) \ldots (2^\mu - 2^{\mu-1}).$$

(*Hint:* Associated with each spanning tree there is a distinct basis in subspace W_Γ, corresponding to the set of fundamental circuits. Therefore, there are at least as many distinct bases in W_Γ as the number of different spanning trees.)

6-19. Sketch a graph G that has the following vectors (among others) in its circuit subspace: (0, 1, 1, 1, 1, 0, 0, 1), (0, 1, 1, 1, 0, 1, 1, 0), (0, 1, 0, 0, 1, 0, 1, 0), (0, 1, 0, 0, 0, 1, 0, 1), (1, 0, 1, 0, 1, 1, 0, 1), (1, 0, 0, 1, 0, 0, 0, 1, 0), (1, 0, 0, 1, 1, 1, 1, 0), and (1, 0, 0, 1, 0, 0, 0, 1).

6-20. Given that a graph is connected and that $W_\Gamma \cap W_S \neq 0$, investigate further the properties of the subgraphs corresponding to the vectors in subspaces (a) $W_\Gamma \cap W_S$ and (b) $W_\Gamma \vee W_S$.

7 MATRIX REPRESENTATION OF GRAPHS

Although a pictorial representation of a graph is very convenient for a visual study, other representations are better for computer processing. A matrix is a convenient and useful way of representing a graph to a computer. Matrices lend themselves easily to mechanical manipulations. Besides, many known results of matrix algebra can be readily applied to study the structural properties of graphs from an algebraic point of view. In many applications of graph theory, such as in electrical network analysis and operations research, matrices also turn out to be the natural way of expressing the problem.

In this chapter we shall consider two most frequently used matrix representations of a graph. Also a correspondence between some graph-theoretic properties and matrix properties will be established. In view of the close tie between matrices and vector spaces, this chapter should, in fact, be looked upon as a continuation of Chapter 6. A rudimentary knowledge of matrix algebra is assumed.

7-1. INCIDENCE MATRIX

Let G be a graph with n vertices, e edges, and no self-loops. Define an n by e matrix $\mathbf{A} = [a_{ij}]$, whose n rows correspond to the n vertices and the e columns correspond to the e edges, as follows:

The matrix element

$$a_{ij} = 1, \quad \text{if } j\text{th edge } e_j \text{ is incident on } i\text{th vertex } v_i, \text{ and}$$
$$\phantom{a_{ij}} = 0, \quad \text{otherwise.}$$

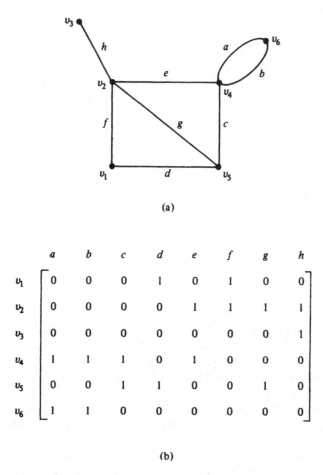

Fig. 7-1 Graph and its incidence matrix.

Such a matrix A is called the *vertex-edge incidence matrix*, or simply *incidence matrix*. Matrix A for a graph G is sometimes also written as A(G). A graph and its incidence matrix are shown in Fig. 7-1.

The incidence matrix contains only two elements, 0 and 1. Such a matrix is called a *binary matrix* or a *(0, 1)-matrix*. Let us stipulate that these two elements are from Galois field modulo 2.† Given any geometric representation of a graph without self-loops, we can readily write its incidence matrix.

†Although matrices are customarily defined over a commutative ring with identity, which need not be a field (such as the ring of integers), we have defined matrix A over a field, GF(2), in keeping with our definition of the vector space W_G in Chapter 6.

On the other hand, if we are given an incidence matrix A(G), we can construct its geometric graph G without ambiguity. The incidence matrix and the geometric graph contain the same information†—they are simply two alternative ways of representing the same (abstract) graph.

The following observations about the incidence matrix A can readily be made:

1. Since every edge is incident on exactly two vertices, each column of A has exactly two 1's.

2. The number of 1's in each row equals the degree of the corresponding vertex.

3. A row with all 0's, therefore, represents an isolated vertex.

4. Parallel edges in a graph produce identical columns in its incidence matrix, for example, columns 1 and 2 in Fig. 7-1.

5. If a graph G is disconnected and consists of two components g_1 and g_2, the incidence matrix A(G) of graph G can be written in a block-diagonal form as

$$A(G) = \begin{bmatrix} A(g_1) & 0 \\ \hline 0 & A(g_2) \end{bmatrix}, \qquad (7\text{-}1)$$

where $A(g_1)$ and $A(g_2)$ are the incidence matrices of components g_1 and g_2. This observation results from the fact that no edge in g_1 is incident on vertices of g_2, and vice versa. Obviously, this remark is also true for a disconnected graph with any number of components.

6. Permutation of any two rows or columns in an incidence matrix simply corresponds to relabeling the vertices and edges of the same graph. This observation leads us to Theorem 7-1.

THEOREM 7-1

Two graphs G_1 and G_2 are isomorphic if and only if their incidence matrices $A(G_1)$ and $A(G_2)$ differ only by permutations of rows and columns.

Rank of the Incidence Matrix: Each row in an incidence matrix A(G) may be regarded as a vector over GF(2) in the vector space of graph G. Let the

†Just as in any two alternative methods of representation, some properties are more evident in one representation than in the other. For example, the fact that the graph is planar is obvious in Fig. 7-1(a), whereas it is not at all obvious from the matrix in Fig. 7-1(b).

vector in the first row be called A_1, in the second row A_2, and so on. Thus

$$A(G) = \begin{bmatrix} A_1 \\ A_2 \\ \cdot \\ \cdot \\ \cdot \\ A_n \end{bmatrix}, \qquad (7\text{-}2)$$

Since there are exactly two 1's in every column of A, the sum of all these vectors is 0 (this being a modulo 2 sum of the corresponding entries). Thus vectors $A_1, A_2, \ldots, A_n$ are not linearly independent. Therefore, the rank of A is less than n; that is, rank $A \leq n - 1$.

Now consider the sum of any m of these n vectors ($m \leq n - 1$). If the graph is connected, $A(G)$ cannot be partitioned, as in Eq. (7-1), such that $A(g_1)$ is with m rows and $A(g_2)$ with $n - m$ rows. In other words, no m by m submatrix of $A(G)$ can be found, for $m \leq n - 1$, such that the modulo 2 sum of those m rows is equal to zero.

Since there are only two constants 0 and 1 in this field, the additions of all vectors taken m at a time for $m = 1, 2, \ldots, n - 1$ exhausts all possible linear combinations of $n - 1$ row vectors. Thus we have just shown that no linear combination of m row vectors of A (for $m \leq n - 1$) can be equal to zero. Therefore, the rank of $A(G)$ must be at least $n - 1$.

Since the rank of $A(G)$ is no more than $n - 1$ and is no less than $n - 1$, it must be exactly equal to $n - 1$. Hence Theorem 7-2.

THEOREM 7-2

If $A(G)$ is an incidence matrix of a connected graph G with n vertices, the rank of $A(G)$ is $n - 1$.

The argument leading to Theorem 7-2 can be extended to prove that the rank of $A(G)$ is $n - k$, if G is a disconnected graph with n vertices and k components (Problem 7-3). This is the reason why the number $n - k$ has been called the rank of a graph with k components.

If we remove any one row from the incidence matrix of a connected graph, the remaining $(n - 1)$ by e submatrix is of rank $n - 1$ (Theorem 7-2). In other words, the remaining $n - 1$ row vectors are linearly independent. Thus we need only $n - 1$ rows of an incidence matrix to specify the corresponding graph completely, for $n - 1$ rows contain the same amount of information as the entire matrix. (This is obvious, since given $n - 1$ rows we can easily reconstitute the missing row, because each column in the matrix has exactly two 1's.)

Such an $(n-1)$ by e submatrix A_f of A is called a *reduced incidence matrix*. The vertex corresponding to the deleted row in A_f is called the *reference vertex*. Clearly, any vertex of a connected graph can be made the reference vertex.

Since a tree is a connected graph with n vertices and $n-1$ edges, its reduced incidence matrix is a square matrix of order and rank $n-1$. In other words,

COROLLARY

The reduced incidence matrix of a tree is nonsingular.

A graph with n vertices and $n-1$ edges that is not a tree is disconnected. The rank of the incidence matrix of such a graph will be less than $n-1$. Therefore, the $(n-1)$ by $(n-1)$ reduced incidence matrix of such a graph will not be nonsingular. In other words, the reduced incidence matrix of a graph is nonsingular if and only if the graph is a tree.

7-2. SUBMATRICES OF A(G)

Let g be a subgraph of a graph G, and let $A(g)$ and $A(G)$ be the incidence matrices of g and G, respectively. Clearly, $A(g)$ is a submatrix of $A(G)$ (possibly with rows or columns permuted). In fact, there is a one-to-one correspondence between each n by k submatrix of $A(G)$ and a subgraph of G with k edges, k being any positive integer less than e and n being the number of vertices in G.

Submatrices of $A(G)$ corresponding to special types of subgraphs, such as circuits, spanning trees, or cut-sets in G, will undoubtedly exhibit special properties. Theorem 7-3 gives one such property.

THEOREM 7-3

Let $A(G)$ be an incidence matrix of a connected graph G with n vertices. An $(n-1)$ by $(n-1)$ submatrix of $A(G)$ is nonsingular if and only if the $n-1$ edges corresponding to the $n-1$ columns of this matrix constitute a spanning tree in G.

Proof: Every square submatrix of order $n-1$ in $A(G)$ is the reduced incidence matrix of the same subgraph in G with $n-1$ edges, and vice versa. From the remarks following Theorem 7-2, it is clear that a square submatrix of $A(G)$ is nonsingular if and only if the corresponding subgraph is a tree. The tree in this case is a spanning tree, because it contains $n-1$ edges of the n-vertex graph. Thus the theorem. ∎

7-3. CIRCUIT MATRIX

Let the number of different circuits in a graph G be q and the number of edges in G be e. Then a *circuit matrix* $\mathbf{B} = [b_{ij}]$ of G is a q by e, (0, 1)-matrix defined as follows:

$$b_{ij} = 1, \quad \text{if } i\text{th circuit includes } j\text{th edge, and}$$
$$= 0, \quad \text{otherwise.}$$

To emphasize the fact that **B** is a circuit matrix of graph G, the circuit matrix may also be written as $\mathbf{B}(G)$.

The graph in Fig. 7-1(a) has four different circuits, $\{a, b\}$, $\{c, e, g\}$, $\{d, f, g\}$, and $\{c, d, f, e\}$. Therefore, its circuit matrix is a 4 by 8, (0, 1)-matrix as shown:

$$\mathbf{B}(G) = \begin{array}{c} \\ 1 \\ 2 \\ 3 \\ 4 \end{array} \begin{array}{c} \begin{array}{cccccccc} a & b & c & d & e & f & g & h \end{array} \\ \left[\begin{array}{cccccccc} 1 & 1 & 0 & 0 & 0 & 0 & 0 & 0 \\ 0 & 0 & 1 & 0 & 1 & 0 & 1 & 0 \\ 0 & 0 & 0 & 1 & 0 & 1 & 1 & 0 \\ 0 & 0 & 1 & 1 & 1 & 1 & 0 & 0 \end{array} \right] \end{array}. \quad (7\text{-}3)$$

The following observations can be made about a circuit matrix $\mathbf{B}(G)$ of a graph G:

1. A column of all zeros corresponds to a noncircuit edge (i.e., an edge that does not belong to any circuit).
2. Each row of $\mathbf{B}(G)$ is a circuit vector.
3. Unlike the incidence matrix, a circuit matrix is capable of representing a self-loop—the corresponding row will have a single 1.
4. The number of 1's in a row is equal to the number of edges in the corresponding circuit.
5. If graph G is separable (or disconnected) and consists of two blocks (or components) g_1 and g_2, the circuit matrix $\mathbf{B}(G)$ can be written in a block-diagonal form as

$$\mathbf{B}(G) = \left[\begin{array}{c|c} \mathbf{B}(g_1) & 0 \\ \hline 0 & \mathbf{B}(g_2) \end{array} \right],$$

where $\mathbf{B}(g_1)$ and $\mathbf{B}(g_2)$ are the circuit matrices of g_1 and g_2. This observation results from the fact that circuits in g_1 have no edges belonging to g_2, and vice versa (Problem 4-14).

6. Permutation of any two rows or columns in a circuit matrix simply corresponds to relabeling the circuits and edges.

7. Two graphs G_1 and G_2 will have the same circuit matrix if and only if G_1 and G_2 are 2-isomorphic (Theorem 4-15). In other words, (unlike an incidence matrix) the circuit matrix does not specify a graph completely. It only specifies the graph within 2-isomorphism. For instance, it can be easily verified that the two graphs in Figs. 4-11(a) and (d) have the same circuit matrix, yet the graphs are not isomorphic.

An important theorem relating the incidence matrix and the circuit matrix of a self-loop-free graph G is

THEOREM 7-4

Let B and A be, respectively, the circuit matrix and the incidence matrix (of a self-loop-free graph) whose columns are arranged using the same order of edges. Then every row of B is orthogonal to every row A; that is,

$$A \cdot B^T = B \cdot A^T = 0 \quad (\text{mod } 2), \tag{7-4}$$

where superscript T denotes the transposed matrix.

Proof: Consider a vertex v and a circuit Γ in the graph G. Either v is in Γ or it is not. If v is not in Γ, there is no edge in the circuit Γ that is incident on v. On the other hand, if v is in Γ, the number of those edges in the circuit Γ that are incident on v is exactly two.

With this remark in mind, consider the ith row in A and the jth row in B. Since the edges are arranged in the same order, the nonzero entries in the corresponding positions occur only if the particular edge is incident on the ith vertex and is also in the jth circuit.

If the ith vertex is not in the jth circuit, there is no such nonzero entry, and the dot product of the two rows is zero. If the ith vertex is in the jth circuit, there will be exactly two 1's in the sum of the products of individual entries. Since $1 + 1 = 0 \pmod 2$, the dot product of the two arbitrary rows—one from A and the other from B—is zero. Hence the theorem. ∎

As an example, let us multiply the incidence matrix and transposed circuit of the graph in Fig. 7-1(a), after making sure that the edges are in the same order in both.

$$A \cdot B^T = \begin{bmatrix} 0 & 0 & 0 & 1 & 0 & 1 & 0 & 0 \\ 0 & 0 & 0 & 0 & 1 & 1 & 1 & 1 \\ 0 & 0 & 0 & 0 & 0 & 0 & 0 & 1 \\ 1 & 1 & 1 & 0 & 1 & 0 & 0 & 0 \\ 0 & 0 & 1 & 1 & 0 & 0 & 1 & 0 \\ 1 & 1 & 0 & 0 & 0 & 0 & 0 & 0 \end{bmatrix} \cdot \begin{bmatrix} 1 & 0 & 0 & 0 \\ 1 & 0 & 0 & 0 \\ 0 & 1 & 0 & 1 \\ 0 & 0 & 1 & 1 \\ 0 & 1 & 0 & 1 \\ 0 & 0 & 1 & 1 \\ 0 & 1 & 1 & 0 \\ 0 & 0 & 0 & 0 \end{bmatrix}$$

$$= \begin{bmatrix} 0 & 0 & 0 & 0 \\ 0 & 0 & 0 & 0 \\ 0 & 0 & 0 & 0 \\ 0 & 0 & 0 & 0 \\ 0 & 0 & 0 & 0 \\ 0 & 0 & 0 & 0 \end{bmatrix} \quad (\text{mod } 2).$$

7-4. FUNDAMENTAL CIRCUIT MATRIX AND RANK OF *B*

A set of fundamental circuits (or basic circuits) with respect to any spanning tree in a connected graph, as discussed in Chapters 3 and 6, are the only independent circuits in a graph. The rest of the circuits can be obtained as ring sums (i.e., linear combinations) of these circuits. Thus, in a circuit matrix, if we retain only those rows that correspond to a set of fundamental circuits and remove all other rows, we would not lose any information. The remaining rows can be reconstituted from the rows corresponding to the set of fundamental circuits. For example, in the circuit matrix in Eq. (7-3), the fourth row is simply the mod 2 sum of the second and third rows.

A submatrix (of a circuit matrix) in which all rows correspond to a set of fundamental circuits is called a *fundamental circuit matrix* B_f. A graph and its fundamental circuit matrix with respect to a spanning tree (indicated by heavy lines) are shown in Fig. 7-2.

As in matrices A and B, permutations of rows (and/or of columns) do not affect B_f. If n is the number of vertices and e the number of edges in a connected graph, then B_f is an $(e - n + 1)$ by e matrix, because the number of fundamental circuits is $e - n + 1$, each fundamental circuit being produced by one chord.

Let us arrange the columns in B_f such that all the $e - n + 1$ chords correspond to the first $e - n + 1$ columns. Furthermore, let us rearrange the rows such that the first row corresponds to the fundamental circuit made

SEC. 7-4 FUNDAMENTAL CIRCUIT MATRIX AND RANK OF B

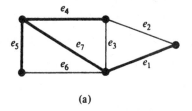

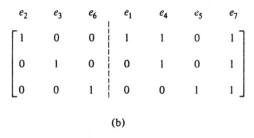

Fig. 7-2 Graph and its fundamental circuit matrix (with respect to the spanning tree shown in heavy lines).

by the chord in the first column, the second row to the fundamental circuit made by the second, and so on. This indeed is how the fundamental circuit matrix is arranged in Fig. 7-2(b).

A matrix B_f thus arranged can be written as

$$B_f = [I_\mu \mid B_t], \qquad (7\text{-}5)$$

where I_μ is an identity matrix of order $\mu = e - n + 1$, and B_t is the remaining μ by $(n - 1)$ submatrix, corresponding to the branches of the spanning tree.

From Eq. (7-5) it is clear that the

$$\text{rank of } B_f = \mu = e - n + 1.$$

Since B_f is a submatrix of the circuit matrix B, the

$$\text{rank of } B \geq e - n + 1.$$

In fact, we can prove Theorem 7-5.

THEOREM 7-5

If B is a circuit matrix of a connected graph G with e edges and n vertices,

$$\text{rank of } B = e - n + 1.$$

Proof: If A is an incidence matrix of G, from Eq. (7-4) we have

$$A \cdot B^T = 0 \pmod{2}.$$

Therefore, according to Sylvester's theorem (Appendix B),

$$\text{rank of A} + \text{rank of B} \leq e;$$

that is,

$$\text{rank of B} \leq e - \text{rank of A}.$$

Since

$$\text{rank of A} = n - 1$$

we have

$$\text{rank of B} \leq e - n + 1.$$

But

$$\text{rank of B} \geq e - n + 1.$$

Therefore, we must have

$$\text{rank of B} = e - n + 1. \blacksquare$$

An Alternative Proof: Theorem 7-5 can also be proved by considering the circuit subspace W_Γ in the vector space W_G of a graph, as discussed in Chapter 6.

Every row in circuit matrix B is a vector in W_Γ, and since the rank of any matrix is equal to the number of linearly independent rows (or columns) in the matrix, we have.

$$\text{rank of matrix B} = \text{number of linearly independent rows in B};$$

but the number of linearly independent rows in B $\leq$ number of linearly independent vectors in W_Γ, and the number of linearly independent vectors in W_Γ = dimension of $W_\Gamma = \mu$. Therefore, rank of B $\leq e - n + 1$. Since we already showed that rank of B $\geq e - n + 1$, Theorem 7-5 follows. $\blacksquare$

Note that in talking of spanning trees of a graph G it is necessary to assume that G is connected. In the case of a disconnected graph, we would have to consider a spanning forest and fundamental circuits with respect to this forest. It is not difficult to show (considering component by component) that if G is a disconnected graph with k components, e edges, and n vertices,

$$\text{rank of B} = \mu = e - n + k.$$

7-5. APPLICATION TO A SWITCHING NETWORK

Suppose you are given a box that contains a switching network consisting of eight switches a, b, c, d, e, f, g, and h. The switches can be turned on or off from outside. You are asked to determine how the switches are connected inside the box, without opening the box, of course.

One way to find the answer is to connect a lamp at the available terminals

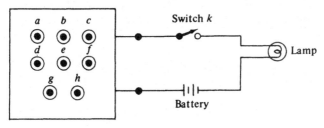

Fig. 7-3 Black box with a switching network.

in series with a battery and an additional switch k, as shown in Fig. 7-3. And then find out which of the various combinations light up the lamp.

In this experiment, suppose you discover that the combinations that turn on the lamp are eight:

(a, b, f, h, k), (a, b, g, k), (a, e, f, g, k), (a, e, h, k),
(b, c, e, h, k), (c, f, h, k), (c, g, k), (d, k).

Solution: Consider the switching network as a graph whose edges represent switches. We can assume that the graph is connected, and has no self-loop. Since a lit lamp implies the formation of a circuit, we can regard the preceding list as a partial list of circuits in the corresponding graph. With this list we form a circuit matrix:

$$B = \begin{array}{c} \\ 1 \\ 2 \\ 3 \\ 4 \\ 5 \\ 6 \\ 7 \\ 8 \end{array} \begin{array}{c} a\ \ b\ \ c\ \ d\ \ e\ \ f\ \ g\ \ h\ \ k \end{array} \begin{bmatrix} 1 & 1 & 0 & 0 & 0 & 1 & 0 & 1 & 1 \\ 1 & 1 & 0 & 0 & 0 & 0 & 1 & 0 & 1 \\ 1 & 0 & 0 & 0 & 1 & 1 & 1 & 0 & 1 \\ 1 & 0 & 0 & 0 & 1 & 0 & 0 & 1 & 1 \\ 0 & 1 & 1 & 0 & 1 & 0 & 0 & 1 & 1 \\ 0 & 0 & 1 & 0 & 0 & 1 & 0 & 1 & 1 \\ 0 & 0 & 1 & 0 & 0 & 0 & 1 & 0 & 1 \\ 0 & 0 & 0 & 1 & 0 & 0 & 0 & 0 & 1 \end{bmatrix}$$

Next, to simplify the matrix, we should remove the obviously redundant circuits. Observe that the following ring sums of circuits give rise to other circuits:

$(a, b, g, k) \oplus (c, f, h, k) \oplus (c, g, k) = (a, b, f, h, k)$,
$(a, b, g, k) \oplus (a, e, h, k) \oplus (c, g, k) = (b, c, e, h, k)$,
$(a, e, h, k) \oplus (c, f, h, k) \oplus (c, g, k) = (a, e, f, g, k)$.

Therefore, we can delete the first, third, and fifth rows from matrix B, without any loss of information. Remaining is a 5 by 9 matrix B_1:

$$B_1 = \begin{matrix} & a & b & c & d & e & f & g & h & k \end{matrix} \\ \begin{bmatrix} 1 & 1 & 0 & 0 & 0 & 0 & 1 & 0 & 1 \\ 1 & 0 & 0 & 0 & 1 & 0 & 0 & 1 & 1 \\ 0 & 0 & 1 & 0 & 0 & 1 & 0 & 1 & 1 \\ 0 & 0 & 1 & 0 & 0 & 0 & 1 & 0 & 1 \\ 0 & 0 & 0 & 1 & 0 & 0 & 0 & 0 & 1 \end{bmatrix}.$$

Our next goal is to bring matrix B_1 to the form of Eq. (7-5). For this we interchange columns to get B_2:

$$B_2 = \begin{matrix} & b & e & f & g & d & a & c & h & k \end{matrix} \\ \begin{bmatrix} 1 & 0 & 0 & 1 & 0 & 1 & 0 & 0 & 1 \\ 0 & 1 & 0 & 0 & 0 & 1 & 0 & 1 & 1 \\ 0 & 0 & 1 & 0 & 0 & 0 & 1 & 1 & 1 \\ 0 & 0 & 0 & 1 & 0 & 0 & 1 & 0 & 1 \\ 0 & 0 & 0 & 0 & 1 & 0 & 0 & 0 & 1 \end{bmatrix}.$$

Adding the fourth row in B_2 to the first, we get B_3.

$$B_3 = \begin{matrix} & b & e & f & g & d & a & c & h & k \end{matrix} \\ \begin{bmatrix} 1 & 0 & 0 & 0 & 0 & 1 & 1 & 0 & 0 \\ 0 & 1 & 0 & 0 & 0 & 1 & 0 & 1 & 1 \\ 0 & 0 & 1 & 0 & 0 & 0 & 1 & 1 & 1 \\ 0 & 0 & 0 & 1 & 0 & 0 & 1 & 0 & 1 \\ 0 & 0 & 0 & 0 & 1 & 0 & 0 & 0 & 1 \end{bmatrix} = [I_5 \mid F].$$

We note that there are no redundant circuits in matrix B_3, and B_3 is a fundamental circuit matrix of the required graph. Since the rank of B_3 is five, and the network was assumed to be connected, we have the following information about the graph:

$$\text{number of edges } e = 9,$$
$$\text{nullity } \mu = 5,$$
$$\text{rank } r = 4,$$
$$\text{number of vertices } n = 5.$$

SEC. 7-5 APPLICATION TO A SWITCHING NETWORK

Constructing a graph from its incidence matrix is simple, but constructing a graph from its fundamental circuit matrix is difficult. We shall, therefore, construct an incidence matrix from B_3.

Since the rows in the incidence matrix are orthogonal to those in B_3—according to Eq. (7-4)—we must first look for a 4 by 9 matrix M, whose rows are linearly independent and are orthogonal to those of B_3.

Since,

$$B_3 = [I_5 \mid F],$$

an orthogonal matrix to B_3 is

$$M = [-F^T \mid I_4]$$
$$= [F^T \mid I_4],$$

because in mod 2 arithmetic $-1 = 1$, [i.e., in GF(2) the additive inverse of 1 is 1].

Thus

$$M = \begin{matrix} & b & e & f & g & d & a & c & h & k \\ & \begin{bmatrix} 1 & 1 & 0 & 0 & 0 & 1 & 0 & 0 & 0 \\ 1 & 0 & 1 & 1 & 0 & 0 & 1 & 0 & 0 \\ 0 & 1 & 1 & 0 & 0 & 0 & 0 & 1 & 0 \\ 0 & 1 & 1 & 1 & 1 & 0 & 0 & 0 & 1 \end{bmatrix} \end{matrix}.$$

Clearly, the rank of M is four, and it is easy to check that

$$B_3 \cdot M^T = 0.$$

Before M can be regarded as a reduced incidence matrix, it must have at most two 1's in each column. This can be achieved by adding (mod 2) the third row to the fourth in M, which gives us M'.

$$M' = \begin{matrix} & b & e & f & g & d & a & c & h & k \\ & \begin{bmatrix} 1 & 1 & 0 & 0 & 0 & 1 & 0 & 0 & 0 \\ 1 & 0 & 1 & 1 & 0 & 0 & 1 & 0 & 0 \\ 0 & 1 & 1 & 0 & 0 & 0 & 0 & 1 & 0 \\ 0 & 0 & 0 & 1 & 1 & 0 & 0 & 1 & 1 \end{bmatrix} \end{matrix}.$$

Matrix M' is the reduced incidence matrix. The incidence matrix A can be obtained by adding a fifth row to M' such that there are exactly two 1's in every column; that is,

$$A = \begin{bmatrix} & b & e & f & g & d & a & c & h & k \\ & 1 & 1 & 0 & 0 & 0 & 1 & 0 & 0 & 0 \\ & 1 & 0 & 1 & 1 & 0 & 0 & 1 & 0 & 0 \\ & 0 & 1 & 1 & 0 & 0 & 0 & 0 & 1 & 0 \\ & 0 & 0 & 0 & 1 & 1 & 0 & 0 & 1 & 1 \\ & 0 & 0 & 0 & 0 & 1 & 1 & 1 & 0 & 1 \end{bmatrix}.$$

From the incidence matrix A we can readily construct the graph and hence the corresponding switching network, as shown in Fig. 7-4.

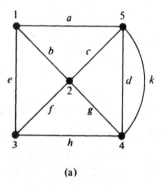

(a)

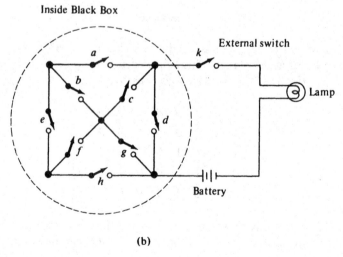

(b)

Fig. 7-4 Graph and the corresponding switching network.

7-6. CUT-SET MATRIX

Analogous to a circuit matrix, we can define a *cut-set matrix* $\mathbf{C} = [c_{ij}]$ in which the rows correspond to the cut-sets and the columns to the edges of the graph, as follows:

$$c_{ij} = 1, \quad \text{if } i\text{th cut-set contains } j\text{th edge, and}$$
$$\phantom{c_{ij}} = 0, \quad \text{otherwise.}$$

For example, a graph and its cut-set matrix are shown in Fig. 7-5.

The following remarks may be made about a cut-set matrix $\mathbf{C}(G)$ of a graph G.

1. As in the case of the incidence matrix, a permutation of rows or columns in a cut-set matrix corresponds simply to a renaming of the cut-sets and edges, respectively.

2. Each row in $\mathbf{C}(G)$ is a cut-set vector.

3. A column with all 0's corresponds to an edge forming a self-loop.

4. Parallel edges produce identical columns in the cut-set matrix (e.g., first two columns in Fig. 7-5).

5. In a nonseparable graph, every set of edges incident on a vertex is a cut-set (Problem 4-8). Therefore, every row of incidence matrix $\mathbf{A}(G)$ is included as a row in the cut-set matrix $\mathbf{C}(G)$. In other words, for a nonseparable graph G, $\mathbf{C}(G)$ contains $\mathbf{A}(G)$. For a separable graph, the incidence matrix of each block is contained in the cut-set matrix. For example, the incidence matrix of the block $\{c, d, e, f, g\}$ in Fig. 7-5 is the 4 by 5 submatrix of $\mathbf{C}$ left after deleting rows a, b, and h and columns 1, 2, 5, and 8.

6. In view of observation 5,

$$\text{rank of } \mathbf{C}(G) \geq \text{rank of } \mathbf{A}(G).$$

Hence, for a connected graph of n vertices,

$$\text{rank of } \mathbf{C}(G) \geq n - 1. \tag{7-6}$$

7. Since the number of edges common to a cut-set and a circuit is always even, every row in $\mathbf{C}$ is orthogonal to every row in $\mathbf{B}$, provided the edges in both $\mathbf{B}$ and $\mathbf{C}$ are arranged in the same order. In other words,

$$\mathbf{B} \cdot \mathbf{C}^T = \mathbf{C} \cdot \mathbf{B}^T = 0 \quad (\text{mod } 2). \tag{7-7}$$

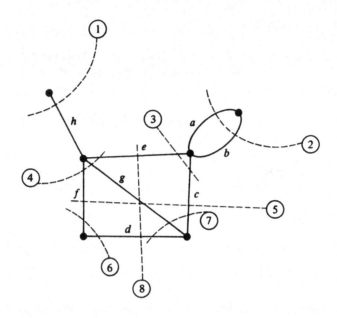

Fig. 7-5 Graph and its cut-set matrix.

On applying Sylvester's theorem to Eq. (7-7),

$$\text{rank of } \mathbf{B} + \text{rank of } \mathbf{C} \leq e.$$

and since for a connected graph

$$\text{rank of } \mathbf{B} = e - n + 1,$$
$$\text{rank of } \mathbf{C} \leq n - 1. \tag{7-8}$$

Combining Eqs. (7-6) and (7-8),

$$\text{rank of } \mathbf{C} = n - 1.$$

Thus we have the following important theorem for a connected graph G.

THEOREM 7-6

The rank of cut-set matrix C(G) is equal to the rank of the incidence matrix A(G), which equals the rank of graph G.

As in the case of the circuit matrix, the cut-set matrix generally has many redundant (or linearly dependent) rows. Therefore, it is convenient to define a fundamental cut-set matrix, $\mathbf{C}_f$, as follows:

A fundamental cut-set matrix $\mathbf{C}_f$ (of a connected graph G with e edges and n vertices) is an $(n - 1)$ by e submatrix of C such that the rows correspond to the set of fundamental cut-sets with respect to some spanning tree.

As in the case of a fundamental circuit matrix, a fundamental cut-set matrix $\mathbf{C}_f$ can also be partitioned into two submatrices, one of which is an identity matrix $\mathbf{I}_{n-1}$ of order $n - 1$. That is,

$$\mathbf{C}_f = [\mathbf{C}_c \mid \mathbf{I}_{n-1}], \tag{7-9}$$

where the last $n - 1$ columns forming the identity matrix correspond to the $n - 1$ branches of the spanning tree, and the first $e - n + 1$ columns forming $\mathbf{C}_c$ correspond to the chords.

A connected graph and a fundamental cut-set matrix with respect to a spanning tree (shown in heavy lines) are given in Fig. 7-6.

Again note that in talking of cut-set matrices we have confined ourselves to connected graphs only. This treatment can be generalized to include disconnected graphs by considering one component at a time.

7-7. RELATIONSHIPS AMONG $\mathbf{A}_f$, $\mathbf{B}_f$, AND $\mathbf{C}_f$

In this section we shall explore the relationships among the reduced incidence matrix $\mathbf{A}_f$, the fundamental circuit matrix $\mathbf{B}_f$, and the fundamental cut-set matrix $\mathbf{C}_f$ of a connected graph.

It has been shown that

$$\mathbf{B}_f = [\mathbf{I}_\mu \mid \mathbf{B}_t], \tag{7-5}$$

$$\mathbf{C}_f = [\mathbf{C}_c \mid \mathbf{I}_{n-1}], \tag{7-9}$$

where subscript t denotes the submatrix corresponding to the branches of a spanning tree, and subscript c denotes the submatrix corresponding to the chords.

Let the spanning tree T in Eqs. (7-5) and (7-9) be the same, and let the

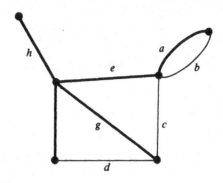

$$C_f = \begin{bmatrix} b & c & d & a & e & f & g & h \\ 1 & 0 & 0 & 1 & 0 & 0 & 0 & 0 \\ 0 & 1 & 0 & 0 & 1 & 0 & 0 & 0 \\ 0 & 0 & 1 & 0 & 0 & 1 & 0 & 0 \\ 0 & 1 & 1 & 0 & 0 & 0 & 1 & 0 \\ 0 & 0 & 0 & 0 & 0 & 0 & 0 & 1 \end{bmatrix} = [C_c \mid I_5]$$

Fig. 7-6 Spanning tree in a graph and the corresponding fundamental cut-set matrix.

order of the edges in both equations be the same. Furthermore, in the reduced incidence matrix A_f—of size $(n - 1)$ by e—let the edges (i.e., the columns) be arranged in the same order as in B_f and C_f. Partition A_f into two submatrices:

$$A_f = [A_c \mid A_t], \qquad (7\text{-}10)$$

where A_t consists of the $n - 1$ columns corresponding to the branches of the spanning tree T, and A_c is the remaining submatrix corresponding to the $e - n + 1$ chords.

Since the columns in A_f and B_f are arranged in the same order, from Eq. (7-4) we have (in mod 2 arithmetic)

$$A_f \cdot B_f^T = 0.$$

That is,
$$[A_c \mid A_t] \cdot \left[-\frac{I_\mu}{B_t^T} \right] = 0,$$

and
$$A_c + A_t \cdot B_t^T = 0. \qquad (7\text{-}11)$$

Since A_t is nonsingular, its inverse A_t^{-1} exists. Premultiplying both sides of Eq. (7-11) by A_t^{-1}, we get

$$A_t^{-1} \cdot A_c = -B_t^T. \tag{7-12}$$

Since in mod 2 arithmetic $-1 = 1$,

$$B_t^T = A_t^{-1} \cdot A_c. \tag{7-13}$$

Similarly, since the columns in B_f and C_f are arranged in the same order, according to Eq. (7-4), we have (in mod 2 arithmetic)

$$C_f \cdot B_f^T = 0.$$

That is, $[C_c \mid I_{n-1}] \cdot \left[-\dfrac{I_\mu}{B_t^T} \right] = 0,$

$$C_c = -B_t^T \tag{7-14}$$
$$= B_t^T \tag{7-15}$$
$$= A_t^{-1} \cdot A_c, \quad \text{from (7-13)}.$$

For example, let us look at the following three matrices for the graph used in Figs. 7-1, 7-5, and 7-6. Using $\{a, e, f, g, h\}$ as the spanning tree, and dropping the sixth row from matrix A in Fig. 7-1 to get A_f, we have

$$A_f = \begin{array}{c} \begin{array}{cccccccc} b & c & d & a & e & f & g & h \end{array} \\ \begin{bmatrix} 0 & 0 & 1 & 0 & 0 & 1 & 0 & 0 \\ 0 & 0 & 0 & 0 & 1 & 1 & 1 & 1 \\ 0 & 0 & 0 & 0 & 0 & 0 & 0 & 1 \\ 1 & 1 & 0 & 1 & 1 & 0 & 0 & 0 \\ 0 & 1 & 1 & 0 & 0 & 0 & 1 & 0 \end{bmatrix} \end{array} = [A_c \mid A_t],$$

$$B_f = \begin{array}{c} \begin{array}{cccccccc} b & c & d & a & e & f & g & h \end{array} \\ \begin{bmatrix} 1 & 0 & 0 & 1 & 0 & 0 & 0 & 0 \\ 0 & 1 & 0 & 0 & 1 & 0 & 1 & 0 \\ 0 & 0 & 1 & 0 & 0 & 1 & 1 & 0 \end{bmatrix} \end{array} = [I_3 \mid B_t],$$

$$C_f = \begin{array}{c} \begin{array}{cccccccc} b & c & d & a & e & f & g & h \end{array} \\ \begin{bmatrix} 1 & 0 & 0 & 1 & 0 & 0 & 0 & 0 \\ 0 & 1 & 0 & 0 & 1 & 0 & 0 & 0 \\ 0 & 0 & 1 & 0 & 0 & 1 & 0 & 0 \\ 0 & 1 & 1 & 0 & 0 & 0 & 1 & 0 \\ 0 & 0 & 0 & 0 & 0 & 0 & 0 & 1 \end{bmatrix} \end{array} = [C_c \mid I_5].$$

$B_t^T = C_c$ is immediate. It can also be readily verified that

$$A_t^{-1} \cdot A_c = B_t^T.$$

This leads to three conclusions:

1. Given A or A_f, we can readily construct B_f and C_f, starting from an arbitrary spanning tree and its subgraph A_t in A_f.
2. Given either B_f or C_f, we can construct the other. Thus since B_f determines a graph within 2-isomorphism, so does C_f.
3. Given either B_f or C_f, A_f in general cannot be determined completely.

7-8. PATH MATRIX

Another (0, 1)-matrix often convenient to use in communication and transportation networks is the *path matrix*. A path matrix is defined for a specific pair of vertices in a graph, say (x, y), and is written as $P(x, y)$. The rows in $P(x, y)$ correspond to different paths between vertices x and y, and the columns correspond to the edges in G. That is, the path matrix for (x, y) vertices is $P(x, y) = [p_{ij}]$, where

$$p_{ij} = 1, \quad \text{if } j\text{th edge lies in } i\text{th path, and}$$
$$= 0, \quad \text{otherwise.}$$

As an illustration, consider all paths between vertices v_3 and v_4 in Fig. 7-1(a). There are three different paths; $\{h, e\}$, $\{h, g, c\}$, and $\{h, f, d, c\}$. Let us number them 1, 2, and 3, respectively. Then we get the 3 by 8 path matrix $P(v_3, v_4)$:

$$P(v_3, v_4) = \begin{matrix} & \begin{matrix} a & b & c & d & e & f & g & h \end{matrix} \\ \begin{matrix} 1 \\ 2 \\ 3 \end{matrix} & \begin{bmatrix} 0 & 0 & 0 & 0 & 1 & 0 & 0 & 1 \\ 0 & 0 & 1 & 0 & 0 & 0 & 1 & 1 \\ 0 & 0 & 1 & 1 & 0 & 1 & 0 & 1 \end{bmatrix} \end{matrix}.$$

Some of the observations one can make at once about a path matrix $P(x, y)$ of a graph G are

1. A column of all 0's corresponds to an edge that does not lie in any path between x and y.
2. A column of all 1's corresponds to an edge that lies in every path between x and y.
3. There is no row with all 0's.
4. The ring sum of any two rows in $P(x, y)$ corresponds to a circuit or an edge-disjoint union of circuits.

THEOREM 7-7

If the edges of a connected graph are arranged in the same order for the columns of the incidence matrix A and the path matrix $P(x, y)$, then the product (mod 2)

$$A \cdot P^T(x, y) = M,$$

where the matrix M has 1's in two rows x and y, and the rest of the $n - 2$ rows are all 0's.

Proof: The proof is left as an exercise for the reader (Problem 7-14).

As an example, multiply the incidence matrix in Fig. 7-1 to the transposed $P(v_3, v_4)$, just discussed.

$$A \cdot P^T(v_3, v_4) = \begin{bmatrix} 0 & 0 & 0 & 1 & 0 & 1 & 0 & 0 \\ 0 & 0 & 0 & 0 & 1 & 1 & 1 & 1 \\ 0 & 0 & 0 & 0 & 0 & 0 & 0 & 1 \\ 1 & 1 & 1 & 0 & 1 & 0 & 0 & 0 \\ 0 & 0 & 1 & 1 & 0 & 0 & 1 & 0 \\ 1 & 1 & 0 & 0 & 0 & 0 & 0 & 0 \end{bmatrix} \begin{bmatrix} 0 & 0 & 0 \\ 0 & 0 & 0 \\ 0 & 1 & 1 \\ 0 & 0 & 1 \\ 1 & 0 & 0 \\ 0 & 0 & 1 \\ 0 & 1 & 0 \\ 1 & 1 & 1 \end{bmatrix}$$

$$= \begin{matrix} & & 1 & 2 & 3 \\ v_1 & \\ v_2 & \\ v_3 & \\ v_4 & \\ v_5 & \\ v_6 & \end{matrix} \begin{bmatrix} 0 & 0 & 0 \\ 0 & 0 & 0 \\ 1 & 1 & 1 \\ 1 & 1 & 1 \\ 0 & 0 & 0 \\ 0 & 0 & 0 \end{bmatrix} \quad (\text{mod } 2).$$

Other properties of the path matrix, such as the rank, are left for the reader to investigate on his own. It should be noted that a path matrix contains less information about the graph in general than any of the matrices A, B, or C does.

7-9. ADJACENCY MATRIX

As an alternative to the incidence matrix, it is sometimes more convenient to represent a graph by its *adjacency matrix* or *connection matrix*. The adjacency matrix of a graph G with n vertices and no parallel edges is an n by n symmetric binary matrix $X = [x_{ij}]$ defined over the ring of integers such that

$$x_{ij} = 1, \quad \text{if there is an edge between } i\text{th and } j\text{th vertices, and}$$
$$= 0, \quad \text{if there is no edge between them.}$$

158   MATRIX REPRESENTATION OF GRAPHS   CHAP. 7

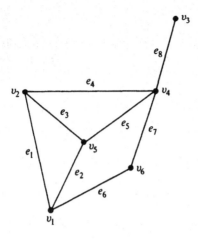

$$X = \begin{array}{c} \\ v_1 \\ v_2 \\ v_3 \\ v_4 \\ v_5 \\ v_6 \end{array} \begin{array}{cccccc} v_1 & v_2 & v_3 & v_4 & v_5 & v_6 \end{array} \\ \left[\begin{array}{cccccc} 0 & 1 & 0 & 0 & 1 & 1 \\ 1 & 0 & 0 & 1 & 1 & 0 \\ 0 & 0 & 0 & 1 & 0 & 0 \\ 0 & 1 & 1 & 0 & 1 & 1 \\ 1 & 1 & 0 & 1 & 0 & 0 \\ 1 & 0 & 0 & 1 & 0 & 0 \end{array} \right]$$

Fig. 7-7 Simple graph and its adjacency matrix.

A simple graph and its adjacency matrix are shown in Fig. 7-7.

Observations that can be made immediately about the adjacency matrix X of a graph G are

1. The entries along the principal diagonal of X are all 0's if and only if the graph has no self-loops. A self-loop at the ith vertex corresponds to $x_{ii} = 1$.

2. The definition of adjacency matrix makes no provision for parallel edges. This is why the adjacency matrix X was defined for graphs without parallel edges.†

---

†Some authors (see Busacker and Saaty [1-2], page 109, for example) define x_{ij} as equal to the number of edges incident on both vertices i and j, and thus take into account parallel edges.

3. If the graph has no self-loops (and no parallel edges, of course), the degree of a vertex equals the number of 1's in the corresponding row or column of X.

4. Permutations of rows and of the corresponding columns imply reordering the vertices. It must be noted, however, that the rows and columns must be arranged in the same order. Thus, if two rows are interchanged in X, the corresponding columns must also be interchanged. Hence two graphs G_1 and G_2 with no parallel edges are isomorphic if and only if their adjacency matrices $X(G_1)$ and $X(G_2)$ are related:

$$X(G_2) = R^{-1} \cdot X(G_1) \cdot R,$$

where R is a permutation matrix.

5. A graph G is disconnected and is in two components g_1 and g_2 if and only if its adjacency matrix $X(G)$ can be partitioned as

$$X(G) = \begin{bmatrix} X(g_1) & 0 \\ 0 & X(g_2) \end{bmatrix},$$

where $X(g_1)$ is the adjacency matrix of the component g_1 and $X(g_2)$ is that of the component g_2.

This partitioning clearly implies that there exists no edge joining any vertex in subgraph g_1 to any vertex in subgraph g_2.

6. Given any square, symmetric, binary matrix Q of order n, one can always construct a graph G of n vertices (and no parallel edges) such that Q is the adjacency matrix of G.

Powers of X: Let us multiply by itself the 6 by 6 adjacency matrix of the simple graph in Fig. 7-7. The result, another 6 by 6 symmetric matrix X^2, is shown below (note that this is ordinary matrix multiplication in the ring of integers and *not* mod 2 multiplication):

$$X^2 = \begin{bmatrix} 3 & 1 & 0 & 3 & 1 & 0 \\ 1 & 3 & 1 & 1 & 2 & 2 \\ 0 & 1 & 1 & 0 & 1 & 1 \\ 3 & 1 & 0 & 4 & 1 & 0 \\ 1 & 2 & 1 & 1 & 3 & 2 \\ 0 & 2 & 1 & 0 & 2 & 2 \end{bmatrix}$$

The value of an off-diagonal entry in X^2, that is, ijth entry ($i \neq j$) in X^2,
= number of 1's in the dot product of ith row and jth column (or jth row) of X.
= number of positions in which both ith and jth rows of X have 1's.
= number of vertices that are adjacent to both ith and jth vertices.
= number of different paths of length two between ith and jth vertices.

Similarly, the ith diagonal entry in X^2 is the number of 1's in the ith row (or column) of matrix X. Thus the value of each diagonal entry in X^2 equals the degree of the corresponding vertex, if the graph has no self-loops.

Since a matrix commutes with matrices that are its own power,

$$X \cdot X^2 = X^2 \cdot X = X^3.$$

And since the product of two square symmetric matrices that commute is also a symmetric matrix, X^3 is a symmetric matrix. (Again note that this is an ordinary product and not mod 2.)

The matrix X^3 for the graph of Fig. 7-7 is

$$X^3 = \begin{bmatrix} 2 & 7 & 3 & 2 & 7 & 6 \\ 7 & 4 & 1 & 8 & 5 & 2 \\ 3 & 1 & 0 & 4 & 1 & 0 \\ 2 & 8 & 4 & 2 & 8 & 7 \\ 7 & 5 & 1 & 8 & 4 & 2 \\ 6 & 2 & 0 & 7 & 2 & 0 \end{bmatrix}.$$

Let us now consider the ijth entry of X^3.

ijth entry of X^3 = dot product of ith row X^2 and jth column (or row) of X.

$= \sum_{k=1}^{n} i k$th entry of $X^2 \cdot kj$th entry of X.

$= \sum_{k=1}^{n}$ number of all different edge sequences† of three edges from ith to jth vertex via kth vertex.

= number of different edge sequences of three edges between ith and jth vertices.

For example, consider how the 1,5th entry on X^3 for the graph of Fig. 7-7 is formed. It is given by the dot product

row 1 of $X^2 \cdot$ row 5 of $X = (3, 1, 0, 3, 1, 0) \cdot (1, 1, 0, 1, 0, 0)$
$= 3 + 1 + 0 + 3 + 0 + 0 = 7.$

These seven different edge sequences of three edges between v_1 and v_5 are

$\{e_1, e_1, e_2\}, \{e_2, e_2, e_2\}, \{e_6, e_6, e_2\}, \{e_2, e_3, e_3\},$
$\{e_6, e_7, e_5\}, \{e_2, e_5, e_5\}, \{e_1, e_4, e_5\}.$

†An edge sequence is a sequence of edges in which each edge (except, of course, the first and the last) has one vertex in common with the edge preceding it and one vertex in common with the edge following it. A path, a circuit, and a walk are all special examples of an edge sequence. An edge may appear more than once in an edge sequence.

Clearly this list includes all the paths of length three between v_1 and v_5, that is, $\{e_6, e_7, e_5\}$ and $\{e_1, e_4, e_5\}$.

It is left as an exercise for the reader to show (Problem 7-19) that the iith entry in X^3 equals twice the number of different circuits of length three (i.e., triangles) in the graph passing through the corresponding vertex v_i.

The general result that includes the properties of X, X^2, and X^3 discussed so far is expressed in Theorem 7-8.

Theorem 7-8

Let X be the adjacency matrix of a simple graph G. Then the ijth entry in X^r is the number of different edge sequences of r edges between vertices v_i and v_j.

Proof: The theorem holds for $r = 1$, and it has been proved for $r = 2$ and 3 also. It can be proved for any positive integer r, by induction.

In other words, assume that it holds for $r - 1$, and then evaluate the ijth entry in X, with the help of the relation

$$X^r = X^{r-1} \cdot X,$$

as was done for X^3.

The rest of the proof is left as an exercise (Problem 7-17).

Corollary A

In a connected graph, the distance between two vertices v_i and v_j (for $i \neq j$) is k, if and only if k is the smallest integer for which the i, jth entry in x^k is nonzero.

This is a useful result in determining the distances between different pairs of vertices.

Corollary B

If X is the adjacency matrix of a graph G with n vertices, and

$$Y = X + X^2 + X^3 + \cdots + X^{n-1}, \quad \text{(in the ring of integers),}$$

then G is disconnected if and only if there exists at least one entry in matrix Y that is zero.

Relationship Between A(G) and X(G): Recall that if a graph G has no self-loops, its incidence matrix A(G) contains all the information about G. Likewise, if G has no parallel edges, its adjacency matrix X(G) contains all the information about G. Therefore, if a graph G has neither self-loops nor parallel edges (i.e., G is a simple graph), both A(G) and X(G) contain the entire information. Thus it is natural to expect that either matrix can be obtained directly from the other, in the case of a simple graph. This relationship is given in Problem 7-23.

SUMMARY

The theory of matrices has been brought to bear upon the theory of graphs. The use of matrices in studying graphs has been amply demonstrated in this chapter.

We have seen that there are several matrices which can be associated with graphs. Two of these, the incidence matrix A and the adjacency matrix X, describe a simple graph completely, that is, up to isomorphism. Two others, the circuit matrix B and the cut-set matrix C, display some important features of the graph and describe the graph only within 2-isomorphism. The path matrix $P(x, y)$ contains even less information than B or C does.

To see further into the structure of the graph, we investigated these matrices, pulled out submatrices A_f, B_f, C_f, I_μ, I_{n-1}, B_t, B_c, C_t, and C_c, and studied them and their interrelationships.

The properties brought out in this chapter do not by any means exhaust the list. Many interesting and useful results are contained in the problems of this chapter.

The converse problem of finding a graph to represent a given matrix has been touched upon lightly in Section 7-5. The problem of realizability, that is, what conditions must a given matrix B satisfy so that a graph can be found whose circuit matrix is B, is very useful and interesting. We shall encounter this problem of realizability again in Chapter 12.

REFERENCES

Some knowledge of elementary matrix algebra was assumed in this chapter. For those not familiar with matrices, dozens of good books are available. Out of these we have listed two, [7-1] and [7-3]. Two somewhat special results, the Binet-Cauchy theorem and Sylvester's law of nullity, are explained and proved in Appendices A and B, respectively.

Most textbooks referred to earlier have some portion devoted to matrices associated with graphs. Particularly recommended readings are Chapter 13 of Harary [1-5], Chapter 4 of Seshu and Reed [1-13], and Chapter 5 of Busacker and Saaty [1-2]. Gould's paper [6-4] referred to in the last chapter is also relevant to this chapter. A survey paper by Harary [7-2] proves some of the results given in this chapter.

7-1. AITKEN, A. C., "Determinants and Matrices," 9th ed., Oliver & Boyd Ltd., Edinburgh, 1956.
7-2. HARARY, F., "Graphs and Matrices," *SIAM Rev.*, Vol. 9, No. 1, Jan., 1967, 83–90.
7-3. HOHN, F. E., *Elementary Matrix Algebra*, The Macmillan Company, New York, 1958.

PROBLEMS

7-1. Write the incidence matrices for the labeled simple graphs shown in Figs. 1-12 and 4-1(b). Put the incidence matrix of the graph of Fig. 4-1(b) in the block-diagonal form of Eq. (7-1).

7-2. Consider the graph in Fig. 4-3. With respect to the spanning tree $\{b, c, e, h, k\}$, write matrices A_f, B_f, and C_f in the forms of Eqs. (7-10), (7-5), and (7-9), respectively. Verify by actual computation Eqs. (7-13) and (7-15).

7-3. Show that for a simple disconnected graph of k components, n vertices, and e edges the ranks of matrices A, B, and C are $n - k$, $e - n + k$, and $n - k$, respectively.

7-4. Label the edges of the graph in Fig. 4-8, and write down its circuit matrix B. Verify observations 1–5 made in Section 7-3 about the properties of matrix B.

7-5. Draw two nonisomorphic, connected, simple, and nonseparable graphs G_1 and G_2, with as small a number of edges as you can; such that the circuit matrices $B(G_1) = B(G_2)$. (*Hint:* G_1 and G_2 are 2-isomorphic, and must be 2-connected.)

7-6. A black box containing a switching network of seven switches—1, 2, 3, 4, 5, 6, and 7—was subjected to the experiment shown in Fig. 7-3. The lamp was lit when each of the following combinations of switches was turned on, in addition to the external switch k, of course: (1, 4, 5), (1, 4, 6, 7), (2, 5, 7), (2, 6), (3, 5), and (3, 6, 7). Show the switching network configuration.

7-7. In Section 7-5 a graph was obtained corresponding to a given fundamental circuit matrix. Similarly, sketch a procedure for obtaining a graph if its fundamental cut-set matrix C_f is given. Can you get two different (nonisomorphic) graphs for the same C_f? If yes, how are these different graphs related?

7-8. Show that you can determine a graph within 2-isomorphism if you were given the set of all spanning trees. (*Hint:* From the set of all spanning trees every cut-set can be determined, using Theorem 4-2. And the set of all cut-sets determines a graph within 2-isomorphism.)

7-9. If the following is the list of all spanning trees of a graph G, determine G.

$\{a, c, d, e\}$, $\{a, c, d, f\}$, $\{b, c, d, e\}$, $\{b, c, d, f\}$,
$\{a, c, e, f\}$, $\{b, c, e, f\}$, $\{a, d, e, f\}$, $\{b, d, e, f\}$,
$\{a, b, d, e\}$, $\{a, b, d, f\}$, $\{a, b, e, f\}$.

7-10. Express the relationship of dualism between two planar, simple graphs in terms of appropriate matrices.

7-11. Characterize simple, self-dual graphs in terms of their circuit and cut-set matrices.

7-12. Prove that

$$B_f = [I_\mu \mid A_c^T \cdot A_t^{-1^T}],$$

$$C_f = A_t^{-1} \cdot A_f.$$

7-13. Write down the path matrix $P(v_1, v_6)$ for the graph in Fig. 4-3. Verify observations 1–4 in Section 7-8 and Theorem 7-7.

7-14. Prove Theorem 7-7.

7-15. Characterize A_f, B_f, C_f, and X matrices of the complete graph of n vertices.

7-16. After having labeled the graph in Fig. 4-8 (as required in Problem 7-4), write its adjacency matrix X. How does the fact that the graph is separable reflect in X? Characterize the adjacency matrix X of a separable graph, in general.

7-17. Complete the proof of Theorem 7-8.

7-18. The diameter of a connected graph is defined (Chapter 3) as the largest distance between two vertices in the graph. Given the adjacency matrix X, how will you determine the diameter of the corresponding graph? (*Hint:* Consider a sum of the powers of X.)

7-19. Show that each diagonal entry in X^3 equals twice the number of triangles passing through the corresponding vertex.

7-20. Prove that the number of spanning trees in a connected graph equals the value of
$$\det(A_f \cdot A_f^T),$$
where A_f is the reduced incidence matrix of the graph, and the arithmetic operations are carried out in the real field and *not* mod 2.

7-21. Similar to the circuit or cut-set matrix, define a *spanning-tree matrix* for a connected graph, and observe some of its properties.

7-22. Let C be the cut-set matrix of a nonseparable graph G, and let $C(x, y)$ be the submatrix of C, containing only those rows of C that represent cut-sets with respect to vertices x and y. Show that $C(x, y)$ contains a fundamental cut-set matrix C_f of G.

7-23. For a labeled graph G of n vertices, define an n by n diagonal matrix D (called the *degree matrix* of G) such that the ith diagonal entry in D equals the degree of the ith vertex in G. Define another matrix E, obtained from the incidence matrix A of G by arbitrarily replacing one of the two 1's in every column by a -1. Show that if G is a simple, connected graph the following holds (the computations are in the ring of integers and *not* mod 2):
(a) $E \cdot E^T = D - X$.
(b) All cofactors of the matrix $D - X$ are equal.
(c) Each cofactor of $D - X$ equals the number of spanning trees in G, where X is as usual the adjacency matrix of G.

7-24. Use the result obtained in Problem 7-23(c) to prove Cayley's formula (Theorem 3-10).

7-25. Let x and y be a pair of vertices in a simple nonseparable graph G, and $P(x, y)$ be the corresponding path matrix of G. Prove that every circuit in G is obtained as a mod 2 sum of two rows of $P(x, y)$. From this result, prove that a path matrix in a simple, nonseparable graph determines the graph within 2-isomorphism. [*Hint:* Every circuit Γ in G falls in one of three categories: (1) Γ passes through both x and y; (2) Γ passes through neither x nor y; or (3) Γ passes through either x or y. Consider all three cases, and use Theorem 4-11.]

7-26. Prove that for a connected, self-loop-free graph G, subspaces W_Γ and W_S are orthogonal complements of W_G over GF(2) if and only if the number of spanning trees in G is odd. [*Hint:* Define a new e by e matrix $M = \begin{bmatrix} C_f \\ B_f \end{bmatrix}$. Compute $\det(MM^T)$, using the identity $C_f B_f^T = B_f C_f^T = 0$, and the Binet–Cauchy theorem (see Appendix A). Show that $\det(MM^T) \equiv 1 \pmod 2$ if and only if G has an odd number of spanning trees.]

8 COLORING, COVERING, AND PARTITIONING

Suppose that you are given a graph G with n vertices and are asked to paint its vertices such that no two adjacent vertices have the same color. What is the minimum number of colors that you would require? This constitutes a coloring problem. Having painted the vertices, you can group them into different sets—one set consisting of all red vertices, another of blue, and so forth. This is a partitioning problem. The coloring and partitioning can, of course, be performed on edges or vertices of a graph. In the case of a planar graph, one may even be interested in coloring the regions. These are the types of questions to be considered in this chapter.

Earlier we came across the subject of partitioning the edges of a given graph into sets with some specified properties. For example, finding a spanning tree in a connected graph is equivalent to partitioning the edges into two sets—one set consisting of the edges included in the spanning tree, and the other consisting of the remaining edges. Identification of a Hamiltonian circuit (if it exists) is another partitioning of the set of edges in a given graph.

The coloring and partitioning of vertices (or edges) is not performed out of mere playfulness, as it may appear from this introduction. Partitioning is applicable to many practical problems, such as coding theory, partitioning of logic in digital computers, and state reduction of sequential machines.

8-1. CHROMATIC NUMBER

Painting all the vertices of a graph with colors such that no two adjacent vertices have the same color is called the *proper coloring* (or sometimes simply *coloring*) of a graph. A graph in which every vertex has been assigned a color

166   COLORING, COVERING, AND PARTITIONING   CHAP. 8

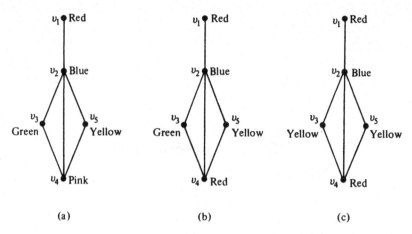

Fig. 8-1 Proper colorings of a graph.

according to a proper coloring is called a *properly colored* graph. Usually a given graph can be properly colored in many different ways. Figure 8-1 shows three different proper colorings of a graph.

The proper coloring which is of interest to us is one that requires the minimum number of colors. A graph G that requires κ different colors for its proper coloring, and no less, is called a *κ-chromatic* graph, and the number κ is called the *chromatic number* of G. You can verify that the graph in Fig. 8-1 is 3-chromatic.

In coloring graphs there is no point in considering disconnected graphs. How we color vertices in one component of a disconnected graph has no effect on the coloring of the other components. Therefore, it is usual to investigate coloring of connected graphs only. All parallel edges between two vertices can be replaced by a single edge without affecting adjacency of vertices. Self-loops must be disregarded. Thus for coloring problems we need to consider only simple, connected graphs.

Some observations that follow directly from the definitions just introduced are

1. A graph consisting of only isolated vertices is 1-chromatic.

2. A graph with one or more edges (not a self-loop, of course) is at least *2-chromatic* (also called *bichromatic*).

3. A complete graph of n vertices is n-chromatic, as all its vertices are adjacent. Hence a graph containing a complete graph of r vertices is at least r-chromatic. For instance, every graph having a triangle is at least 3-chromatic.

4. A graph consisting of simply one circuit with $n \geq 3$ vertices is 2-

chromatic if n is even and 3-chromatic if n is odd. (This can be seen by numbering vertices 1, 2, ..., n in sequence and assigning one color to odd vertices and another to even. If n is even, no adjacent vertices will have the same color. If n is odd, the nth and first vertex will be adjacent and will have the same color, thus requiring a third color for proper coloring.)

Proper coloring of a given graph is simple enough, but a proper coloring with the minimum number of colors is, in general, a difficult task. In fact, there has not yet been found a simple way of characterizing a κ-chromatic graph. (The brute-force method of using all possible combinations can, of course, always be applied, as in any combinatorial problem. But brute force is highly unsatisfactory, because it gets out of hand as soon as the size of the graph increases beyond a few vertices.) Chromatic numbers of some specific types of graphs will be discussed in the rest of this section.

THEOREM 8-1

Every tree with two or more vertices is 2-chromatic.

Proof: Select any vertex v in the given tree T. Consider T as a rooted tree at vertex v. Paint v with color 1. Paint all vertices adjacent to v with color 2. Next, paint the vertices adjacent to these (those that just have been colored with 2) using color 1. Continue this process till every vertex in T has been painted. (See Fig. 8-2). Now in T we find that all vertices at odd distances from v have color 2, while v and vertices at even distances from v have color 1.

Now along any path in T the vertices are of alternating colors. Since there is one and only one path between any two vertices in a tree, no two adjacent vertices have the same color. Thus T has been properly colored with two colors. One color would not have been enough (observation 2 in this section). ∎

Though a tree is 2-chromatic, not every 2-chromatic graph is a tree. (The utilities graph, for instance, is not a tree.) What then is the characterization

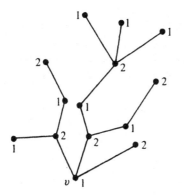

Fig. 8-2 Proper coloring of a tree.

of a 2-chromatic graph? Theorem 8-2 (due to König) characterizes all 2-chromatic graphs.

THEOREM 8-2

A graph with at least one edge is 2-chromatic if and only if it has no circuits of odd length.

Proof: Let G be a connected graph with circuits of only even lengths. Consider a spanning tree T in G. Using the coloring procedure and the result of Theorem 8-1, let us properly color T with two colors. Now add the chords to T one by one. Since G had no circuits of odd length, the end vertices of every chord being replaced are differently colored in T. Thus G is colored with two colors, with no adjacent vertices having the same color. That is, G is 2-chromatic.

Conversely, if G has a circuit of odd length, we would need at least three colors just for that circuit (observation 4 in this section). Thus the theorem. ∎

An upper limit on the chromatic number of a graph is given by Theorem 8-3, whose proof is left as an exercise (Problem 8-1).

THEOREM 8-3

If d_{max} is the maximum degree of the vertices in a graph G,

$$\text{chromatic number of } G \leq 1 + d_{max}.$$

Brooks [8-1] showed that this upper bound can be improved by 1 if G has no complete graph of $d_{max} + 1$ vertices. In that case

$$\text{chromatic number of } G \leq d_{max}.$$

A graph G is called *bipartite* if its vertex set V can be decomposed into two disjoint subsets V_1 and V_2 such that every edge in G joins a vertex in V_1 with a vertex in V_2. Thus every tree is a bipartite graph. So are the graphs in Figs. 8-6 and 8-8. Obviously, a bipartite graph can have no self-loop. A set of parallel edges between a pair of vertices can all be replaced with one edge without affecting bipartiteness of a graph.

Clearly, every 2-chromatic graph is bipartite because the coloring partitions the vertex set into two subsets V_1 and V_2 such that no two vertices in V_1 (or V_2) are adjacent. Similarly, every bipartite graph is 2-chromatic, with one trivial exception; a graph of two or more isolated vertices and with no edges is bipartite but is 1-chromatic.

In generalizing this concept, a graph G is called p-partite if its vertex set can be decomposed into p disjoint subsets $V_1, V_2, \ldots, V_p$, such that no edge in G joins the vertices in the same subset. Clearly, a κ-chromatic graph is p-partite if and only if

$$\kappa \leq p.$$

With this qualification, the results of this section on κ-chromatic graphs are applicable to κ-partite graphs also.

8-2. CHROMATIC PARTITIONING

A proper coloring of a graph naturally induces a partitioning of the vertices into different subsets. For example, the coloring in Fig. 8-1(c) produces the partitioning

$$\{v_1, v_4\}, \quad \{v_2\}, \quad \text{and} \quad \{v_3, v_5\}.$$

No two vertices in any of these three subsets are adjacent. Such a subset of vertices is called an independent set; more formally:

A set of vertices in a graph is said to be an *independent set* of vertices or simply an *independent set* (or an *internally stable set*) if no two vertices in the set are adjacent. For example, in Fig. 8-3, $\{a, c, d\}$ is an independent set. A single vertex in any graph constitutes an independent set.

A *maximal independent set* (or *maximal internally stable set*) is an independent set to which no other vertex can be added without destroying its independence property. The set $\{a, c, d, f\}$ in Fig. 8-3 is a maximal independent set. The set $\{b, f\}$ is another maximal independent set. The set $\{b, g\}$ is a third one. From the preceding example, it is clear that a graph, in general, has many maximal independent sets; and they may be of different sizes. Among all maximal independent sets, one with the largest number of vertices is often of particular interest.

Suppose that the graph in Fig. 8-3 describes the following problem. Each of the seven vertices of the graph is a possible code word to be used in some communication. Some words are so close (say, in sound) to others that they might be confused for each other. Pairs of such words that may be mistaken for one another are joined by edges. Find a largest set of code words for a reliable communication. This is a problem of finding a maximal independent set with largest number of vertices. In this simple example, $\{a, c, d, f\}$ is an answer.

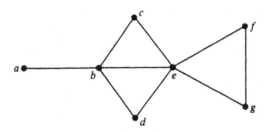

Fig. 8-3

The number of vertices in the largest independent set of a graph G is called the *independence number* (or *coefficient of internal stability*), $\beta(G)$.

Consider a κ-chromatic graph G of n vertices properly colored with κ different colors. Since the largest number of vertices in G with the same color cannot exceed the independence number $\beta(G)$, we have the inequality

$$\beta(G) \geq \frac{n}{\kappa}.$$

Finding a Maximal Independent Set: A reasonable method of finding a maximal independent set in a graph G will be to start with any vertex v of G in the set. Add more vertices to the set, selecting at each stage a vertex that is not adjacent to any of those already selected. This procedure will ultimately produce a maximal independent set. This set, however, is not necessarily a maximal independent set with a largest number of vertices.

Finding All Maximal Independent Sets: A reasonable (but not very efficient for large graphs) method for obtaining all maximal independent sets in any graph can be developed using Boolean arithmetic on the vertices. Let each vertex in the graph be treated as a Boolean variable. Let the logical (or Boolean) sum $a + b$ denote the operation of including vertex a or b or both; let the logical multiplication ab denote the operation of including both vertices a and b, and let the Boolean complement a' denote that vertex a is not included.

For a given graph G we must find a maximal subset of vertices that does not include the two end vertices of any edge in G. Let us express an edge (x, y) as a Boolean product, xy, of its end vertices x and y, and let us sum all such products in G to get a Boolean expression

$$\varphi = \Sigma\, xy \quad \text{for all } (x, y) \text{ in } G.$$

Let us further take the Boolean complement φ' of this expression, and express it as a sum of Boolean products:

$$\varphi' = f_1 + f_2 + \cdots + f_k.$$

A vertex set is a maximal independent set if and only if $\varphi = 0$ (logically false), which is possible if and only if $\varphi' = 1$ (true), which is possible if and only if at least one $f_i = 1$, which is possible if and only if each vertex appearing in f_i (in complemented form) is excluded from the vertex set of G. Thus each f_i will yield a maximal independent set, and every maximal independent set will be produced by this method. This procedure can be best explained by an example. For the graph G in Fig. 8-3,

$$\varphi = ab + bc + bd + be + ce + de + ef + eg + fg,$$
$$\varphi' = (a' + b')(b' + c')(b' + d')(b' + e')(c' + e')(d' + e')$$
$$(e' + f')(e' + g')(f' + g').$$

Multiplying these out and employing the usual identities of Boolean arithmetic, such as

$$aa = a,$$
$$a + a = a,$$
$$a + ab = a,$$

we get

$$\varphi' = b'e'f' + b'e'g' + a'c'd'e'f' + a'c'd'e'g' + b'c'd'f'g'.$$

Now if we exclude from the vertex set of G vertices appearing in any one of these five terms, we get a maximal independent set. The five maximal independent sets are

$$acdf, \quad acdg, \quad bg, \quad bf, \quad \text{and} \quad ae.$$

These are all the maximal independent sets of the graph.

Finding Independence and Chromatic Numbers: Once all the maximal independent sets of G have been obtained, we find the size of the one with the largest number of vertices to get the independence number $\beta(G)$. The independence number of the graph in Fig. 8-3 is four.

To find the chromatic number of G, we must find the minimum number of these (maximal independent) sets, which collectively include all the vertices of G. For the graph in Fig. 8-3, sets $\{a, c, d, f\}$, $\{b, g\}$, and $\{a, e\}$, for example, satisfy this condition. Thus the graph is 3-chromatic.

Chromatic Partitioning: Given a simple, connected graph G, partition all vertices of G into the smallest possible number of disjoint, independent sets. This problem, known as the *chromatic partitioning* of graphs, is perhaps the most important problem in partitioning of graphs.

By enumerating all maximal independent sets and then selecting the smallest number of sets that include all vertices of the graph, we just solved this problem. The following four are some chromatic partitions of the graph in Fig. 8-3, for example.

$$\{(a, c, d, f), (b, g), (e)\},$$
$$\{(a, c, d, g), (b, f), (e)\},$$
$$\{(c, d, f), (b, g), (a, e)\},$$
$$\{(c, d, g), (b, f), (a, e)\}.$$

172 COLORING, COVERING, AND PARTITIONING CHAP. 8

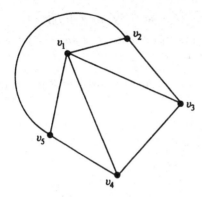

Fig. 8-4 A 3-chromatic graph.

This method of chromatic partitioning (requiring enumeration of all maximal independent sets) is inefficient and needs prohibitively large amounts of computer memory. A more efficient method for computer implementation is proposed in [8-6].

Uniquely Colorable Graphs: A graph that has only one chromatic partition is called a *uniquely colorable* graph. The graph in Fig. 8-3 is *not* a uniquely colorable graph, but the one in Fig. 8-4 is (Problem 8-2). For some interesting properties of uniquely colorable graphs, the reader is referred to Chapter 12 of [1-5].

A concept related to that of the independent set and chromatic partitioning is the dominating set, to be discussed next.

Dominating Sets: A *dominating set* (or an *externally stable set*) in a graph G is a set of vertices that dominates every vertex v in G in the following sense: Either v is included in the dominating set or is adjacent to one or more vertices included in the dominating set. For instance, the vertex set $\{b, g\}$ is a dominating set in Fig. 8-3. So is the set $\{a, b, c, d, f\}$ a dominating set. A dominating set need not be independent. For example, the set of all its vertices is trivially a dominating set in every graph.

In many applications one is interested in finding minimal dominating sets defined as follows:

A *minimal dominating set* is a dominating set from which no vertex can be removed without destroying its dominance property. For example, in Fig. 8-3, $\{b, e\}$ is a minimal dominating set. And so is $\{a, c, d, f\}$. Observations that follow from these definitions are

1. Any one vertex in a complete graph constitutes a minimal dominating set.
2. Every dominating set contains at least one minimal dominating set.
3. A graph may have many minimal dominating sets, and of different

sizes. [The number of vertices in the smallest minimal dominating set of a graph G is called the *domination number*, $\alpha(G)$.]

4. A minimal dominating set may or may not be independent.
5. Every maximal independent set is a dominating set. For if an independent set does not dominate the graph, there is at least one vertex that is neither in the set nor adjacent to any vertex in the set. Such a vertex can be added to the independent set without destroying its independence. But then the independent set could not have been maximal.
6. An independent set has the dominance property only if it is a maximal independent set. Thus an *independent dominating set* is the same as a maximal independent set.
7. In any graph G,

$$\alpha(G) \leq \beta(G).$$

Finding Minimal Dominating Sets: A method for obtaining all minimal dominating sets in a graph will now be developed. The method, like the one for finding all maximal independent sets, also uses Boolean arithmetic.

To dominate a vertex v_i we must either include v_i or any of the vertices adjacent to v_i. A minimum set satisfying this condition for every vertex v_i is a desired set. Therefore, for every vertex v_i in G let us form a Boolean product of sums $(v_i + v_{i_1} + v_{i_2} + \cdots + v_{i_d})$, where $v_{i_1}, v_{i_2}, \ldots, v_{i_d}$ are the vertices adjacent to v_i, and d is the degree of v_i:

$$\theta = \prod (v_i + v_{i_1} + v_{i_2} + \cdots + v_{i_d}) \quad \text{for all } v_i \text{ in } G.$$

When θ is expressed as a sum of products, each term in it will represent a minimal dominating set. Let us illustrate this algorithm using the graph of Fig. 8-3:

Consider the following expression θ for Fig. 8-3:

$$\theta = (a + b)(b + c + d + e + a)(c + b + e)(d + b + e)$$
$$(e + b + c + d + f + g)(f + e + g)(g + e + f).$$

Since in Boolean arithmetic $(x + y)x = x$,

$$\theta = (a + b)(b + c + e)(b + d + e)(e + f + g)$$
$$= ae + be + bf + bg + acdf + acdg.$$

Each of the six terms in the preceding expression represents a minimal dominating set. Clearly, $\alpha(G) = 2$, for this example.

8-3. CHROMATIC POLYNOMIAL

In general, a given graph G of n vertices can be properly colored in many different ways using a sufficiently large number of colors. This property of a graph is expressed elegantly by means of a polynomial. This polynomial is called the *chromatic polynomial* of G and is defined as follows:

The value of the chromatic polynomial $P_n(\lambda)$ of a graph with n vertices gives the number of ways of properly coloring the graph, using λ or fewer colors.

Let c_i be the different ways of properly coloring G using exactly i different colors. Since i colors can be chosen out of λ colors in

$$\binom{\lambda}{i} \quad \text{different ways,}$$

there are $c_i \binom{\lambda}{i}$ different ways of properly coloring G using exactly i colors out of λ colors.

Since i can be any positive integer from 1 to n (it is not possible to use more than n colors on n vertices), the chromatic polynomial is a sum of these terms; that is,

$$P_n(\lambda) = \sum_{i=1}^{n} c_i \binom{\lambda}{i}$$
$$= c_1 \frac{\lambda}{1!} + c_2 \frac{\lambda(\lambda-1)}{2!} + c_3 \frac{\lambda(\lambda-1)(\lambda-2)}{3!} + \cdots$$
$$+ c_n \frac{\lambda(\lambda-1)(\lambda-2)\cdots(\lambda-n+1)}{n!}.$$

Each c_i has to be evaluated individually for the given graph. For example, any graph with even one edge requires at least two colors for proper coloring, and therefore

$$c_1 = 0.$$

A graph with n vertices and using n different colors can be properly colored in $n!$ ways; that is,

$$c_n = n!.$$

As an illustration, let us find the chromatic polynomial of the graph given in Fig. 8-4.

$$P_5(\lambda) = c_1 \lambda + c_2 \frac{\lambda(\lambda-1)}{2} + c_3 \frac{\lambda(\lambda-1)(\lambda-2)}{3!}$$
$$+ c_4 \frac{\lambda(\lambda-1)(\lambda-2)(\lambda-3)}{4!} + c_5 \frac{\lambda(\lambda-1)(\lambda-2)(\lambda-3)(\lambda-4)}{5!}.$$

Since the graph in Fig. 8-4 has a triangle, it will require at least three different colors for proper coloring. Therefore,

$$c_1 = c_2 = 0 \quad \text{and} \quad c_5 = 5!.$$

Moreover, to evaluate c_3, suppose that we have three colors x, y, and z. These three colors can be assigned properly to vertices v_1, v_2, and v_3 in $3! = 6$ different ways. Having done that, we have no more choices left, because vertex v_5 must have the same color as v_3, and v_4 must have the same color as v_2. Therefore,

$$c_3 = 6.$$

Similarly, with four colors, v_1, v_2, and v_3 can be properly colored in $4 \cdot 6 = 24$ different ways. The fourth color can be assigned to v_4 or v_5, thus providing two choices. The fifth vertex provides no additional choice. Therefore,

$$c_4 = 24 \cdot 2 = 48.$$

Substituting these coefficients in $P_5(\lambda)$, we get, for the graph in Fig. 8-4,

$$\begin{aligned} P_5(\lambda) &= \lambda(\lambda - 1)(\lambda - 2) + 2\lambda(\lambda - 1)(\lambda - 2)(\lambda - 3) \\ &\quad + \lambda(\lambda - 1)(\lambda - 2)(\lambda - 3)(\lambda - 4) \\ &= \lambda(\lambda - 1)(\lambda - 2)(\lambda^2 - 5\lambda + 7). \end{aligned}$$

The presence of factors $\lambda - 1$ and $\lambda - 2$ indicates that G is at least 3-chromatic.

Chromatic polynomials have been studied in great detail in the literature. The interested reader is referred to [8-5] for a more thorough discussion of their properties. Theorems 8-4, 8-5, and 8-6 should provide a glimpse into the colorful world of chromatic polynomials.

THEOREM 8-4

A graph of n vertices is a complete graph if and only if its chromatic polynomial is

$$P_n(\lambda) = \lambda(\lambda - 1)(\lambda - 2) \ldots (\lambda - n + 1).$$

Proof: With λ colors, there are λ different ways of coloring any selected vertex of a graph. A second vertex can be colored properly in exactly $\lambda - 1$ ways, the third in $\lambda - 2$ ways, the fourth in $\lambda - 3$ ways, ..., and the nth in $\lambda - n + 1$ ways if and only if every vertex is adjacent to every other. That is, if and only if the graph is complete. ∎

THEOREM 8-5

An n-vertex graph is a tree if and only if its chromatic polynomial

$$P_n(\lambda) = \lambda(\lambda - 1)^{n-1}.$$

176   COLORING, COVERING, AND PARTITIONING   CHAP. 8

Proof: That the theorem holds for $n = 1, 2$ is immediately evident. It is left as an exercise to prove the theorem by induction (Problem 8-9).

THEOREM 8-6

Let a and b be two nonadjacent vertices in a graph G. Let G' be a graph obtained by adding an edge between a and b. Let G'' be a simple graph obtained from G by fusing the vertices a and b together and replacing sets of parallel edges with single edges. Then

$$P_n(\lambda) \text{ of } G = P_n(\lambda) \text{ of } G' + P_{n-1}(\lambda) \text{ of } G''.$$

Proof: The number of ways of properly coloring G can be grouped into two cases, one such that vertices a and b are of the same color and the other such that a and b are of different colors. Since the number of ways of properly coloring G such that a and b have different colors = number of ways of properly coloring G', and
number of ways of properly coloring G such that a and b have the same color

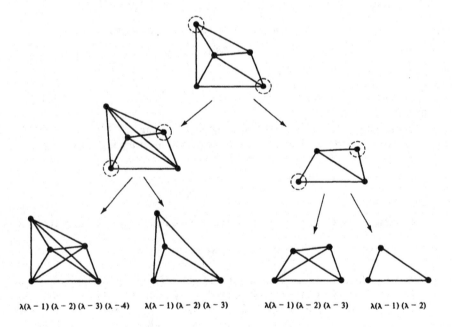

$\lambda(\lambda-1)(\lambda-2)(\lambda-3)(\lambda-4)$ $\lambda(\lambda-1)(\lambda-2)(\lambda-3)$ $\lambda(\lambda-1)(\lambda-2)(\lambda-3)$ $\lambda(\lambda-1)(\lambda-2)$

$$P_5(\lambda) \text{ of } G = \lambda(\lambda - 1)(\lambda - 2) + 2\lambda(\lambda - 1)(\lambda - 2)(\lambda - 3)$$
$$+ \lambda(\lambda - 1)(\lambda - 2)(\lambda - 3)(\lambda - 4)$$
$$= \lambda(\lambda - 1)(\lambda - 2)(\lambda^2 - 5\lambda + 7)$$

Fig. 8-5 Evaluation of a chromatic polynomial.

= number of ways of properly coloring G'',

$$P_n(\lambda) \text{ of } G = P_n(\lambda) \text{ of } G' + P_{n-1}(\lambda) \text{ of } G''. \blacksquare$$

Theorem 8-6 is often used in evaluating the chromatic polynomial of a graph. For example, Fig. 8-5 illustrates how the chromatic polynomial of a graph G is expressed as a sum of the chromatic polynomials of four complete graphs. The pair of nonadjacent vertices shown enclosed in circles is the one used for reduction at that stage.

In the last three sections we have been concerned with proper coloring of the vertices in a graph. Suppose that we are interested in coloring the edges rather than the vertices. It is reasonable to call two edges *adjacent* if they have one end vertex in common (but are not parallel). A proper coloring of edges then requires that adjacent edges should be of different colors. Some results on proper coloring of edges, similar to the results given in Sections 8-1 and 8-2, can be derived (Problem 8-19).

Moreover, a set of edges in which no two are adjacent is similar to an independent set of vertices. Such a set of edges is called a *matching*, the subject of the next section.

8-4. MATCHINGS

Suppose that four applicants a_1, a_2, a_3, and a_4 are available to fill six vacant positions p_1, p_2, p_3, p_4, p_5, and p_6. Applicant a_1 is qualified to fill position p_2 or p_5. Applicant a_2 can fill p_2 or p_5. Applicant a_3 is qualified for p_1, p_2, p_3, p_4, or p_6. Applicant a_4 can fill jobs p_2 or p_5. This situation is represented by the graph in Fig. 8-6. The vacant positions and applicants are represented by vertices. The edges represent the qualifications of each applicant for filling

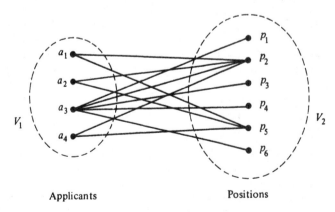

Applicants   Positions

Fig. 8-6 Bipartite graph.

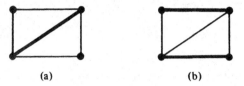

Fig. 8-7 Graph and two of its maximal matchings.

different positions. The graph clearly is bipartite, the vertices falling into two sets $V_1 = \{a_1, a_2, a_3, a_4\}$ and $V_2 = \{p_1, p_2, p_3, p_4, p_5, p_6\}$.

The questions one is most likely to ask in this situation are: Is it possible to hire all the applicants and assign each a position for which he is suitable? If the answer is no, what is the maximum number of positions that can be filled from the given set of applicants?

This is a problem of *matching* (or *assignment*) of one set of vertices into another. More formally, a *matching* in a graph is a subset of edges in which no two edges are adjacent. A single edge in a graph is obviously a matching.

A *maximal matching* is a matching to which no edge in the graph can be added. For example, in a complete graph of three vertices (i.e., a triangle) any single edge is a maximal matching. The edges shown by heavy lines in Fig. 8-7 are two maximal matchings. Clearly, a graph may have many different maximal matchings, and of different sizes. Among these, the maximal matchings with the largest number of edges are called the *largest maximal matchings*. In Fig. 8-7(b), a largest maximal matching is shown in heavy lines. The number of edges in a largest maximal matching is called the *matching number* of the graph.

Although matching is defined for any graph, it is mostly studied in the context of bipartite graphs, as suggested by the introduction to this section. In a bipartite graph having a vertex partition V_1 and V_2, a *complete matching* of vertices in set V_1 into those in V_2 is a matching in which there is one edge incident with every vertex in V_1. In other words, every vertex in V_1 is matched against some vertex in V_2. Clearly, a complete matching (if it exists) is a largest maximal matching, whereas the converse is not necessarily true.

For the existence of a complete matching of set V_1 into set V_2, first we must have at least as many vertices in V_2 as there are in V_1. In other words, there must be at least as many vacant positions as the number of applicants if all the applicants are to be hired. This condition, however, is not sufficient. For example, in Fig. 8-6, although there are six positions and four applicants, a complete matching does not exist. Of the three applicants a_1, a_2, and a_4, each qualifies for the same two positions p_2 and p_5, and therefore one of the three applicants cannot be matched.

This leads us to another necessary condition for a complete matching: Every subset of r vertices in V_1 must collectively be adjacent to at least r vertices in V_2, for all values of $r = 1, 2, \ldots, |V_1|$. This condition is not satisfied in Fig. 8-6. The subset $\{a_1, a_2, a_4\}$ of three vertices has only two vertices p_2 and p_5 adjacent to them. That this condition is also sufficient for existence of a complete matching is indeed surprising. Theorem 8-7 is a formal statement and proof of this result.

THEOREM 8-7

A complete matching of V_1 into V_2 in a bipartite graph exists if and only if every subset of r vertices in V_1 is collectively adjacent to r or more vertices in V_2 for all values of r.

Proof: The "only if" part (i.e., the necessity of a subset of r applicants collectively qualifying for at least r jobs) is immediate and has already been pointed out. The sufficiency (i.e., the "if" part) can be proved by induction on r, as the theorem trivially holds for $r = 1$. For a complete proof, the student is referred to Theorem 11-1 in [8-3], Theorem 5-19 in [4-5], or Chapter 4 in [1-9].

Let us illustrate this important theorem with an example.

Problem of Distinct Representatives: Five senators s_1, s_2, s_3, s_4, and s_5 are members of three committees, c_1, c_2, and c_3. The membership is shown in Fig. 8-8. One member from each committee is to be represented in a supercommittee. Is it possible to send one distinct representative from each of the committees†?

This problem is one of finding a complete matching of a set V_1 into set V_2 in a bipartite graph. Let us use Theorem 8-7 and check if r vertices from V_1 are collectively adjacent to at least r vertices from V_2, for all values of r. The result is shown in Table 8-1 (ignore the last column for the time being).

Thus for this example the condition for the existence of a complete matching is satisfied as stated in Theorem 8-7. Hence it is possible to form the supercommittee with one distinct representative from each committee.

The problem of distinct representatives just solved was a small one. A

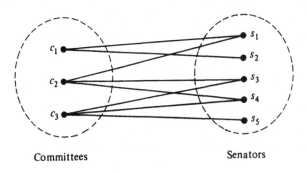

Committees   Senators

Fig. 8-8 Membership of committees.

†This problem, known as the problem of *distinct representatives*, was first formulated and studied by the English mathematician, Philip Hall, in 1935.

	V_1	V_2	$r - q$
$r = 1$	$\{c_1\}$	$\{s_1, s_2\}$	-1
	$\{c_2\}$	$\{s_1, s_3, s_4\}$	-2
	$\{c_3\}$	$\{s_3, s_4, s_5\}$	-2
$r = 2$	$\{c_1, c_2\}$	$\{s_1, s_2, s_3, s_4\}$	-2
	$\{c_2, c_3\}$	$\{s_1, s_3, s_4, s_5\}$	-2
	$\{c_3, c_1\}$	$\{s_1, s_2, s_3, s_4, s_5\}$	-3
$r = 3$	$\{c_1, c_2, c_3\}$	$\{s_1, s_2, s_3, s_4, s_5\}$	-2

Table 8-1

larger problem would have become unwieldy. If there are M vertices in V_1, Theorem 8-7 requires that we take all $2^M - 1$ nonempty subsets of V_1 and find the number of vertices of V_2 adjacent collectively to each of these. In most cases, however, the following simplified version of Theorem 8-7 will suffice for detection of a complete matching in any large graph.

THEOREM 8-8

In a bipartite graph a complete matching of V_1 into V_2 exists if (but not only if) there is a positive integer m for which the following condition is satisfied:

degree of every vertex in $V_1 \geq m \geq$ degree of every vertex in V_2.

Proof: Consider a subset of r vertices in V_1. These r vertices have at least $m \cdot r$ edges incident on them. Each $m \cdot r$ edge is incident to some vertex in V_2. Since the degree of every vertex in set V_2 is no greater than m, these $m \cdot r$ edges are incident on at least $(m \cdot r)/m = r$ vertices in V_2.

Thus any subset of r vertices in V_1 is collectively adjacent to r or more vertices in V_2. Therefore, according to Theorem 8-7, there exists a complete matching of V_1 into V_2. ∎

In the bipartite graph of Fig. 8-8,

degree of every vertex in $V_1 \geq 2 \geq$ degree of every vertex in V_2.

Therefore, there exists a complete matching.

In the bipartite graph of Fig. 8-6 no such number is found, because the degree of $p_2 = 4 >$ degree of a_1.

It must be emphasized that the condition of Theorem 8-8 is a sufficient condition and not necessary for the existence of a complete matching. It will be instructive for the reader to sketch a bipartite graph that does not satisfy Theorem 8-8 and yet has a complete matching (Problem 8-15).

The matching problem or the problem of distinct representatives is also called the *marriage problem* (whose solution, unfortunately, is of little use to those with real marital problems!) See Problem 8-16.

If one fails to find a complete matching, he is most likely to be interested in finding a maximal matching, that is, to pair off as many vertices of V_1 with those in V_2 as possible. For this purpose, let us define a new term called *deficiency*, $\delta(G)$, of a bipartite graph G.

A set of r vertices in V_1 is collectively incident on, say, q vertices of V_2. Then the maximum value of the number $r - q$ taken over all values of $r = 1, 2, \ldots$ and all subsets of V_1 is called the deficiency $\delta(G)$ of the bipartite graph G.

Theorem 8-7, expressed in terms of the deficiency, states that a complete matching in a bipartite graph G exists if and only if

$$\delta(G) \leq 0.$$

For example, the deficiency of the bipartite graph in Fig. 8-7 is -1 (the largest number in the last column of Table 8-1). It is suggested that you prepare a table for the graph of Fig. 8-6, similar to Table 8-1, and verify that the deficiency is $+1$ for this graph (Problem 8-17).

Theorem 8-9 gives the size of the maximal matching for a bipartite graph with a positive deficiency.

THEOREM 8-9

The maximal number of vertices in set V_1 that can be matched into V_2 is equal to

number of vertices in $V_1 - \delta(G)$.

The proof of Theorem 8-9 can be found in [8-3], page 288. The size of a maximal matching in Fig. 8-6, using Theorem 8-9, is obtained as follows:

number of vertices in $V_1 - \delta(G) = 4 - 1 = 3$.

Matching and Adjacency Matrix: Consider a bipartite graph G with non-adjacent sets of vertices V_1 and V_2, having number of vertices n_1 and n_2, respectively, and let $n_1 \leq n_2$, $n_1 + n_2 = n$, the number of vertices in G. The adjacency matrix $X(G)$ of G can be written in the form

$$X(G) = \begin{bmatrix} 0 & X_{12} \\ X_{12}^T & 0 \end{bmatrix},$$

where the submatrix X_{12} is the n_1 by n_2, (0, 1)-matrix containing the information as to which of the n_1 vertices of V_1 are connected to which of the n_2 vertices of V_2. Matrix X_{12}^T is the transpose of X_{12}.

Clearly, all the information about the bipartite graph G is contained in its X_{12} matrix.

A matching V_1 into V_2 corresponds to a selection of the 1's in the matrix X_{12} such that no line (i.e., a row or a column) has more than one 1.

The matching is complete if the n_1 by n_2 matrix made of selected 1's has exactly one 1 in every row. For example, the X_{12} matrix for Fig. 8-8 is

$$X_{12} = \begin{array}{c} \\ c_1 \\ c_2 \\ c_3 \end{array} \begin{array}{ccccc} s_1 & s_2 & s_3 & s_4 & s_5 \\ \begin{bmatrix} 1 & 1 & 0 & 0 & 0 \\ 1 & 0 & 1 & 1 & 0 \\ 0 & 0 & 1 & 1 & 1 \end{bmatrix} \end{array},$$

$n_1 = 3, \quad n_2 = 5, \quad n = 8, \quad \text{and} \quad n_1 \leq n_2,$
$V_1 = \{c_1, c_2, c_3\}$
$V_2 = \{s_1, s_2, s_3, s_4, s_5\}.$

A complete matching of V_1 into V_2 is given by

$$M = \begin{array}{c} \\ c_1 \\ c_2 \\ c_3 \end{array} \begin{array}{ccccc} s_1 & s_2 & s_3 & s_4 & s_5 \\ \begin{bmatrix} 0 & 1 & 0 & 0 & 0 \\ 1 & 0 & 0 & 0 & 0 \\ 0 & 0 & 0 & 0 & 1 \end{bmatrix} \end{array}.$$

A maximal matching corresponds to the selection of a largest possible number of 1's from X_{12} such that no row in it has more than one 1. Therefore, according to Theorem 8-9, in matrix X_{12} the largest number of 1's, no two of which are in one row, is equal to

number of vertices in $V_1 - \delta(G)$.

Matching problems in bipartite graphs can also be formulated in terms of the flow problem (see Section 14-5). All edges are assumed to be of unit capacity, and the problem of finding a maximal matching is reduced to the problem of maximizing flow from the source to the sink (also see [8-3]).

8-5. COVERINGS

In a graph G, a set g of edges is said to *cover* G if every vertex in G is incident on at least one edge in g. A set of edges that covers a graph G is said to be an *edge covering*, a *covering subgraph*, or simply a *covering* of G. For example, a graph G is trivially its own covering. A spanning tree in a connected graph (or a spanning forest in an unconnected graph) is another covering. A Hamiltonian circuit (if it exists) in a graph is also a covering.

Just any covering is too general to be of much interest. We have already dealt with some coverings with specific properties, such as spanning trees and

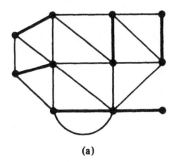 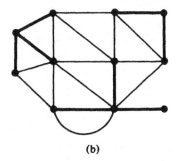

Fig. 8-9 Graph and two of its minimal coverings.

Hamiltonian circuits. In this section we shall investigate the *minimal covering*—a covering from which no edge can be removed without destroying its ability to cover the graph. In Fig. 8-9 a graph and two of its minimal coverings are shown in heavy lines.

The following observations should be made:

1. A covering exists for a graph if and only if the graph has no isolated vertex.
2. A covering of an n-vertex graph will have at least $\lceil n/2 \rceil$ edges. ($\lceil x \rceil$ denotes the smallest integer not less than x.)
3. Every pendant edge in a graph is included in every covering of the graph.
4. Every covering contains a minimal covering.
5. If we denote the remaining edges of a graph by $(G - g)$, the set of edges g is a covering if and only if, for every vertex v, the degree of vertex in $(G - g) \leq$ (degree of vertex v in G) $-$ 1.
6. No minimal covering can contain a circuit, for we can always remove an edge from a circuit without leaving any of the vertices in the circuit uncovered. Therefore, a minimal covering of an n-vertex graph can contain no more than $n - 1$ edges.
7. A graph, in general, has many minimal coverings, and they may be of different sizes (i.e., consisting of different numbers of edges). The number of edges in a minimal covering of the smallest size is called the *covering number* of the graph.

THEOREM 8-10

A covering g of a graph is minimal if and only if g contains no paths of length three or more.

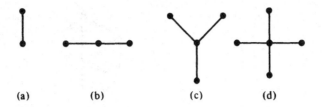

(a) (b) (c) (d)

Fig. 8-10 Star graphs of one, two, three, and four edges.

Proof: Suppose that a covering g contains a path of length three, and it is

$$v_1 e_1 v_2 e_2 v_3 e_3 v_4.$$

Edge e_2 can be removed without leaving its end vertices v_2 and v_3 uncovered. Therefore, g is not a minimal covering.

Conversely, if a covering g contains no path of length three or more, all its components must be *star graphs* (i.e., graphs in the shape of stars; see Fig. 8-10). From a star graph no edge can be removed without leaving a vertex uncovered. That is, g must be a minimal covering. ∎

Suppose that the graph in Fig. 8-9 represents the street map of a part of a city. Each of the vertices is a potential trouble spot and must be kept under the surveillance of a patrol car. How will you assign a minimum number of patrol cars to keep every vertex covered?

The answer is a smallest minimal covering. The covering shown in Fig. 8-9(a) is an answer, and it requires six patrol cars. Clearly, since there are 11 vertices and no edge can cover more than two, less than six edges cannot cover the graph.

Minimization of Switching Functions†: An important step in the logical design of a digital machine is to minimize Boolean functions before implementing them. Suppose we are interested in building a logical circuit that gives the following function F of four Boolean variables w, x, y, and z.

$$F = \bar{w}\bar{x}\bar{y}\bar{z} + \bar{w}\bar{x}y\bar{z} + w\bar{x}\bar{y}\bar{z} + \bar{w}\bar{x}yz + \bar{w}xy\bar{z} + \bar{w}xyz + wxyz,$$

where $+$ denotes logical OR, xy denotes x AND y, and $\bar{x}$ denotes NOT x.

Let us represent each of the seven terms in F by a vertex, and join every pair of vertices that differ only in one variable. Such a graph is shown in Fig. 8-11.

An edge between two vertices represents a term with three variables.

A minimal cover of this graph will represent a simplified form of F, performing the same function as F, but with less logic hardware.

The pendant edges 1 and 7 must be included in every covering of the

---

†Those not familiar with switching functions may skip this subsection.

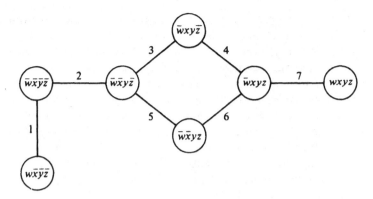

Fig. 8-11 Graph representation of a Boolean function.

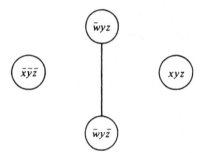

Fig. 8-12

graph. Therefore, the terms

$$\bar{x}\bar{y}\bar{z} \quad \text{and} \quad xyz \quad \text{are essential.}$$

Two additional edges 3 and 6 (or 4 and 5 or 3 and 5) will cover the remainder. Thus a simplified version of F is

$$F = \bar{x}\bar{y}\bar{z} + xyz + \bar{w}y\bar{z} + \bar{w}yz.$$

This expression can again be represented by a graph of four vertices, as shown in Fig. 8-12.

The essential terms $\bar{x}\bar{y}\bar{z}$ and xyz cannot be covered by any edge, and hence cannot be minimized further. One edge will cover the remaining two vertices in Fig. 8-12. Thus the minimized Boolean expression is

$$F = \bar{x}\bar{y}\bar{z} + xyz + \bar{w}y.$$

Dimer Problem: In crystal physics, a crystal is represented by a three-dimensional lattice. Each vertex in the lattice represents an atom, and an edge between vertices represents the bond between the two atoms. In the study of the surface properties of crystals, one is interested in two-dimensional lattices, such as the two shown in Fig. 1-10.

To obtain an analytic expression for certain surface properties of crystals consisting of diatomic molecules (also called *dimers*), one is required to find the number of ways in which all atoms on a two-dimensional lattice can be paired off as molecules (each consisting of two atoms). The problem is equivalent to finding all different coverings of a given graph such that every vertex in the covering is of degree one. Such a covering in which every vertex is of degree one is called a *dimer covering* or a *1-factor*. A dimer covering is obviously a matching because no two edges in it are adjacent. Moreover, a dimer covering is a maximal matching. This is why a dimer covering is often referred to as a *perfect matching*.

Two different dimer coverings are shown in heavy lines in the graph in Fig. 8-13.

Clearly, a graph must have an even number of vertices to have a dimer covering. This condition, however, is not enough (Problem 8-21).

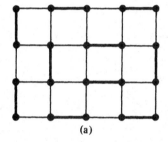

(a)

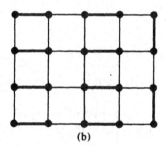

(b)

Fig. 8-13 Two dimer coverings of a graph.

8-6. FOUR-COLOR PROBLEM

So far we have considered proper coloring of vertices and proper coloring of edges. Let us briefly consider the *proper coloring of regions* in a planar graph (embedded on a plane or sphere). Just as in coloring of vertices and edges, the regions of a planar graph are said to be properly colored if no two *contiguous* or *adjacent regions* have the same color. (Two regions are said to be adjacent if they have a common edge between them. Note that one or more vertices in common does not make two regions adjacent.) The proper coloring of regions is also called *map coloring*, referring to the fact that in an atlas different countries are colored such that countries with common boundaries are shown in different colors.

Once again we are not interested in just properly coloring the regions of a given graph. We are interested in a coloring that uses the minimum number of colors. This leads us to the most famous conjecture in graph theory. The conjecture is that every map (i.e., a planar graph) can be properly colored with four colors. The *four-color conjecture*, already referred to in Chapter 1, has been worked on by many famous mathematicians for the past 100 years. No one has yet been able to either prove the theorem or come up with a map (in a plane) that requires more than four colors.

That at least four colors are necessary to properly color a graph is immediate from Fig. 8-14, and that five colors will suffice for any planar graph will be shown shortly.

Two remarks may be made here in passing. Paradoxically, for surfaces more complicated than the plane (or sphere) corresponding theorems have been proved. For example, it has been proved that seven colors are necessary and sufficient for properly coloring maps on the surface of a torus.† Second, it has been proved that all maps containing less than 40 regions can be properly colored with four colors. Therefore, if in general the four-color conjecture is false, the counterexample has to be a very complicated and large one.

Vertex Coloring Versus Region Coloring: From Chapter 5 we know that a graph has a dual if and only if it is planar. Therefore, coloring the regions of a planar graph G is equivalent to coloring the vertices of its dual G^*, and

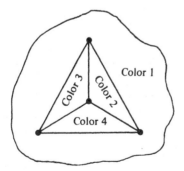

Fig. 8-14 Necessity of four colors.

---

†In fact, the Heawood map-coloring theorem gives the exact number of colors required for every orientable surface more complicated than that of a sphere. See page 136, [1-5], or page 94, [1-2].

188 COLORING, COVERING, AND PARTITIONING   CHAP. 8

vice versa. Thus the four-color conjecture can be restated as follows: *Every planar graph has a chromatic number of four or less.*

Five-Color Theorem: We shall now show that every planar map can be properly colored with five colors:

THEOREM 8-11

The vertices of every planar graph can be properly colored with five colors.

Proof: The theorem will be proved by induction. Since the vertices of all graphs (self-loop-free, of course†) with 1, 2, 3, 4, or 5 vertices can be properly colored with five colors, let us assume that vertices of every planar graph with $n - 1$ vertices can be properly colored with five colors. Then, if we prove that any planar graph G with n vertices will require no more than five colors, we shall have proved the theorem.

Consider the planar graph G with n vertices. Since G is planar, it must have at least one vertex with degree five or less (Problem 5-4). Let this vertex be v.

Let G' be a graph (of $n - 1$ vertices) obtained from G by deleting vertex v (i.e., v and all edges incident on v). Graph G' requires no more than five colors, according to the induction hypothesis. Suppose that the vertices in G' have been properly colored, and now we add to it v and all edges incident on v. If the degree of v is 1, 2, 3, or 4, we have no difficulty in assigning a proper color to v.

This leaves only the case in which the degree of v is five, and all the five colors have been used in coloring the vertices adjacent to v, as shown in Fig. 8-15(a). (Note that Fig. 8-15 is part of a planar representation of graph G'.)

Suppose that there is a path in G' between vertices a and c colored alternately with colors 1 and 3, as shown in Fig. 8-15(b). Then a similar path between b and d, colored alternately with colors 2 and 4, cannot exist; otherwise, these two paths

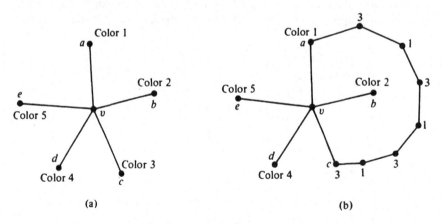

Fig. 8-15 Reassigning of colors.

†See "Regularization of a Planar Graph" in this section.

will intersect and cause G to be nonplanar. (This is a consequence of the Jordan curve theorem, used in Section 5-3, also.)

If there is no path between b and d colored alternately with colors 2 and 4, starting from vertex b we can interchange colors 2 and 4 of all vertices connected to b through vertices of alternating colors 2 and 4. This interchange will paint vertex b with color 4 and yet keep G' properly colored. Since vertex d is still with color 4, we have color 2 left over with which to paint vertex v.

Had we assumed that there was no path between a and c of vertices painted alternately with colors 1 and 3, we would have released color 1 or 3 instead of color 2. And thus the theorem. ∎

Regularization of a Planar Graph: Removing every vertex of degree one (together with the pendant edge) from the graph G does not affect the regions of a planar graph. Nor does the elimination of every vertex of degree two, by merging the two edges in series (Fig. 5-6), have any effect on the regions of a planar graph.

Now consider a typical vertex v of degree four or more in a planar graph. Let us replace vertex v by a small circle with as many vertices as there were incidences on v. This results in a number of vertices each of degree three (see Fig. 8-16).

Performing this transformation on every vertex of degree four or more in a planar graph G will produce another planar graph H in which every vertex is of degree three. When the regions of H have been properly colored, a proper coloring of the regions of G can be obtained simply by shrinking each of the new regions back to the original vertex.

Such a transformation may be called *regularization* of a planar graph, because it converts a planar graph G into a regular planar graph H of degree three. Clearly, if H can be colored with four colors, so can G. Thus, for map-coloring problems, it is sufficient to confine oneself to (connected) planar, regular graphs of degree three. And the four-color conjecture may be restated as follows:

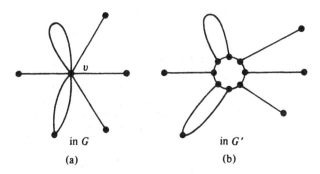

in G in G'
(a)   (b)

Fig. 8-16 Regularization of a graph.

The regions of every planar, regular graph of degree three can be colored properly with four colors.

If, in a planar graph G, every vertex is of degree three, its dual G^* is a planar graph in which every region is bounded by three edges; that is, G^* is a triangular graph. Thus the four-color conjecture may again be restated as follows: The chromatic number of every triangular, planar graph is four or less.

SUMMARY

In the first three sections of this chapter, we were concerned with proper coloring of the vertices of a graph. This led us to the chromatic partitioning of the vertices. In the process we also developed the concept of an independent set of vertices and a dominating set of vertices. Associated in a natural way with these sets we found maximal independent sets, largest maximal independent sets, and the independence number; minimal dominating sets, smallest minimal dominating sets, and the domination number. The chromatic polynomial, studied in Section 8-3, was also a direct consequence of proper coloring of vertices.

Sections 8-4 and 8-5 contain developments parallel to those in Sections 8-1 and 8-2, except that Sections 8-4 and 8-5 are concerned with proper coloring of the *edges* of a graph rather than the vertices. A matching is an independent set of edges, that is, a set of edges no two of which are adjacent. A maximal matching is a maximal set of independent edges. An edge covering is somewhat similar to a dominating set of edges in the sense that every edge in the graph is either in a covering or is adjacent to it. A dimer covering is a perfect matching. An independent set is also a dominating set if and only if it is maximal. Likewise, a matching is also a covering if and only if it is perfect. The sketch in Fig. 8-17 summarizes the relationships among these concepts. The arrows indicate the direction of increasing restriction.

The last section deals with proper coloring of regions in a planar graph rather than vertices or edges.

REFERENCES

A good deal of research has been done and published on coloring, covering, matching, and partitioning of the vertices as well as edges of graphs. The survey paper of Mirsky and Perfect [8-4] is an excellent source of material and references for most of this chapter. Read's survey paper [8-5] on chromatic polynomials is very readable. Wilkov and Kim [8-6] present an efficient algorithm for chromatic partitioning of graphs. Among textbooks, recommended readings are Chapters 4, 10, 11, and 18 of Berge [1-1]; Chapters 10 and 12 of Harary [1-5]; and Chapters 9 and 11 of Liu [8-3]. Rouse Ball [1-12] gives the interesting

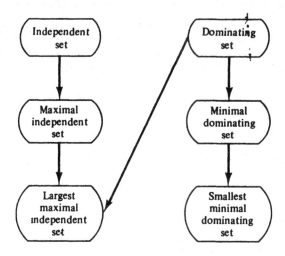

(a) Vertex Subsets

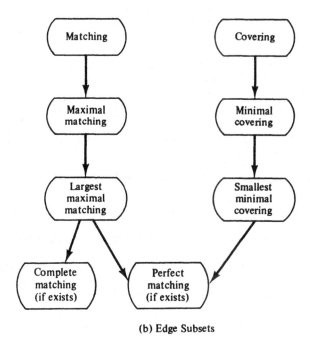

(b) Edge Subsets

Fig. 8-17 Structure of terms.

192 COLORING, COVERING, AND PARTITIONING CHAP. 8

early history of the four-color problem. For an entire book devoted to this unsolved problem, see Ore [1-11].

8-1. BROOKS, R. L., "On Coloring the Nodes of a Network," *Proc. Cambridge Phil. Soc.*, Vol., 37, 1941, 194–197.
8-2. KASTELEYN, P. W., "Graph Theory and Crystal Physics," in *Graph Theory and Theoretical Physics* (F. Harary, ed.), New York, Academic Press, Inc., 1967, 43–110.
8-3. LIU, C. L., *Introduction to Combinatorial Mathematics*, McGraw-Hill Book Company, New York, 1968.
8-4. MIRSKY, L., and H. PERFFCT, "Systems of Representatives," *J. Math. Anal. Appl.*, Vol. 3, 1966, 520–568.
8-5. READ, R. C., "An Introduction to Chromatic Polynomials," *J. Combinatorial Theory*, Vol. 4, No. 1, 1968, 52–71.
8-6. WILKOV, R. S., and W. H. KIM, "A Practical Approach to the Chromatic Partition Problem," *J. Franklin Inst.*, Vol. 289, No. 5, May 1970, 333–349.

PROBLEMS

8-1. Prove that the chromatic number of a graph will not exceed by more than one the maximum degree of the vertices in a graph.
8-2. Show that the graph in Fig. 8-4 has only one chromatic partition. What is it?
8-3. Show that the chromatic number of a graph G cannot exceed the diameter (i.e., the length of the longest path) of G by more than one.
8-4. Show that a simple graph with n vertices and more than $\lfloor n^2/4 \rfloor$ edges cannot be a bipartite graph.
8-5. A bipartite graph is said to be a *complete bipartite* graph if there is one edge between every vertex of set V_1 to every vertex of set V_2. Show that the maximum number of edges in a complete bipartite graph of n vertices is $\lfloor n^2/4 \rfloor$.
8-6. Show that if a bipartite graph has any circuits, they must all be of even lengths.
8-7. In a chessboard, show the positions of
   (a) The minimum number of queens that collectively dominate all 64 squares (an example of a minimal dominating with smallest number of vertices).
   (b) The maximum number of queens such that none of them can take another (an example of a maximal independent set with largest number of vertices).
8-8. Find the chromatic polynomial of the graph in Fig. 8-7.
8-9. Using induction on n, prove Theorem 8-5. (*Hint:* Use a technique similar to one used in proving Theorem 3-3.)
8-10. Show that the chromatic polynomial of a graph of n vertices satisfies the inequality
$$P_n(\lambda) \leq \lambda(\lambda - 1)^{n-1}.$$
(*Hint:* Use Theorem 8-5.)
8-11. Show that the chromatic polynomial of a graph consisting of a single circuit of length n (i.e., an n-gon) is
$$P_n(\lambda) = (\lambda - 1)^n + (\lambda - 1)(-1)^n.$$
8-12. Show that the absolute value of the second coefficient of λ^{n-1} in the chromatic polynomial $P_n(\lambda)$ of a graph equals the number of edges in the graph.
8-13. Sketch two different (i.e., nonisomorphic) graphs that have the same chromatic polynomial.

8-14. Suppose that you are required to make a class schedule in a university. There are a total of n courses to be taught in m available hours of the week. There are pairs of courses that cannot be taught at the same time because some students might like to take both. Explain how you will make the schedule. State the condition when it will be impossible to make a compatible schedule. (*Hint:* Try properly coloring n vertices with m available colors.)

8-15. Sketch a bipartite graph that does not satisfy the condition in Theorem 8-8 and yet has a complete matching.

8-16. In a village there are an equal number of boys and girls of marriageable age. Each boy dates a certain number of girls and each girl dates a certain number of boys. Under what condition is it possible that every boy and girl gets married to one of their dates? (Polygamy and polyandry not allowed.)

8-17. Make a complete table (like Table 8-1) for the graph of Fig. 8-6 to determine whether or not a complete matching exists. Find the deficiency number from this table.

8-18. Show that a nonnull graph is 2-chromatic if and only if, for all odd r, every diagonal entry in matrix X^r is zero. The matrix X is the adjacency matrix of the graph. (*Hint:* Use Theorem 8-2.)

8-19. Just as with an independent set of vertices (or simply *independent set*), define an *independent set of edges* in a graph as a set of nonadjacent edges (not incident on a common vertex). Make some observations parallel to those in Section 8-2. Observe that matching is an independent set of edges. What are complete matchings, maximal matchings, and so on?

8-20. Explore how the covering number of a graph G with n vertices is related to the diameter of G.

8-21. Sketch a graph with an even number of vertices that has no dimer covering.

8-22. Show that the regions of a simple planar graph G can be colored properly with two colors if and only if every vertex in G is of even degree. (*Hint:* Use Theorem 8-2 and Problem 4-28.)

8-23. From v distinct objects one can select $v!/[k!(v-k)!]$ combinations of k objects. Two such combinations are m-related if they have m or more objects in common, $m \leq k$. This relationship can be expressed by means of a graph with $v!/[k!(v-k)!]$ vertices and with edges between every pair of m-related combinations. Make observations on the properties of such a graph. Give conditions for which this graph is (a) a null graph; (b) a complete graph. How will you select a largest set of combinations that are not m-related? Illustrate your method by sketching the graph for $v = 6$, $k = 3$, and $m = 2$.

8-24. An N by N square in which objects $a_1, a_2, \ldots, a_N$ are arranged in such a way that each object appears exactly once in each row and exactly once in each column is called a *Latin square*. A Latin square can also be represented by a complete bipartite graph of $2N$ vertices. What is the total number of different matchings in such a graph? How many of these matchings are edge disjoint?

8-25. Call a subset of vertices that includes at least one vertex incident on every edge of G a *vertex cover of G*. Show that the number of vertices in the smallest vertex cover is equal to or less than the domination number of G.

9 DIRECTED GRAPHS

The graphs studied in this book so far have been undirected graphs. No direction was assigned to the edges in a graph. An edge e_k between vertices v_i and v_j could be considered as going from vertex v_i to vertex v_j or from v_j to v_i. In this chapter we shall consider directed graphs—graphs in which edges have directions.

Many physical situations require directed graphs. The street map of a city with one-way streets, flow networks with valves in the pipes, and electrical networks, for example, are represented by directed graphs. Directed graphs are employed in abstract representations of computer programs, where the vertices stand for the program instructions and the edges specify the execution sequence. The directed graph is an invaluable tool in the study of sequential machines. Directed graphs in the form of signal-flow graphs are used for system analysis in control theory.

Most of the concepts and terminology of undirected graphs are also applicable to directed graphs. For example, the planarity of a graph does not depend on whether the graph is directed or undirected, and therefore Chapter 5 is applicable to both directed and undirected graphs. The same is true for most other topics covered so far. It would be wasteful to devote another eight chapters to the study of directed graphs, mostly repeating, with minor changes, what has already been said. In this chapter, therefore, we shall mainly bring out those properties of directed graphs that are not shared by undirected graphs.

9-1. WHAT IS A DIRECTED GRAPH?

A *directed graph* (or a *digraph* for short) G consists of a set of vertices $V = \{v_1, v_2, \ldots\}$, a set of edges $E = \{e_1, e_2, \ldots\}$, and a mapping Ψ that maps

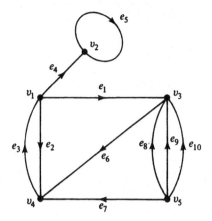

Fig. 9-1 Directed graph with 5 vertices and 10 edges.

every edge onto some *ordered* pair of vertices (v_i, v_j). As in the case of undirected graphs, a vertex is represented by a point and an edge by a line segment between v_i and v_j with an arrow directed from v_i to v_j. For example, Fig. 9-1 shows a digraph with five vertices and ten edges. A digraph is also referred to as an *oriented graph*.†

In a digraph an edge is not only incident on a vertex, but is also *incident out of* a vertex and *incident into* a vertex. The vertex v_i, which edge e_k is incident out of, is called the *initial vertex* of e_k. The vertex v_j, which e_k is incident into, is called the *terminal vertex* of e_k. In Fig. 9-1, v_5 is the initial vertex and v_4 is the terminal vertex of edge e_7. An edge for which the initial and terminal vertices are the same forms a *self-loop*, such as e_5. (Some authors reserve the term *arc* for an oriented or directed edge. We use the term edge to mean either an undirected or a directed edge. Whenever there is a possibility of confusion, we shall explicitly state directed or undirected edge.)

The number of edges incident out of a vertex v_i is called the *out-degree* (or *out-valence* or *outward demidegree*) of v_i and is written $d^+(v_i)$. The number of edges incident into v_i is called the *in-degree* (or *in-valence* or *inward demidegree*) of v_i and is written as $d^-(v_i)$. In Fig. 9-1, for example,

$$d^+(v_1) = 3, \quad d^-(v_1) = 1,$$
$$d^+(v_2) = 1, \quad d^-(v_2) = 2,$$
$$d^+(v_5) = 4, \quad d^-(v_5) = 0.$$

It is not difficult to prove (Problem 9-1) that in any digraph G the sum of all in-degrees is equal to the sum of all out-degrees, each sum being equal to

†Some authors make a distinction between the terms "oriented graph" and "directed graph" by reserving the former for only those digraphs which have at most one directed edge between a pair of vertices. This often leads to confusion; therefore, we use these two terms synonymously.

the number of edges in G; that is,

$$\sum_{i=1}^{n} d^+(v_i) = \sum_{i=1}^{n} d^-(v_i).$$

An *isolated vertex* is a vertex in which the in-degree and the out-degree are both equal to zero. A vertex v in a digraph is called *pendant* if it is of degree one, that is, if

$$d^+(v) + d^-(v) = 1.$$

Two directed edges are said to be *parallel* if they are mapped onto the same ordered pair of vertices. That is, in addition to being parallel in the sense of undirected edges, parallel directed edges must also agree in the direction of their arrows. In Fig. 9-1, edges e_8, e_9, and e_{10} are parallel, whereas edges e_2 and e_3 are not.

Since many properties of directed graphs are the same as those of undirected ones, it is often convenient to disregard the orientations of edges in a digraph. Such an undirected graph obtained from a directed graph G will be called the *undirected graph corresponding to G*.

On the other hand, given an undirected graph H, we can assign each edge of H some arbitrary direction. The resulting digraph, designated by $\vec{H}$ is called an *orientation of H* (or a *digraph associated with H*). Note that while a given digraph has a unique (within isomorphism) undirected graph corresponding to it, a given undirected graph may have "different" orientations possible. This is why we say *the* undirected graph corresponding to a digraph, but *an* orientation of a graph.

This brings us to the question: When are two digraphs considered to be the same or isomorphic?

Isomorphic Digraphs: Isomorphic graphs were defined such that they have identical behavior in terms of graph properties. In other words, if their labels are removed, two isomorphic graphs are indistinguishable. For two digraphs

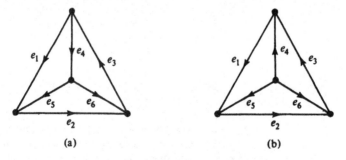

Fig. 9-2 Two nonisomorphic digraphs.

to be isomorphic not only must their corresponding undirected graphs be isomorphic, but the directions of the corresponding edges must also agree. For example, Fig. 9-2 shows two digraphs that are not isomorphic, although they are orientations of the same undirected graph.

Figure 9-2 immediately suggests a problem. What is the number of distinct (i.e., nonisomorphic) orientations of a given undirected graph? The problem was solved by F. Harary and E. M. Palmer in 1966. Some specific cases are left as an exercise (Problem 9-3).

9-2. SOME TYPES OF DIGRAPHS

Like their undirected sisters, digraphs come in many varieties. In fact, due to the choice of assigning a direction to each edge, directed graphs have more varieties than undirected ones.

Simple Digraphs: A digraph that has no self-loop or parallel edges is called a simple digraph (Figs. 9-2 and 9-3, for example).

Asymmetric Digraphs: Digraphs that have at most one directed edge between a pair of vertices, but are allowed to have self-loops, are called *asymmetric* or *antisymmetric*.

Symmetric Digraphs: Digraphs in which for every edge (a, b) (i.e., from vertex a to b) there is also an edge (b, a).

A digraph that is both simple and symmetric is called a *simple symmetric digraph*. Similarly, a digraph that is both simple and asymmetric is *simple asymmetric*. The reason for the terms symmetric and asymmetric will be apparent in the context of binary relations in Section 9-3.

Complete Digraphs: A complete undirected graph was defined as a simple graph in which every vertex is joined to every other vertex exactly by one edge. For digraphs we have two types of complete graphs. A *complete symmetric digraph* is a simple digraph in which there is exactly one edge directed from every vertex to every other vertex (Fig. 9-3), and a *complete asymmetric digraph* is an asymmetric digraph in which there is exactly one edge between every pair of vertices (Fig. 9-2).

A complete asymmetric digraph of n vertices contains $n(n-1)/2$ edges, but a complete symmetric digraph of n vertices contains $n(n-1)$ edges. A complete asymmetric digraph is also called a *tournament* or a *complete tournament* (the reason for this term will be made clear in Section 9-10).

A digraph is said to be *balanced* if for every vertex v_i the in-degree equals the out-degree; that is, $d^+(v_i) = d^-(v_i)$. (A balanced digraph is also referred to as a *pseudosymmetric* digraph, or an *isograph*.) A balanced digraph is said to be *regular* if every vertex has the same in-degree and out-degree as every other vertex.

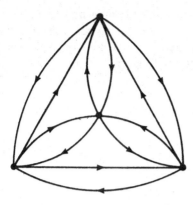

Fig. 9-3 Complete symmetric digraph of four vertices.

9-3. DIGRAPHS AND BINARY RELATIONS

The theory of graphs and the calculus of binary relations are closely related (so much so that some people often mistakenly come to regard graph theory as a branch of the theory of relations).

In a set of objects, X, where

$$X = \{x_1, x_2, \ldots\},$$

a *binary relation* R between pairs (x_i, x_j) may exist. In which case, we write

$$x_i R x_j$$

and say that *x_i has relation R to x_j*.

Relation R may for instance be "is parallel to," "is orthogonal to," or "is congruent to" in geometry. It may be "is greater than," "is a factor of," "is equal to," and so on, in the case when X consists of numbers. On the other hand, if the set X is composed of people, the relation R may be "is son of," "is spouse of," "is friend of," and so forth. Each of these relations is defined only on pairs of numbers of the set, and this is why the name *binary relation*. Although there are relations other than binary (x_i "is a product of" x_j and x_k, for example, will be a tertiary relation), binary relations are the most important in mathematics, and the word "relation" implies a binary relation.

A digraph is the most natural way of representing a binary relation on a set X. Each $x_i \in X$ is represented by a vertex x_i. If x_i has the specified relation R to x_j, a directed edge is drawn from vertex x_i to x_j, for every pair (x_i, x_j). For example, the digraph in Fig. 9-4 represents the relation "is greater than" on a set consisting of five numbers {3, 4, 7, 5, 8}.

Clearly, every binary relation on a finite set can be represented by a digraph without parallel edges. Conversely, every digraph without parallel edges defines a binary relation on the set of its vertices.

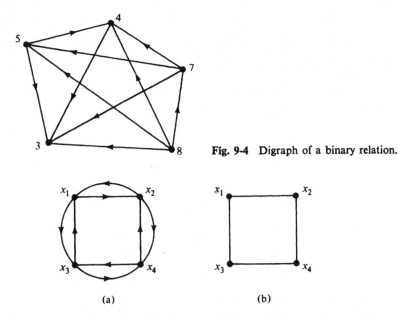

Fig. 9-4 Digraph of a binary relation.

Fig. 9-5 Graphs of symmetric binary relation.

Reflexive Relation: For some relation R it may happen that every element is in relation R to itself. For example, a number is always equal to itself, or a line is always parallel to itself. Such a relation R on set X that satisfies

$$x_i R x_i$$

for every $x_i \in X$ is called a *reflexive* relation. The digraph of a reflexive relation will have a self-loop at every vertex. Such a digraph representing a reflexive binary relation on its vertex set may be called a *reflexive digraph*. A digraph in which no vertex has a self-loop is called an *irreflexive digraph*.

Symmetric Relation: For some relation R it may happen that for all x_i and x_j, if

$$x_i R x_j \text{ holds, then } x_j R x_i \text{ also holds.}$$

Such a relation is called a *symmetric relation*. "Is spouse of" is a symmetric but irreflexive relation. "Is equal to" is both symmetric and reflexive.

The digraph of a symmetric relation is a *symmetric digraph* because for every directed edge from vertex x_i to x_j there is a directed edge from x_j to x_i. Figure 9-5(a) shows the graph of an irreflexive, symmetric binary relation on a set of four elements. The same relation can also be represented by drawing just one undirected edge between every pair of vertices that are related, as in Fig. 9-5(b). Thus every undirected graph is a representation of some symmetric binary relation (on the set of its vertices). Furthermore, every undirected graph with e edges can be thought of as a symmetric digraph with $2e$ directed edges. (A two-way street is equivalent to two one-way streets pointed in opposite directions.)

Transitive Relation: A relation R is said to be *transitive* if for any three elements x_i, x_j, and x_k in the set,

$$x_i R x_j \quad \text{and} \quad x_j R x_k$$

always imply

$$x_i R x_k.$$

The binary relation "is greater than," for example, is a transitive relation. If $x_i > x_j$ and $x_j > x_k$, clearly $x_i > x_k$. "Is descendent of" is another example of a transitive relation.

The digraph of a transitive (but irreflexive and asymmetric) binary relation is shown in Fig. 9-4. Note the triangular subgraphs. A digraph representing a transitive relation (on its vertex set) is called a *transitive directed graph*.

Equivalence Relation: A binary relation is called an *equivalence relation* if it is reflexive, symmetric, and transitive. Some examples of equivalence relations are "is parallel to," "is equal to," "is congruent to," "is equal to modulo m," and "is isomorphic to."

The graph representing an equivalence relation may be called an *equivalence graph*. What does an equivalence graph look like? Let us look at an example, consisting of the equivalence relation "is congruent to modulo 3" defined on the set of 11 integers, 10 through 20. The graph is shown in Fig. 9-6. (Recall that each undirected edge in Fig. 9-6 represents two parallel but oppositely directed edges.)

In Fig. 9-6 we see that the vertex set of the graph is divided into three disjoint classes, each in a separate component. Each component is an undirected subgraph (due to symmetry) with a self-loop at each vertex (due to reflexivity). Furthermore, in each component every vertex is related to (i.e., joined by an edge to) every other vertex.

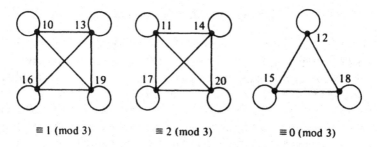

Fig. 9-6 Equivalence graph.

In general, an equivalence relation on a set partitions the elements of the set into classes (called *equivalence classes*) such that two elements are in the same class if and only if they are related. Symmetry ensures that there is no ambiguity regarding membership in the equivalence class; otherwise, x_i may have been related to x_j but not vice versa. Transitivity ensures that in each component every vertex is joined to every other vertex, because if a is related to b and b is related to c, a is also related to c. Transitivity also guarantees that no element can be in more than one class. Reflexivity allows an element to be in a class by itself, if it is not related to any other element in the set.

Relation Matrices: A binary relation R on a set can also be represented by a matrix, called a *relation matrix*. It is a (0, 1), n by n matrix, where n is the number of elements in the set. The i, jth entry in the matrix is 1 if $x_i R x_j$ is true, and is 0, otherwise. For example, the relation matrix of the relation "is greater than" on the set of integers $\{3, 4, 7, 5, 8\}$ is

$$\begin{array}{c c} & \begin{array}{c c c c c} 3 & 4 & 7 & 5 & 8 \end{array} \\ \begin{array}{c} 3 \\ 4 \\ 7 \\ 5 \\ 8 \end{array} & \left[\begin{array}{c c c c c} 0 & 0 & 0 & 0 & 0 \\ 1 & 0 & 0 & 0 & 0 \\ 1 & 1 & 0 & 1 & 0 \\ 1 & 1 & 0 & 0 & 0 \\ 1 & 1 & 1 & 1 & 0 \end{array} \right] \end{array}.$$

We shall see in Section 9-8 that this is precisely the adjacency matrix of the digraph representing the binary relation.

9-4. DIRECTED PATHS AND CONNECTEDNESS

Walks, paths, and circuits in a directed graph, in addition to being what they are in the corresponding undirected graph, have the added consideration of orientation. For example, in Fig. 9-1, the sequence of vertices and edges $v_5 \, e_8 \, v_3 \, e_6 \, v_4 \, e_3 \, v_1$ is a path "directed" from v_5 to v_1, whereas $v_5 \, e_7 \, v_4 \, e_6 \, v_3 \, e_1 \, v_1$ (although a path in the corresponding undirected graph) has no such consistent direction from v_5 to v_1. A distinction must be made between these two types of paths. It is natural to call the first one a *directed path* from v_5 to v_1, and the second one a *semipath*. The word "path" in a digraph could mean either a directed path or a semipath, and similarly for walks, circuits, and cutsets. More precisely:

A *directed walk* from a vertex v_i to v_j is an alternating sequence of vertices and edges, beginning with v_i and ending with v_j, such that each edge is oriented from the vertex preceding it to the vertex following it. Of course, no edge in a directed walk appears more than once, but a vertex may appear more than once, just as in the case of undirected graphs. A *semiwalk* in a directed graph is a walk in the corresponding undirected graph, but is *not* a *directed walk*. A *walk* in a digraph can mean either a directed walk or a semiwalk.

The definitions of *circuit*, *semicircuit*, and *directed circuit* can be written similarly. Let us turn to Fig. 9-1 once more. The set of edges $\{e_1, e_6, e_3\}$ is a directed circuit. But $\{e_1, e_6, e_2\}$ is a semicircuit. Both of them are circuits.

Connected Digraphs: In Chapter 2 a graph (i.e., undirected graph) was defined as connected if there was at least one path between every pair of vertices. In a digraph there are two different types of paths. Consequently, we have two different types of connectedness in digraphs. A digraph G is said to be *strongly connected* if there is at least one directed path from every vertex to every other vertex. A digraph G is said to be *weakly connected* if its corresponding undirected graph is connected but G is not strongly connected. In Fig. 9-2 one of the digraphs is strongly connected, and the other one is weakly connected. The statement that a digraph G is connected simply means that its corresponding undirected graph is connected; and thus G may be strongly or weakly connected. A directed graph that is not connected is dubbed as disconnected.

Since there are two types of connectedness in a digraph, we can define two types of components also. Each maximal connected (weakly or strongly) subgraph of a digraph G will still be called a *component* of G. But within each component of G the maximal strongly connected subgraphs will be called the *fragments* (or *strongly connected fragments*) of G.

For example, the digraph in Fig. 9-7 consists of two components. The component g_1 contains three fragments $\{e_1, e_2\}$, $\{e_5, e_6, e_7, e_8\}$, and $\{e_{10}\}$. Observe that e_3, e_4, and e_9 do not appear in any fragment of g_1.

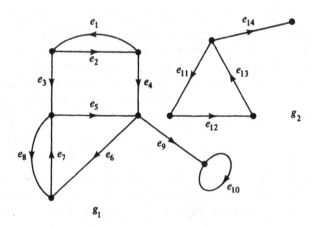

Fig. 9-7 Disconnected digraph with two components.

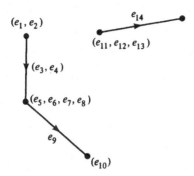

Fig. 9-8 Condensation of Fig. 9-7.

Condensation: The *condensation* G_c of a digraph G is a digraph in which each strongly connected fragment is replaced by a vertex, and all directed edges from one strongly connected component to another are replaced by a single directed edge. The condensation of the digraph G in Fig. 9-7 is shown in Fig. 9-8.

Two observations can be made from the definition:

1. The condensation of a strongly connected digraph is simply a vertex.

2. The condensation of a digraph has no directed circuit.

Accessibility: In a digraph a vertex b is said to be *accessible* (or *reachable*) from vertex a if there is a directed path from a to b. Clearly, a digraph G is strongly connected if and only if every vertex in G is accessible from every other vertex.

9-5. EULER DIGRAPHS

The notion of the Euler graph can be extended to digraphs also. In a digraph G a closed directed walk (i.e., a directed walk that starts and ends at the same vertex) which traverses every edge of G exactly once is called a *directed Euler line.* A digraph containing a directed Euler line is called an *Euler digraph.* The graph in Fig. 9-9 is an Euler digraph, in which the walk $a\ b\ c\ d\ e\ f$ is an Euler line.

When is a digraph an Euler digraph? Clearly, the digraph must be connected, with the possible exception of isolated vertices; otherwise, every edge cannot be traversed in one walk. In fact, an Euler digraph must be strongly connected, although every strongly connected digraph need not be an Euler digraph (Problem 9-13). Theorem 9-1 (whose proof follows the proof of Theorem 2-4 almost verbatim) provides a very simple test for determining whether or not a digraph has an Euler line.

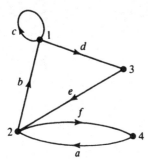

Fig. 9-9 Euler digraph.

THEOREM 9-1

A digraph G is an Euler digraph if and only if G is connected and is balanced [i.e., $d^-(v) = d^+(v)$ for every vertex v in G].

Let us now consider an application of the Euler digraph for solving an important problem in communication theory. The problem, which is often referred to as the *teleprinter's problem*, was solved in 1940 by I. G. Good using the digraph, and was presented in a classic paper in 1946 [9-2].

Teleprinter's Problem: How long is a longest circular (or cyclic) sequence of 1's and 0's such that no subsequence of r bits appears more than once in the sequence? Construct one such longest sequence.

Solution: Since there are 2^r distinct r-tuples formed from 0 and 1, the sequence can be no longer than 2^r bits long. Using Theorem 9-1, we shall construct a circular sequence 2^r bits long with the required property that no subsequence of r bits be repeated.

Construct a digraph G whose vertices are all $(r-1)$-tuples of 0's and 1's. Clearly, there are 2^{r-1} vertices in G. Let a typical vertex be

$$\alpha_1 \alpha_2 \ldots \alpha_{r-1}, \quad \text{where } \alpha_i = 0 \text{ or } 1.$$

Draw an edge directed from this vertex $(\alpha_1 \alpha_2 \ldots \alpha_{r-1})$ to each of two vertices $(\alpha_2 \alpha_3 \ldots \alpha_{r-1} 0)$ and $(\alpha_2 \alpha_3 \ldots \alpha_{r-1} 1)$; label these directed edges $\alpha_1 \alpha_2 \ldots \alpha_{r-1} 0$ and $\alpha_1 \alpha_2 \ldots \alpha_{r-1} 1$, respectively. Draw two such edges directed from each of the 2^{r-1} vertices. (A self-loop will result in each of the two cases when $\alpha_1 = \alpha_2 = \cdots = \alpha_{r-1} = 0$ or 1.)

The resulting digraph is an Euler digraph because for each vertex the in-degree equals the out-degree (each being equal to two). A directed Euler line in G consists of the 2^r edges, each with a distinct r-bit label. The labels of any two consecutive edges in the Euler line are of the form $\alpha_1 \alpha_2 \ldots \alpha_{r-1} \alpha_r$; $\alpha_2 \alpha_3 \ldots \alpha_r \alpha_{r+1}$; that is, the $r-1$ trailing bits of the first edge are identical to the $r-1$ leading bits of the second edge. Thus in the sequence of 2^r bits, made of the first bit of each of the edges in the Euler line, every possible

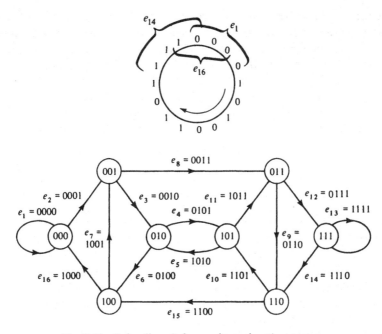

Fig. 9-10 Euler digraph for maximum-length sequence.

subsequence of r bits occurs as the label of an edge; and since no two edges have the same label, no subsequence occurs more than once. The circular arrangement is achieved by joining the two ends of the sequence.)

For $r = 4$, the graph in Fig. 9-10 illustrates the procedure of obtaining such a maximum-length sequence. One such sequence is 0000101001101111 corresponding to the walk $e_1\ e_2\ e_3\ e_4\ e_5\ e_6\ e_7\ e_8\ e_9\ e_{10}\ e_{11}\ e_{12}\ e_{13}\ e_{14}\ e_{15}\ e_{16}$.

This problem is also encountered in locating the position of a rotating drum (Problem 9-19) with the surface carrying two different types of marks. The problem of the rotating drum is similar to that of a feedback shift register (to be discussed in Chapter 12). The quest for a word in which each arrangement of r letters of a given alphabet appears exactly once, encountered in cryptography, is also the same problem (Problem 9-20).

It is not difficult to see that the alphabet size need not be 2. It could be any number m. In that case, the maximum-length sequence is m^r symbols long. The in-degree and out-degree of each vertex in the corresponding Euler digraph equals m, rather than two.

Number of Euler Lines: In Fig. 9-10 there is more than one Euler line. In fact, the digraph has 16 distinct Euler lines. (Note that rotations of the same sequence of edges are not considered distinct.) Finding the number of distinct directed Euler lines in a given Euler digraph is also of interest in many applications. This problem of enumeration was solved by N. G. deBruijn in

1946 for those regular Euler digraphs in which the in-degree and out-degree of every vertex were exactly two, the digraph in Fig. 9-10, for example. In 1951 van Aardenne-Ehrenfest and deBruijn [9-9] solved the more general problem of counting the number of distinct directed Euler lines in any Euler digraph.

The number of distinct Euler lines in a balanced, connected digraph G can be obtained by counting certain types of spanning trees in G, to be covered in the next section.

9-6. TREES WITH DIRECTED EDGES

A tree (for undirected graphs) was defined as a connected graph without any circuit. The basic concept as well as the term "tree" remains the same for digraphs. A *tree* is a connected digraph that has no circuit—neither a directed circuit nor a semicircuit. A tree of n vertices contains $n - 1$ directed edges and has properties similar to those with undirected edges. Trees with directed edges are of great importance in many applications, such as electrical network analysis, game theory, theory of languages, computer programming, and counting problems, to name a few.

In addition to being trees in the undirected sense, trees in digraphs have additional properties and variations resulting from the relative orientations of the edges. One such particularly useful type of rooted tree with directed edges is called an arborescence and is defined as follows:

Arborescence: A digraph G is said to be an arborescence if

1. G contains no circuit—neither directed nor semicircuit.

2. In G there is precisely one vertex v of zero in-degree.

This vertex v is called the *root of the arborescence*. An arborescence is shown in Fig. 9-11.

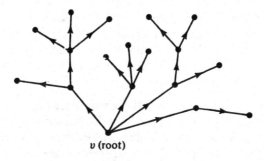

Fig. 9-11 Arborescence.

Theorem 9-2

An arborescence is a tree in which every vertex other than the root has an in-degree of exactly one.

Proof: An arborescence with n vertices can have at most $n - 1$ edges because of condition 1. Therefore, the sum of in-degrees of all vertices in G

$$d^-(v_1) + d^-(v_2) + \cdots + d^-(v_n) \leq n - 1.$$

Of the n terms on the left-hand side of this equation, only one is zero because of condition 2; others must all be positive integers. Therefore, they must all be 1's. Now, since there are exactly $n - 1$ vertices of in-degree one and one vertex of in-degree zero, digraph G has exactly $n - 1$ edges. Since G is also circuitless, it must be connected, and hence a tree. ∎

An arborescence is in a sense a tree directed out of the root. Therefore, an arborescence is sometimes referred to as an *out-tree*. (Reversing the direction of every edge in an arborescence will produce what may be called an *in-tree*.)

Theorem 9-3

In an arborescence there is a directed path from the root R to every other vertex. Conversely, a circuitless digraph G is an arborescence if there is a vertex v in G such that every other vertex is accessible from v, and v is not accessible from any other vertex.

Proof: (a) In an arborescence consider a directed path P starting from the root R and continuing as far as possible. P can end only at a pendant vertex; otherwise, we get a vertex whose in-degree is two or more. A contradiction.

Since an arborescence is connected, every vertex lies on some directed path from the root R to each of the pendant vertices.

(b) Conversely, since every vertex in G is accessible from v, and G has no circuit, G is a tree. Moreover, since v is not accessible from any other vertex, $d^-(v) = 0$. Every other vertex is accessible from v, and therefore the in-degree of each of these vertices must be at least one. The in-degree cannot be greater than one because there are only $n - 1$ edges in G (n being the number of vertices in G). ∎

The following is an important application of arborescences to the theory of computer algorithms.

Polish Notation: Consider the arithmetic expression

$$a + b - \frac{c \cdot d}{g^x - f}. \tag{9-1}$$

In a procedural language (such as FORTRAN or ALGOL) this expression might be written as

$$a + b - c \cdot d \div (g \uparrow x - f), \qquad (9\text{-}2)$$

where $\uparrow$ denotes exponentiation.

In evaluating this expression the computer must perform the arithmetic operations in a certain order; otherwise, it will produce a wrong result. Let us number the operations in this expression in the order in which they might be performed.

$$a \underset{\underset{\textcircled{6}}{\uparrow}}{\,} + b \underset{\underset{\textcircled{5}}{\uparrow}}{\,} - c \underset{\underset{\textcircled{3}}{\uparrow}}{\cdot} d \underset{\underset{\textcircled{4}}{\uparrow}}{\div} (g \underset{\underset{\textcircled{1}}{\uparrow}}{\uparrow} x \underset{\underset{\textcircled{2}}{\uparrow}}{\,} - f). \qquad (9\text{-}3)$$

To evaluate such an expression, the machine will have to scan the expression back and forth to find the sequence of operations to be performed.

To avoid scanning back and forth, the computer makes a preliminary translation of expressions such as (9-2) into the *Polish notation* (invented by the Polish logician, Lukasiewicz). Polish notation is also called *parenthesis-free notation*, because it contains no parentheses. This notation has the advantage that the operations appear exactly in the same order as they are performed.

The basic idea in Polish notation is that a binary operator appears just to the left of the two operands rather than in between the two operands. Thus $x + y$ is written as $+ xy$. The translation of expressions from procedural language into Polish notation is extremely important in compiling and can be accomplished by first representing the given expression by means of an arborescence as follows: Each variable (or constant) appearing in the expression is represented by a pendant vertex. Each internal vertex represents a binary operator having the two subarborescences as its operands. An arborescence for expression (9-2) is shown in Fig. 9-12.

To obtain the expression in Polish notation, we traverse the arborescence

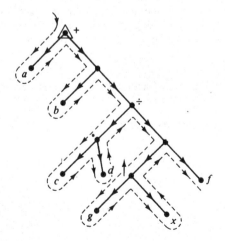

Fig. 9-12 Arborescence for
$a + b - c \cdot d \div (g \uparrow x - f)$.

starting from the root from left to right and from top to bottom, as indicated by the dotted line in Fig. 9-12. Each time we come across a vertex that has not been traversed before, we append its label to the existing string. This process in Fig. 9-12 yields

$$+ a - b \div \cdot cd - \uparrow gxf. \tag{9-4}$$

An expression in Polish notation is evaluated as follows: We start at the right extreme and move to the left. Whenever an operator is encountered the operation is performed between the two operands immediately to the right of it. After an operation is performed, the resultant is regarded as one operand for the next operation. You can verify that under this procedure expression (9-4) is equal to (9-2). The advantage of expression (9-4) over (9-2) is that in (9-4) there are no parentheses and the operators appear in the order (from right to left) in which they are to be acted upon. Therefore, no back and forth scanning is required during the computation.

Ordered Trees: You must have noticed that in the expression arborescence of Fig. 9-12 the relative positions of the vertices in the plane of the paper—left or right, up or down—are important and must be preserved. In this sense it is a "rigid" graph, and a graph isomorphic to it may not preserve its properties. This in fact is not purely a graph-theoretic problem; and this is the first and only time in this book we shall consider such a structure.

In computer literature a tree in which the relative order of subtrees meeting at each vertex must be preserved is called an *ordered tree* or a *planar tree* (because the tree can be visualized as rigidly embedded in the plane of the paper). In computer science the term tree usually means an ordered tree, and by convention a tree is drawn hanging down with the root at the top.

Spanning Arborescence: A spanning tree in an n-vertex connected digraph, analogous to a spanning tree in an undirected graph, consists of $n - 1$ directed edges. A *spanning arborescence* in a connected digraph is a spanning tree that is an arborescence. For example, a spanning arborescence in Fig. 9-13 is $\{f, b, d\}$. There is a striking relationship between a spanning arborescence and an Euler line. This is brought out by Theorems 9-4 and 9-5.

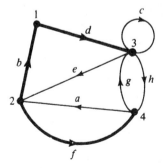

Fig. 9-13 Euler digraph.

Theorem 9-4

In a connected, balanced digraph G of n vertices and m edges, let $W = (e_1, e_2, \ldots, e_m)$ be an Euler line, which starts and ends at a vertex v (i.e., v is the initial vertex of e_1 and the terminal vertex of e_m). Among the m edges in W there are $n-1$ edges that "enter" each of $n-1$ vertices, other than v, for the first time. The subdigraph g of these $n-1$ directed edges together with the n vertices is a spanning arborescence of G, rooted at vertex v.

Illustration: Before proving the theorem, let us look at an example. In Fig. 9-13, $W = (b\ d\ c\ e\ f\ g\ h\ a)$ is an Euler line, starting and ending at vertex 2. The subdigraph $\{b, d, f\}$ is a spanning arborescence rooted at vertex 2.

Proof: In the subdigraph g, vertex v is of in-degree zero, and every other vertex is of in-degree one; for g includes exactly one edge going to each of the $n-1$ vertices, and no edge going to v. Moreover, the way g is defined in W, g is connected and contains $n-1$ directed edges. Therefore, g is a spanning arborescence in G and is rooted at v. ∎

Theorem 9-4 provides a method of obtaining a spanning arborescence rooted at any specified vertex, provided the digraph is Eulerian. Conversely, given a spanning arborescence in an Euler digraph, an Euler line can be constructed using Theorem 9-5. This important result discovered by T. van Aardenne-Ehrenfest and N. G. deBruijn in 1951 is used in counting the number of distinct Euler lines.

For the sake of traversing the edges along with rather than opposite to the direction of edges, it is better to express Theorem 9-5 in terms of an in-tree, that is, an arborescence in which the direction of every edge has been inverted. In Fig. 9-14 the subdigraph $\{e_2, e_3, e_7, e_{10}, e_{11}\}$ is a spanning in-tree.

Theorem 9-5

Let G be an Euler digraph and T be a spanning in-tree in G, rooted at a vertex R. Let e_1 be an edge in G incident out of the vertex R. Then a directed walk $W = (e_1, e_2, \ldots, e_m)$ is a directed Euler line, if it is constructed as follows:

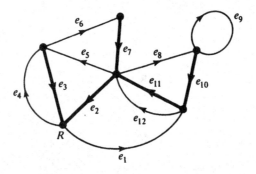

Fig. 9-14 Spanning in-tree rooted at R.

1. No edge is included in W more than once.
2. In exiting a vertex the one edge belonging to T is not used until all other outgoing edges have been traversed.
3. The walk is terminated only when a vertex is reached from which there is no edge left on which to exit.

Proof: The walk W must terminate at R, because all vertices must have been entered as often as they have been left (because G is balanced). Now suppose there is an edge a in G that has not been included in W. Let v be the terminal vertex of a. Since G is balanced, v must also be the initial vertex of some edge b not included in W. Edge b going out of vertex v must be in T, according to rule 1. This omitted edge leads to another omitted edge c in T, and so on. Ultimately, we arrive at R, and find an outgoing edge there not included in W. This contradicts rule 3. ■

The number of distinct Euler lines formed from a given in-tree T, and starting with edge e_1 at R, can be computed by considering all the choices available at each vertex, after starting with e_1.

Since there is exactly one outgoing edge in T at each vertex and this edge is to be selected last (rule 2), the remaining $d^+(v_i) - 1$ edges at vertex v_i can be chosen in

$$[d^+(v_i) - 1]! \text{ ways.}$$

And since these are independent choices, we have altogether

$$\prod_{i=1}^{n} [d^+(v_i) - 1]! \qquad (9\text{-}5)$$

different Euler lines that meet the three rules in Theorem 9-5.

Let us apply these three rules to obtain different directed Euler lines in Fig. 9-14, from the in-tree $\{e_2, e_3, e_7, e_{10}, e_{11}\}$, starting with edge e_1. We get the following two directed Euler lines:

$$(e_1\ e_{12}\ e_5\ e_6\ e_7\ e_8\ e_9\ e_{10}\ e_{11}\ e_2\ e_4\ e_3),$$

and

$$(e_1\ e_{12}\ e_8\ e_9\ e_{10}\ e_{11}\ e_5\ e_6\ e_7\ e_2\ e_4\ e_3).$$

The value of expression (9-5) for Fig. 9-14 is 2. Note that these are not all the directed Euler lines in the digraph, but only those that are generated by the specific in-tree in accordance with the rules in Theorem 9-5.

The result in Theorem 9-5 may seem contrived at first sight, but it is a very natural step in counting the number of distinct directed Euler lines in a digraph, which will be undertaken in Section 9-9.

9-7. FUNDAMENTAL CIRCUITS IN DIGRAPHS

The edges of a connected digraph not included in a specified spanning tree T are also called chords with respect to T. Just as in the case of undirected graphs, every chord c_i when added to the spanning tree T produces a fundamental circuit (which may be a directed circuit or a semicircuit).

A cut-set in a connected digraph G (just as in an undirected graph) induces a partitioning of the vertices of G into two disjoint subsets V_1 and V_2 such that the cut-set consists of all those edges that have one end vertex in V_1 and the other in V_2. All edges in the cut-set may be directed from V_1 to V_2, or vice versa, or some edges may be directed from V_1 to V_2 and others from V_2 to V_1.†

The concepts of spanning trees, fundamental circuits, and fundamental cut-sets are illustrated in Fig. 9-15. A spanning tree is shown in heavy lines. Observe that some of the fundamental circuits are directed circuits and others are semicircuits. The five fundamental cut-sets, each corresponding to an edge in the spanning tree, are also shown.

Ring Sum of Circuits: Just as in undirected graphs, we can define operations between subgraphs of a digraph. In particular, the ring sum of two subdigraphs $g_1 \oplus g_2$ is another subdigraph consisting of edges that are either in g_1 or in g_2 but not in both.

As in undirected graphs, the ring sum of two circuits (directed or semicircuit) in a digraph is either a third circuit or a union of edge-disjoint circuits. For if we disregard the directions of edges in the circuits, the earlier results from undirected graphs are also applicable to digraphs. Under the ring-sum operation, $\oplus$, circuits and unions of edge-disjoint circuits form a group. Every element of this group can be expressed as a ring sum of some of the fundamental circuits with respect to a spanning tree. The same holds for cut-sets also.

Set of Directed Circuits Only: The most important property of a set of $\mu \, (= e - n + 1)$ fundamental circuits in a connected graph G (directed or undirected) is that these circuits form a basis for all circuits in G. Any circuit (directed or semicircuit in a digraph) can be obtained as a ring sum of some of these circuits. But in many applications we are interested only in the set of directed circuits in a digraph. Is there a similar basis for all directed circuits?

The answer, unfortunately, is no. The ring sum of two directed circuits is not necessarily another directed circuit or edge-disjoint union of directed circuits. For example, in Fig. 9-15 the ring sum of directed circuits $a\,c\,b \oplus a\,d\,g\,b = d\,g\,c$, a semicircuit. In fact, it can be shown that there exists no binary operation under which all directed circuits (and edge-disjoint unions of directed circuits) form a group, let alone a vector space.

†A cut-set in which all edges are oriented in the same direction is called a *directed cut-set*.

SEC. 9-8 MATRICES A, B, AND C OF DIGRAPHS

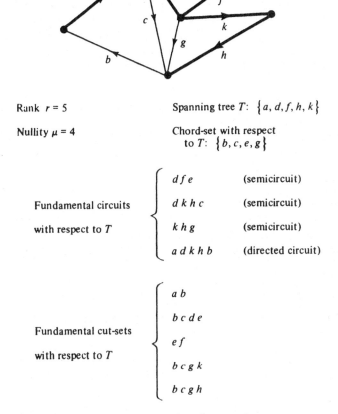

Rank $r = 5$

Nullity $\mu = 4$

Spanning tree T: $\{a, d, f, h, k\}$

Chord-set with respect to T: $\{b, c, e, g\}$

Fundamental circuits with respect to T
$$\begin{cases} d\,f\,e & \text{(semicircuit)} \\ d\,k\,h\,c & \text{(semicircuit)} \\ k\,h\,g & \text{(semicircuit)} \\ a\,d\,k\,h\,b & \text{(directed circuit)} \end{cases}$$

Fundamental cut-sets with respect to T
$$\begin{cases} a\,b \\ b\,c\,d\,e \\ e\,f \\ b\,c\,g\,k \\ b\,c\,g\,h \end{cases}$$

Fig. 9-15 Directed graph and a spanning tree.

The practical significance of this situation is to be discussed in connection with acyclic digraphs, the topic of Section 9-11.

9-8. MATRICES A, B, AND C OF DIGRAPHS

The matrices associated with a digraph are quite similar to those discussed in Chapter 7 for an undirected graph, with the following basic difference. In order to account for the orientation of the edges, the incidence, circuits, and cut-set matrices consist of $+1, 0, -1$ (instead of only 0 and 1 for undirected graphs). The numbers $+1, 0, -1$ are regarded as ordinary real numbers. Their addition and multiplication are interpreted as in ordinary arithmetic (and not modulo 2 arithmetic as in the case of undirected graphs). Consequently, the vectors and vector spaces associated with a digraph and its subdigraphs are over the field of all real numbers, and not GF(2).

Incidence Matrix: The *incidence matrix of a digraph* with n vertices, e edges, and no self-loops is an n by n matrix $A = [a_{ij}]$, whose rows correspond to vertices and columns correspond to edges, such that

$$a_{ij} = 1, \quad \text{if } j\text{th edge is incident out of } i\text{th vertex,}$$
$$= -1, \quad \text{if } j\text{th edge is incident into } i\text{th vertex,}$$
$$= 0, \quad \text{if } j\text{th edge is not incident on } i\text{th vertex.}$$

A digraph and its incidence matrix are shown in Fig. 9-16. Observe that if we disregard the orientations of the edges and correspondingly change -1 to 1 in the incidence matrix, Fig. 9-16 becomes identical to Fig. 7-1.

Observations 1–6 made in Section 7-1 on the properties of the incidence matrix of an undirected graph, with minor changes, also hold for digraphs.

Since the sum (in the real field) of each column is zero, the rank of the incidence matrix of a digraph of n vertices is less than n. In fact, we have the following theorem, identical to Theorem 7-2, which can be proved along similar lines.

THEOREM 9-6

If $A(G)$ is the incidence matrix of a connected digraph of n vertices, the rank of $A(G) = n - 1$.

Deleting any one row from A we get A_f, the $(n - 1)$ by e reduced incidence matrix. The vertex corresponding to the deleted row is called the reference vertex.

Unimodularity of A: It was observed in Chapter 7 that if A is the incidence matrix of an undirected graph, the determinant of every square submatrix of A is either 0 or 1. This was a result of the fact that the determinant was defined in modulo 2 arithmetic and, therefore, could have no other value.

In the case of digraphs, the incidence matrix A is in the real field, and on first sight it would appear that the determinants of its square submatrices could acquire any integral value. This, however, is not the case, as shown in the following important theorem.

THEOREM 9-7

The determinant of every square submatrix of A, the incidence matrix of a digraph, is 1, -1, or 0.

Proof: The theorem can be proved directly by expanding the determinant of a square submatrix of A. Consider a k by k submatrix M of A. If M has any column or row consisting of all zeros, det M is clearly zero. Also det $M = 0$ if every column of M contains the two nonzero entries, a 1 and a -1.

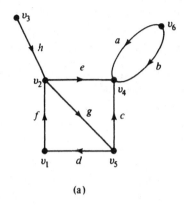

	a	b	c	d	e	f	g	h
v_1	0	0	0	-1	0	1	0	0
v_2	0	0	0	0	1	-1	1	-1
v_3	0	0	0	0	0	0	0	1
v_4	-1	-1	-1	0	-1	0	0	0
v_5	0	0	1	1	0	0	-1	0
v_6	1	1	0	0	0	0	0	0

(b)

Fig. 9-16 Digraph and its incidence matrix.

Now if det $M \neq 0$ (i.e., M is nonsingular), then the sum of entries in each column of M cannot be zero. Therefore, M must have a column in which there is a single nonzero element that is either $+1$ or -1. Let this single element be in the (i,j)th position in M. Thus

$$\det M = \pm 1 \cdot \det M_{ij},$$

where M_{ij} is the submatrix of M with its ith row and jth column deleted. The $(k-1)$ by $(k-1)$ submatrix M_{ij} is also nonsingular (because M is nonsingular); therefore, it too must have at least one column with a single nonzero entry, say in the (p,q)th position. Expanding det M_{ij} about this element in the (p,q)th position,

$$\det M_{ij} = \pm[\text{determinant of a nonsingular} \\ (k-2) \text{ by } (k-2) \text{ submatrix of M}].$$

Repeated application of this procedure yields

$$\det M = \pm 1.$$

Hence the theorem. ∎

Any matrix with every square submatrix having a determinant of 1, −1, or 0 is called a *unimodular matrix*. (Unimodular matrices also play an important role in linear programming.)

Circuit Matrix of a Digraph: Let G be a digraph with e edges and q circuits (directed circuits or semicircuits). An arbitrary orientation (clockwise or counterclockwise) is assigned to each of the q circuits. Then a circuit matrix $B = [b_{ij}]$ of the digraph G is a q by e matrix defined as

$b_{ij} = 1,$ if ith circuit includes jth edge, and the orientations of the edge and circuit coincide,

$= -1,$ if ith circuit includes jth edge, but the orientations of the two are opposite,

$= 0,$ if ith circuit does not include the jth edge.

For example, a circuit matrix of the digraph in Fig. 9-16 is

$$\begin{array}{cccccccc} a & b & c & d & e & f & g & h \end{array}$$
$$\begin{bmatrix} 0 & 0 & 0 & 1 & 0 & 1 & 1 & 0 \\ 0 & 0 & 1 & 0 & -1 & 0 & 1 & 0 \\ 0 & 0 & 1 & -1 & -1 & -1 & 0 & 0 \\ -1 & 1 & 0 & 0 & 0 & 0 & 0 & 0 \end{bmatrix}.$$

Note that the orientation assigned to each of the four circuits is entirely arbitrary. The circuit in the first row is assigned clockwise orientation, in the second row counterclockwise, the third counterclockwise, and the fourth clockwise. Changing the orientation of any circuit will simply change the sign of every nonzero entry in the corresponding row. Also note that if we subtract the first row from the second, we get the third row. Thus the rows are not all linearly independent (in the real field, of course).

Observations 1–7 made in Section 7-3 about the circuit matrix of an undirected graph are applicable to the circuit matrix of a digraph also—with some obvious minor changes. Just as for undirected graphs, the rows of the circuit matrix are orthogonal to the rows of the incidence matrix (this time in the real field), as proved in Theorem 9-8.

THEOREM 9-8

Let B and A be, respectively, the circuit matrix and incidence matrix of a self-loop-free digraph such that the columns in B and A are arranged using the same order of edges. Then

$$A \cdot B^T = B \cdot A^T = 0,$$

where superscript T denotes the transposed matrix.

Proof: Consider the mth row in B and the kth row in A. If the circuit m does not include any edge incident on vertex k, the product of the two rows is clearly zero. If, on the other hand, vertex k is in circuit m, there are exactly two edges (say x and y) incident on k that are also in circuit m. This situation can occur in only four different ways, as shown in Fig. 9-17. (The other four cases with the orientation of m reversed are identical to these when x and y are interchanged.)

The possible entries in row k of A and row m of B in column positions x and y are tabulated for each of these four cases.

Case	Row k		Row m		Dot Product
	Column x	Column y	Column x	Column y	Row $k \cdot$ Row m
(a)	-1	1	1	1	0
(b)	1	-1	-1	-1	0
(c)	-1	-1	1	-1	0
(d)	1	1	-1	1	0

In each case, the dot product is zero. Therefore, the theorem. ∎

Using Sylvester's theorem (Appendix B) and Theorem 9-8, it can be shown

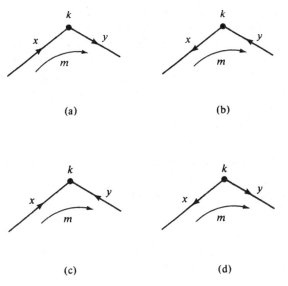

Fig. 9-17 Vertex k in circuit m.

that in a digraph with e edges

$$\text{rank of } \mathbf{B} + \text{rank of } \mathbf{A} = e.$$

Moreover, for a connected graph

$$\text{rank of } \mathbf{A} = n - 1,$$

and therefore

$$\text{rank of } \mathbf{B} = e - n + 1.$$

The following two important properties of matrices $\mathbf{A}$ and $\mathbf{B}$ hold for digraphs also, and can be proved as was done for undirected graphs in Chapter 7 (except that here we are working in ordinary real arithmetic and not in modulo 2 arithmetic).

1. The nonsingular submatrices of order $n - 1$ of $\mathbf{A}$ are in one-to-one correspondence with the spanning trees of the connected digraph of n vertices.
2. The nonsingular submatrices of $\mathbf{B}$ of order $\mu \ (= e - n + 1)$ are in one-to-one correspondence with the chord set (complement of the spanning tree) of the connected digraph of n vertices and e edges.

Sign of a Spanning Tree: For a digraph the determinant of the nonsingular submatrix of $\mathbf{A}$ corresponding to a spanning tree T can assume either a value of $+1$ or -1. This is referred to as the *sign* of T.

As we shall see in Chapter 13, in the analysis of a certain class of electrical networks it is necessary to know the signs of the spanning trees. Note that the sign of a spanning tree is defined only for a particular ordering of vertices and edges in $\mathbf{A}$ (because interchanging two rows or columns in a matrix changes the sign of its determinant). Thus the sign of a spanning tree is relative. Once the sign of one spanning tree is arbitrarily chosen, the sign of every other spanning tree is determined as positive or negative with respect to this spanning tree.

Number of Spanning Trees: We have Theorem 9-9 for determining the number of spanning trees in a connected digraph. (An identical result for undirected graphs was given in Problem 7-20.)

THEOREM 9-9

Let $\mathbf{A}_f$ be the reduced incidence matrix of a connected digraph. Then the number of spanning trees in the graph equals the value of

$$\det(\mathbf{A}_f \cdot \mathbf{A}_f^T).$$

SEC. 9-8   MATRICES A, B, AND C OF DIGRAPHS

Proof: According to the Binet–Cauchy theorem (Appendix A)

$$\det(A_f \cdot A_f^T) = \text{sum of the products of all corresponding majors of } A_f \text{ and } A_f^T.$$

Every major of A_f or A_f^T is zero unless it corresponds to a spanning tree, in which case its value is 1 or -1. Since both majors of A_f and A_f^T have the same value $+1$ or -1, the product is $+1$ for each spanning tree. ∎

Fundamental Circuit Matrix: The μ fundamental circuits each made by a chord (with respect to some specified spanning tree) define a fundamental circuit matrix B_f for a digraph. The orientation assigned to each of the fundamental circuits is chosen to coincide with that of the chord. Therefore, B_f, a μ by e matrix, can be expressed exactly in the same form as in the case of an undirected graph in Section 7-4:

$$B_f = [I_\mu \mid B_t],$$

where I_μ is the identity matrix of order μ, and the columns of B_t correspond to the edges in a spanning tree. This is illustrated in Fig. 9-18.

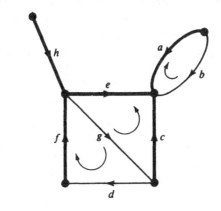

$$B_f = \begin{bmatrix} & b & d & g & a & c & e & f & h \\ & 1 & 0 & 0 & -1 & 0 & 0 & 0 & 0 \\ & 0 & 1 & 0 & 0 & -1 & 1 & 1 & 0 \\ & 0 & 0 & 1 & 0 & 1 & -1 & 0 & 0 \end{bmatrix}$$

$$= [I_\mu \mid B_t]$$

Fig. 9-18 Digraph and its fundamental circuit matrix.

The cut-set matrix C of a digraph G is also similarly defined. And so is its submatrix C_f, the fundamental cut-set matrix with respect to a given spanning tree in G.

9-9. ADJACENCY MATRIX OF A DIGRAPH

Another important matrix used in the representation and study of digraphs is the *adjacency matrix* defined as follows: Let G be a digraph with n vertices, containing no parallel edges. Then the adjacency matrix $\mathbf{X} = [x_{ij}]$ of the digraph G is an n by n (0, 1)-matrix whose element

$x_{ij} = 1,$ if there is an edge directed from ith vertex to jth vertex,
$\phantom{x_{ij}} = 0,$ otherwise.

A digraph and its adjacency matrix are shown in Fig. 9-19.

The adjacency matrix occurs in many different disciplines, and therefore has different names. In the theory of sequential machines it is called the *transition matrix*. In the calculus of relations it is called the *relation matrix*. (Observe that the relation matrix defined in Section 9-3 is the same as the adjacency matrix of the corresponding digraph.) In network flows it is called the *connection matrix*. It is also known as the *precedence matrix* or *preference matrix* in some sociological applications. In scheduling and critical-path analysis the adjacency matrix is known as the *predecessor matrix*.

Let us make the following observations on the properties of the adjacency matrix $\mathbf{X}$ of a digraph G.

1. $\mathbf{X}$ is a symmetric matrix if and only if G is a symmetric digraph.
2. Every nonzero element on the main diagonal represents a self-loop at the corresponding vertex.

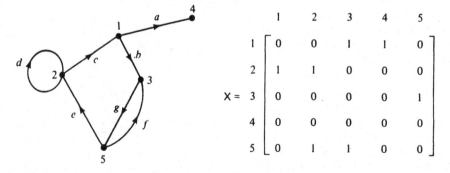

Fig. 9-19 Digraph and its adjacency matrix.

3. There is no way of showing parallel edges in X. This is why the adjacency matrix is defined only for a digraph without parallel edges.
4. The sum of each row equals the out-degree of the corresponding vertex, and the sum of each column equals the in-degree of the corresponding vertex. The number of nonzero entries in X equals the number of edges in G.
5. Permutation of any two rows accompanied by a permutation of the corresponding columns does not alter the graph. The permutation merely corresponds to a reordering of the vertices. Thus two digraphs are isomorphic if and only if their adjacency matrices differ only by such permutations.
6. If X is the adjacency matrix of a digraph G, then the transposed matrix X^T is the adjacency matrix of a digraph G^R obtained by reversing the direction of every edge in G.
7. For any square (0, 1)-matrix Q of order n, there exists a unique digraph G of n vertices, such that Q is the adjacency matrix of G.

The adjacency matrix is used as a tool to investigate the properties of a digraph, specially by means of a digital computer. For example, the connectedness of a digraph is reflected in its adjacency matrix in the following fashion.

Connectedness and the Adjacency Matrix: A digraph is disconnected if and only if its vertices can be ordered in such a way that its adjacency matrix X can be expressed as the direct sum of two square submatrices X_1 and X_2 as follows:

$$X = \begin{bmatrix} X_1 & 0 \\ 0 & X_2 \end{bmatrix}. \tag{9-6}$$

Such a partitioning is possible if and only if the vertices in the submatrix X_1 have no edge going to or coming from the vertex set in X_2.

Similarly, a digraph is weakly connected if and only if its vertices can be ordered in such a way that its adjacency matrix X can be expressed in the form (9-7) or (9-8):

$$X = \begin{bmatrix} X_1 & 0 \\ X_{21} & X_2 \end{bmatrix}, \tag{9-7}$$

$$X = \begin{bmatrix} X_1 & X_{12} \\ 0 & X_2 \end{bmatrix}, \tag{9-8}$$

where X_1 and X_2 are square submatrices. Form (9-7) represents the case when there is no directed edge going from the subdigraph corresponding to X_1 to the one corresponding to X_2. Form (9-8) represents the case when there is no directed edge going to the subdigraph corresponding to X_1.

A digraph that is neither unconnected nor weakly connected is strongly connected. Therefore, we conclude that a digraph G is strongly connected if and only if the vertices of G cannot be ordered such that its adjacency matrix X is expressible in the form (9-6), (9-7), or (9-8).

Number of Edge Sequences: We have the following result, similar to Theorem 7-8, concerning the powers of the adjacency matrix X of a digraph G.

THEOREM 9-10

The (i, j)th entry in X^r equals the number of different, directed edge sequences of r edges from the ith vertex to the jth.

Proof: (By induction) The theorem is trivially true for $r = 1$. As the inductive hypothesis, assume that the theorem holds for X^{r-1}. The (i, j)th entry in $X^r (= X^{r-1} \cdot X)$

$$= \sum_{k=1}^{n} [(i, k)\text{th entry in } X^{r-1}] \cdot x_{kj}$$

$$= \sum_{k=1}^{n} (\text{number of all directed edge sequences} \quad (9\text{-}9)$$
$$\text{of length } r - 1 \text{ from vertex } i \text{ to } k) \cdot x_{kj},$$

according to the induction hypothesis. In (9-9), $x_{kj} = 1$ or 0, depending on whether or not there is a directed edge from k to j. Thus a term in the sum (9-9) is nonzero if and only if there is a directed edge sequence of length r from i to j, whose last edge is from k to j. If the term is nonzero, its value equals the number of such edge sequences from i to j via k. This holds for every vertex k, $1 \leq k \leq n$. Therefore, (9-9) is equal to the number of all possible directed edge sequences from i to j. ∎

As in the case of Theorem 7-7, it must be kept in mind that the (i, j)th entry in X^r gives the number of all directed edge sequences from vertex i to j. These edge sequences fall in three different categories:

1. Directed paths from i to j: Those directed edge sequences in which no vertex is traversed more than once.

2. Directed walks from i to j: Those directed edge sequences in which a vertex may be traversed more than once, but no edge is traversed more than once.

3. Those directed edge sequences in which an edge may also be traversed more than once.

Unfortunately, there is no easy way of separating these, say, category 1 from 2 and 3. This is why this simple method cannot be employed for enumerating directed paths or directed circuits of a specified length.

For example, examine the fourth power of the adjacency matrix of the digraph in Fig. 9-19:

$$X^4 = \begin{bmatrix} 1 & 1 & 0 & 0 & 1 \\ 1 & 2 & 2 & 1 & 1 \\ 1 & 2 & 2 & 1 & 0 \\ 0 & 0 & 0 & 0 & 0 \\ 2 & 2 & 1 & 1 & 2 \end{bmatrix}.$$

The entry in the second row and third column represents two directed edge sequences of length four: $c\,b\,g\,f$ (a directed walk from 2 to 3) and $d\,d\,c\,b$ (not a walk). The third diagonal entry represents two directed edge sequences of length four beginning and ending at vertex 3: $g\,e\,c\,b$ (a directed circuit) and $g\,f\,g\,f$ (not a directed circuit). The reader should also examine the remaining entries in X^4.

Number of Arborescences: A method of counting the number of spanning trees in a labeled, undirected graph was suggested in Problem 7-23. There is a similar formula for counting the number of spanning arborescences in a labeled, connected, simple digraph. (Counting of the spanning arborescences in any connected digraph is a trivial extension of counting them in a simple, connected digraph. The self-loops can be discarded right away, and addition of a parallel edge b to an existing edge a simply doubles the number of the arborescences—repeating the same arborescences with a replaced by b.)

In preparation for the arborescence counting formula, let us define, for a simple digraph G of n vertices, an n by n matrix called the *Kirchhoff matrix* $K(G)$ or $K = [k_{ij}]$:

$k_{ii} = d^-(v_i),$ in-degree of the ith vertex,

$k_{ij} = -x_{ij},$ (i, j)th entry in the adjacency matrix, with a negative sign.

For example, a digraph and its K matrix are shown in Fig. 9-20.

The sum of the entries in each column in a K matrix is equal to zero, which means that the n rows are linearly dependent; therefore,

$$\det K = 0.$$

Next we explore the special property of the K matrix of an arborescence.

THEOREM 9-11

A simple digraph G of n vertices and $n-1$ directed edges is an arborescence rooted at v_1 if and only if the (1, 1) cofactor of $K(G)$ is equal to 1.

Proof: (a) Let G be an arborescence with n vertices and rooted at vertex v_1. Relabel the vertices as $v_1, v_2, \ldots, v_n$ such that vertices along every directed path from the root v_1 have increasing indices. Permute the rows and columns of K(G) to conform with this relabeling.

Since the in-degree of v_1 equals zero, the first column contains only zeros. Other entries in K(G) are

$$k_{ij} = 0, \quad i > j,$$
$$k_{ij} = -x_{ij}, \quad i < j,$$
$$k_{ii} = 1, \quad i > 1.$$

Then the K matrix of an arborescence rooted at v_1 is of the form

$$K(G) = \begin{bmatrix} 0 & -x_{12} & -x_{13} & -x_{14} & \cdots & -x_{1n} \\ 0 & 1 & -x_{23} & -x_{24} & \cdots & -x_{2n} \\ 0 & 0 & 1 & -x_{34} & \cdots & -x_{3n} \\ 0 & 0 & 0 & 1 & \cdots & \cdot \\ \cdot & \cdot & \cdot & \cdot & \cdots & \cdot \\ \cdot & \cdot & \cdot & \cdot & \cdots & \cdot \\ 0 & 0 & 0 & 0 & \cdots & 1 \end{bmatrix}$$

Clearly, the cofactor of the (1, 1) entry is 1; that is, det $K_{11} = 1$.

(b) Conversely, let G be a simple digraph of n vertices and $n - 1$ edges, and let the (1, 1) cofactor of its K matrix be equal to 1; that is, det $K_{11} = 1$.

Since det $K_{11} \neq 0$, every column in K_{11} has at least one nonzero entry. Therefore,

$$d^-(v_i) \geq 1, \quad \text{for } i = 2, 3, \ldots, n.$$

There are only $n - 1$ edges to go around. Therefore,

$$d^-(v_i) = 1, \quad \text{for } i = 2, 3, \ldots, n,$$

and

$$d^-(v_1) = 0.$$

Now since no vertex in G has an in-degree of more than one, if G can have any circuit at all, it has to be a directed circuit. Suppose that such a directed circuit exists, which passes through vertices $v_{i_1}, v_{i_2}, \ldots, v_{i_r}$. Then the sum of the columns $i_1, i_2, \ldots, i_r$ in K_{11} is zero. (This is because each of these columns contains exactly two nonzero entries, a 1 on the main diagonal, and a -1 for the incoming edge from the vertex preceding it in the directed circuit.) Thus these r columns in K_{11} are linearly dependent. Hence det $K_{11} = 0$, a contradiction. Therefore, G has no circuits.

If G has $n - 1$ edges and no circuits, it must be a tree. Since in this tree

$$d^-(v_1) = 0,$$

and

$$d^-(v_i) = 1, \quad \text{for } i = 2, 3, \ldots, n,$$

G must be an arborescence rooted at vertex v_1.

The arguments in (a) and (b) are valid for an arborescence rooted at any vertex v_q. Any reordering of the vertices in G corresponds to identical permutations of rows and columns in $K(G)$. Such permutations do not alter the value or sign of the determinant. ∎

Next we come to an important result, which was first discovered by R. Bott and J. P. Mayberry and was proved by W. T. Tutte.

THEOREM 9-12

Let $K(G)$ be the Kirchhoff matrix of a simple digraph G. Then the value of the (q, q) cofactor of $K(G)$ is equal to the number of arborescences in G rooted at the vertex v_q.

Proof: The proof depends on the result of Theorem 9-11 and on the fact that the determinant of a square matrix is a linear function of its columns. Specifically, if P is a square matrix consisting of n column vectors, each of dimension n, that is,

$$P = [p_1, p_2, \ldots, (p_i + p_i'), \ldots, p_n]$$

then

$$\det P = \det[p_1, p_2, \ldots, p_i, \ldots, p_n] \\ + \det[p_1, p_2, \ldots, p_i', \ldots, p_n]. \quad (9\text{-}10)$$

In graph G suppose that vertex v_j has in-degree of d_j. The jth column of $K(G)$ can be regarded as the sum of d_j different columns, each corresponding to a graph in which v_j has in-degree one. And then (9-10) can be repeatedly applied. After this, splitting of columns can be carried out for each j, $j \neq q$, and $\det K_{qq}(G)$ can be expressed as a sum of determinants of subgraphs; that is,

$$\det K_{qq}(G) = \sum_g \det K_{qq}(g), \quad (9\text{-}11)$$

where g is a subgraph of G, with the following properties:
1. Every vertex in g has an in-degree of exactly one, except v_q.
2. g has $n - 1$ vertices, and hence $n - 1$ edges.

From Theorem 9-11,

$$\det K_{qq}(g) = 1, \quad \text{if and only if } g \text{ is an arborescence rooted at } q,$$
$$= 0, \quad \text{otherwise.}$$

Thus the summation in (9-11) carried over all g's equals the number of arborescences rooted at v_q. ∎

Theorem 9-12 is illustrated in Fig. 9-20. The cofactor of every entry in the second row of the K matrix is 3. The digraph does indeed have three arborescences rooted at vertex 2.

For an Euler digraph G, all cofactors of $K(G)$ are equal, because the sum of each row and the sum of each column equals zero. Let this common value

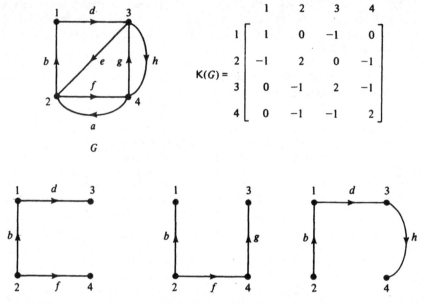

Fig. 9-20 Digraph G, its K matrix, and all arborescences rooted at 2.

of all cofactors of K(G) be σ. This σ is the number of different arborescences rooted at any given vertex in G. The number of different Euler lines associated with each of these distinct arborescences is given by Eq. (9-5). Therefore, Theorem 9-13 is obtained.

THEOREM 9-13

In an Euler digraph the number of Euler lines is

$$\sigma \cdot \prod_{i=1}^{n} [d^{-}(v_i) - 1]!.$$

From this theorem we can compute the number of Euler lines in any connected balanced digraph. As an example, let us compute the number of Euler lines in Fig. 9-10. Its K matrix is

$$\begin{bmatrix} 1 & -1 & 0 & 0 & 0 & 0 & 0 & 0 \\ 0 & 2 & -1 & 0 & -1 & 0 & 0 & 0 \\ 0 & 0 & 2 & -1 & 0 & 0 & 0 & -1 \\ 0 & 0 & -1 & 2 & -1 & 0 & 0 & 0 \\ 0 & 0 & 0 & 0 & 2 & -1 & -1 & 0 \\ 0 & 0 & 0 & 0 & 0 & 1 & -1 & 0 \\ 0 & 0 & 0 & -1 & 0 & 0 & 2 & -1 \\ -1 & -1 & 0 & 0 & 0 & 0 & 0 & 2 \end{bmatrix}.$$

(In this matrix, vertices appear in the order as they do in the directed Hamiltonian path $e_2\, e_3\, e_4\, e_{11}\, e_{12}\, e_{14}\, e_{15}$.)

The cofactor of any term in this matrix is 16, and therefore $\sigma = 16$ in Theorem 9-13. Since $d^-(v_i) = 2$ for each v_i in Fig. 9-10,

$$\prod_{i=1}^{8} [d^-(v_i) - 1]! = 1.$$

Therefore, the number of Euler lines in Fig. 9-10 is 16.

However, for a regular Euler digraph, such as the one in Fig. 9-10, it is often easier to compute the number of Euler lines by other methods (Problem 9-18).

9-10. PAIRED COMPARISONS AND TOURNAMENTS

In many experiments, specially in the social sciences, one is required to rank a number of given objects by comparing only two at a time. This is called the *method of paired comparisons*, and is used in situations where a numerical measurement is difficult, for example, individual preference for pieces of music. The items are presented two at a time to a subject and he is asked to state his preference. After having noted the results of all possible $n(n-1)/2$ paired comparisons of the n objects, the experimenter ranks the n objects in order of preference.

A digraph is a natural way of representing the results of a paired-comparison experiment. The results of a classic experiment of Kendall [9-5] are shown in Fig. 9-21. Six different dog foods $\{1, 2, \ldots, 6\}$ were to be ranked. Each day two of the six delicacies were served to a dog, and the dog established preference for one food over the other according to which plate he finished first. The experiment was conducted for 15 days, so that all possible pairs could be tried. In the graph representation, an edge is drawn from the preferred dish to the less preferred. For example, 1 was preferred to 2 in Fig. 9-21. Such a graph is called a *preference graph*.

Establishing a rank from a given preference graph is, in general, not easy. In Fig. 9-21, for example, due to some canine inconsistency, the dog preferred food 1 over 2, 2 over 4, and then 4 over 1. So which of the three is the best?

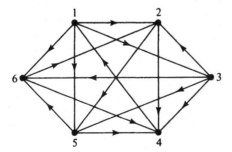

Fig. 9-21 Results of a paired-comparison experiment.

On Tournaments: A similar situation is encountered in tournaments. The results of a round-robin tournament in which every player has played against every other may also be represented by a digraph in which an edge directed from vertex a to b represents the victory of player a over player b. This is why a complete asymmetric digraph was called a tournament or a complete tournament in Section 9-2. The digraph in Fig. 9-21 can also be viewed as the result of a six-player tournament. The problem of ranking players in a tournament is identical to that of ranking in a paired-comparison experiment.

Ranking by Score: A straightforward method of ranking, and the one that has been traditionally used in round-robin tournaments, is to rank each player by his *score*. The score is the number of games the player has won. In terms of the dog food, the number of times the particular dish was preferred is its score. The score of a player in a tournament equals the out-degree of the corresponding vertex in the digraph.

Thus if we use the scores for ranking, we would rank the six dog foods as

$$(1, 3), \quad (2, 5, 6), \quad \text{and} \quad 4.$$

That is, foods 1 and 3 are tied for the first rank; there is a three-way tie for the second rank; and food 4 is the least preferred.

Ranking the vertices according to their out-degrees is not always a satisfactory method, although it is the easiest. In particular, this method loses significance if the tournament is incomplete (that is, the players do not compete in the same number of games).

Ranking by Hamiltonian Path: Another method sometimes used is to rank the players in a directed Hamiltonian path, such that each player has defeated his successor. One such ranking in Fig. 9-21 is 1 3 2 5 6 4. In this context, let us prove the following result regarding Hamiltonian paths in a tournament.

THEOREM 9-14

Every complete tournament has a directed Hamiltonian path.

Proof: The theorem will be proved by induction on the number of vertices. By actual sketching, the theorem can be shown to hold for all complete tournaments of 1, 2, 3, and 4 vertices. Let us make the inductive assumption that the theorem is true for all complete tournaments of n vertices, and then prove that it also holds for all tournaments of $n + 1$ vertices.

Let G be any complete tournament of $n + 1$ vertices. Let g be an n-vertex complete subtournament of G. By inductive assumption, g has a directed Hamiltonian path. Let that path be $v_1 v_2 \ldots v_n$. Let the vertex present in G but not in g be called v_{n+1}.

Since G is a complete tournament of $n + 1$ vertices, the vertex v_{n+1} in G has a directed edge either to or from each of the other vertices $v_1, v_2, \ldots, v_n$. The following three cases are possible.

Case 1: The edge between v_{n+1} and v_1 is directed toward v_1. Then we have

SEC. 9-10 PAIRED COMPARISONS AND TOURNAMENTS 229

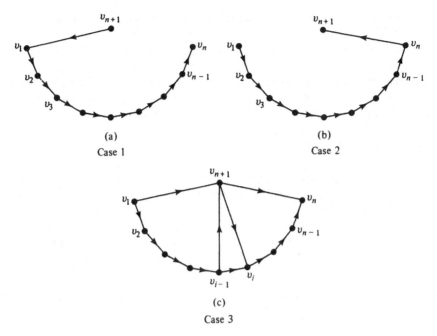

Fig. 9-22 Three cases of Theorem 9-14.

a Hamiltonian path $v_{n+1}\, v_1\, v_2 \ldots v_n$ in G, and the proof is complete [Fig. 9-22(a)].

Case 2: There is an edge directed from v_n to v_{n+1}. Then also we have a Hamiltonian path in G, which is $v_1\, v_2 \ldots v_n\, v_{n+1}$, and the proof is complete [Fig. 9-22(b)].

Case 3: Instead, both these edges are directed from v_1 to v_{n+1} and from v_{n+1} to v_n. In this case, as we move from v_1 to v_n, we encounter a reversal of direction in the edges incident on v_{n+1}. This reversal must occur because edge (v_1, v_{n+1}) is directed toward v_{n+1}, but edge (v_n, v_{n+1}) is directed away from v_{n+1}. Call the vertex at which the first such reversal occurs v_i (v_i may be v_n itself). Then edge (v_{i-1}, v_{n+1}) must be directed toward v_{n+1}. See Fig. 9-22(c). In this case we have a directed Hamiltonian path $v_1\, v_2 \ldots v_{i-1}\, v_{n+1}\, v_i\, v_{i+1} \ldots v_n$ in G. Therefore, the theorem. ∎

Coming back to the original problem of ranking the vertices, we now know that if the digraph is a complete tournament, at least one Hamiltonian ranking is always possible.

However, this method of ranking also suffers from some drawbacks. For one, there may be discrepancies between such a ranking and the scores of the players. Second, a tournament may have more than one directed Hamiltonian path, and therefore several different rankings are possible. In Fig. 9-21, for instance, 1 3 2 5 6 4 and 1 3 5 6 2 4 are two different Hamiltonian rankings.

Ranking with Minimum Violations: For a given ranking of the n vertices in any tournament (complete or incomplete), a *violation* is defined as an edge directed from v_i to v_j if v_j precedes v_i in the ranking. For example, in Fig.

9-21 the order 1 3 2 5 6 4 has the following two violations—edges 4 to 1 and 6 to 2. The order 3 2 5 6 4 1 has five violations, edges 1 to 3, 1 to 2, 6 to 2, 1 to 5, and 1 to 6.

Ranking with the minimum number of violations represents the fewest possible upsets for a given tournament. It can be shown that a ranking with minimum violations automatically includes the ranking according to scores, as well as a Hamiltonian ranking. Moreover, a minimum-violation ranking is also meaningful for incomplete tournaments. Thus this may be considered the best method of ranking.

However, out of all $n!$ possible orders of n vertices, to find one with minimum violations is computationally difficult. A method using dynamic programming has been used and is the best available so far, but it is computationally slow and cumbersome.

A minimum number of violations among all $n!$ rankings represents a smallest set of edges whose removal from the digraph will eliminate all directed circuits, that is, make the digraph acyclic. Acyclic digraphs are discussed in the next section.

9-11. ACYCLIC DIGRAPHS AND DECYCLIZATION

In many situations semicircuits are of no significance, and one is concerned only with whether or not a given digraph has a directed circuit. A digraph that has no directed circuit is called *acyclic*. Let us make the following observations about acyclic digraphs:

1. Every tree (with directed edges) is an acyclic digraph, but the converse is not true. For example, the digraph in Fig. 9-4 is acyclic, but it is not a tree.

2. An acyclic digraph cannot be condensed. That is, the condensation G_c of an acyclic digraph G is G itself. The converse is also true, because if $G_c = G$, obviously G has no directed circuit.

3. An acyclic digraph represents an irreflexive, asymmetric relation. But the digraph of an irreflexive, asymmetric relation is not necessarily acyclic. (Why?)

4. A digraph G is acyclic if and only if every directed walk in G is also a directed path.

5. Observation 4 has a significant implication: If a digraph is acyclic, the (i,j)th entry in X^k gives the number of distinct directed paths of length k from the ith vertex to the jth vertex.

THEOREM 9-15

Every acyclic digraph G has at least one vertex with zero in-degree and at least one vertex with zero out-degree.

Proof: Consider any maximal directed path (i.e., a path whose length cannot be increased by an edge at either end) P in G. Let v be the vertex where P starts and w be the vertex where it ends. Since G is acyclic, v and w must be distinct. Now the vertices in G can be divided into two classes: The set V_1 of vertices that are on P, and the set V_2 of the remaining vertices.

There is no edge incident into vertex v from any vertex in V_1. Otherwise, G would have a directed circuit. Also, there can be no edge incident into v from any vertex in V_2; otherwise, the length of P could have been increased by adding this edge. Thus the in-degree of v, $d^-(v) = 0$. Similarly, vertex w has no edge incident out of it; that is, $d^+(w) = 0$. ∎

THEOREM 9-16

A digraph G is acyclic if and only if its vertices can be ordered such that the adjacency matrix X is an upper (or lower) triangular matrix.

Proof: (a) Let us assume that X is upper triangular; that is,

$$x_{ij} = 0 \quad \text{for } i \geq j.$$

It can be seen by direct multiplication that X^2 is also upper triangular, and so are $X^3, X^4, \ldots$, all powers of X. Since every diagonal entry in all powers of X is zero, G has no directed circuit. That is, G is acyclic.

(b) For the second part of the theorem, assume that G is acyclic and then reorder the vertices of G, as follows: According to Theorem 9-15, there is at least one vertex in G whose in-degree is zero. In the reordering of the vertices let this be the first vertex v_1. Now, remove v_1 and edges incident on v_1 from G. The remaining digraph $G - v_1$ must also be acyclic, because G was acyclic. Therefore, $G - v_1$ has also at least one vertex, whose in-degree in $G - v_1$ is zero. Let v_2 be this second vertex in the reordering. Next remove v_2 from $G - v_1$. By continuing this process all the vertices are reordered† as $v_1, v_2, \ldots$.

Now consider the adjacency matrix X of G with the vertices appearing in this order. The first column (corresponding to v_1) has all zeros. The second column below the first row represents vertex v_2 in $G - v_1$, and therefore contains all zeros. And so forth.

Thus the adjacency matrix is upper triangular. This proves part (b) of the theorem.

The lower triangular portion of the theorem can also be proved either by reordering of the vertices with zero out-degrees, or by considering X^T and G^R. ∎

Given the adjacency matrix X of a digraph G, the following result is quite useful in finding out whether or not G is acyclic.

THEOREM 9-17

Digraph G is acyclic if and only if $\det(I - X)$ is not equal to zero, where I is the identity matrix of the same size as X.

†This is called a *topological sorting*. See Section 14-8 also.

Proof: Det$(I - X) \neq 0$ if and only if the inverse $(I - X)^{-1}$ exists. But

$$(I - X)^{-1} = I + X + X^2 + \cdots + X^n + \cdots. \qquad (9\text{-}13)\dagger$$

This inverse $(I - X)^{-1}$ exists if and only if the infinite series (9-13) converges; that is, $X^k = 0$ for all $k \geq$ some N (because X contains only nonnegative entries).

However, $X^k = 0$ for all $k \geq N$ if and only if G contains no directed edge sequence of length N or larger. And this is possible if and only if G contains no cycle of any length. ∎

Decyclization: Acyclic digraphs are of enormous importance in many applications. It was pointed out that directed circuits represent inconsistencies in ranking by paired comparisons. Directed circuits may represent undesirable feedback paths in an electrical network. In the project graph of a CPM (critical path method) or PERT (program evaluation and review technique) a directed circuit represents a serious error, and must be eliminated. This is because a directed circuit, say *abca*, implies that activity *a* must be completed before activity *b*, and *b* before *c*, and *c* before *a*. Obviously, this is an impossible situation and nothing will get done. A similar situation in computer programming often arises and is justifiably known as the deadly encounter or the deadly embrace (Problem 9-26).

In deductive logic (where vertices represent axioms or statements and directed edges represent the theorems or derivation of one statement from others), a directed circuit implies circular reasoning and hence a fallacy.

Thus it is important to know how to break these vicious cycles with a minimum of effort. In other words, find a smallest set of directed edges whose removal will render the given digraph G acyclic.

Consider, for example, Fig. 9-2(a). The digraph contains several directed circuits: $e_1 e_2 e_3$, $e_4 e_6 e_3$, and $e_4 e_5 e_2 e_3$. In this simple case, one can tell by inspection that the removal of edge e_3 will eliminate all directed circuits. This is *the* smallest set of edges whose removal makes the remaining digraph acyclic. Such a smallest set of edges whose removal destroys all directed circuit in a digraph G is known as a *minimum-feedback arc set* in electrical engineering. In general, a digraph may possess several minimum-feedback arc sets. Obtaining one such smallest set of edges may be called *minimal decyclization* of a digraph.

Minimal decyclization of an arbitrary directed graph is at best a tedious affair. No simple method has been found so far. One method proposed in the literature (in 1969) uses Theorem 9-16, as follows: Make the adjacency matrix X upper triangular as much as possible by interchanging rows (and correspondingly columns). The 1's remaining below (and on) the principal diagonal

†This identity can be seen by premultiplying both sides of (9-13) with the matrix $(I - X)$.

represent a minimum-feedback arc set. Another method can be

1. Obtain all directed circuits in the given digraph G (using the result of Problem 9-11, say).
2. Express each directed circuit as a Boolean sum of its edges.
3. Take the Boolean product of all directed circuit expressions obtained in step 2. (The absorption laws of Boolean algebra are applied, such as $a \cdot a = a$, $a + a = a$, and $a + ab = a$.)
4. Each of the resulting terms in the sum of the products represents a set of edges whose removal will destroy all directed circuits. Pick a term that consists of the smallest number of edges; this is a minimum-feedback arc set.

To illustrate the procedure, let us consider the digraph in Fig. 9-2(a). All the directed circuits are

$$e_1 e_2 e_3, \quad e_3 e_4 e_6, \quad \text{and} \quad e_2 e_3 e_4 e_5.$$

Expressing these as a product of Boolean sums and multiplying out and simplifying, we get

$$(e_1 + e_2 + e_3)(e_3 + e_4 + e_6)(e_2 + e_3 + e_4 + e_5)$$
$$= e_1 e_4 + e_1 e_6 e_5 + e_2 e_4 + e_2 e_6 + e_3.$$

Clearly, any one of these terms represents the set of edges whose removal would break all directed circuits in Fig. 9-2(a). The set with the smallest number of edges, $\{e_3\}$, is the answer we were seeking. Both these methods are impractical for large digraphs.

SUMMARY

Most of the important and fundamental features of directed graphs were investigated in this chapter. We saw that there are two different aspects of digraphs: one in which their properties are similar to those of undirected graphs, such as planarity, thickness, spanning trees, fundamental circuits, and cut-sets; in their second aspect, digraphs have properties altogether different from those of undirected graphs, such as strong connectedness, arborescence, decyclization, and so on.

The close relationship between binary relations and digraphs was explored. Applications of digraphs are virtually unlimited. Some important ones, such as in sequence generation in telecommunications and paired comparisons, were dealt with in detail. Others were simply mentioned.

Undoubtedly, a great deal more remains to be said. Additional properties of digraphs are presented in the form of problems at the end of this chapter. For the rest the reader must explore on his own, using the tools and results presented in the chapter.

REFERENCES

An entire 400-page book has been written by Harary, Norman, and Cartwright [9-3] on the theory of digraphs. It is a textbook written specially for those with little mathematical background. This book is recommended reading for many of the topics not covered in this chapter. Specializing even further, a 100-page monograph was written by Moon [9-7] on tournaments (complete asymmetric digraphs) alone. This is a compactly written definitive book and is highly recommended for those wishing to know all about complete, asymmetric digraphs. For applications of directed graphs in operations research, Kaufmann's book [9-4] is a good source. In particular, Chapter 4 contains an excellent presentation of the properties of digraphs. (Be prepared for slightly different terminology.) For methods of paired comparison, Kendall's book [9-5] is recommended.

Other books recommended are Chapters 5–7 in Ore [1-10] as very introductory reading; Chapters 13–17 of Berge [1-1], intermediate-level reading; Chapters 8–10 of Ore [1-9]; and Chapter 16 of Harary [1-5] as a more advanced level of reading on digraphs.

The classic works of Good, Kendall, van Aardenne-Ehrenfest, and deBruijn, and Tutte have already been referred to. Chen and Wing [9-1] give some properties and interesting applications of acyclic digraphs. Minimal decyclization of a digraph was the subject of the doctoral thesis by Lempel [9-6].

9-1. CHEN, Y. C., and O. WING, "Some Properties of Cycle-Free Directed Graphs and the Identification of the Longest Path," *J. Franklin Inst.*, Vol. 281, No. 4, April 1966, 293–301.

9-2. GOOD, I. G., "Normal Recurring Decimals," *J. London Math. Soc.*, Vol. 21 (part 3), 1946, 167–172.

9-3. HARARY, F., R. Z. NORMAN, and D. CARTWRIGHT, *Structural Models: An Introduction to the Theory of Directed Graphs*, John Wiley & Sons, Inc., New York, 1965.

9-4. KAUFMANN, A., *Graphs, Dynamic Programming and Finite Games*, Academic Press, Inc., New York, 1967. (Originally published in French in 1964, Dunod Editeur, Paris.)

9-5. KENDALL, M. G., *Rank Correlation Methods*, Charles Griffin and Co., London, 1948; 3rd ed., Hafner Publishing Company, Inc., New York, 1962.

9-6. LEMPEL, A., "Minimum Feedback Arc and Vertex Sets of a Directed Graph," *IEEE Trans. Circuit Theory*, Vol. CT-13, No. 4, Dec. 1966, 399–403.

9-7. MOON, J. W., *Topics on Tournaments*, Holt, Rinehart and Winston, Inc., New York, 1968.

9-8. TUTTE, W. T., "The Dissection of Equilateral Triangles into Equilateral Triangles," *Proc. Cambridge Phil. Soc.*, Vol. 44, 1948, 463–482.

9-9. VAN AARDENNE-EHRENFEST, T., and N. G. DEBRUIJN, "Circuits and Trees in Oriented Graphs," *Simon Stevin*, Vol. 28, 1951, 203–217.

PROBLEMS

9-1. Prove that in any digraph the sum of the in-degrees of all vertices is equal to the sum of their out-degrees; and this sum is equal to the number of edges in the digraph.

9-2. Sketch all different (nonisomorphic) simple digraphs with 1, 2, and 3 vertices.

9-3. Sketch all distinct (nonisomorphic) orientations of a complete graph of four vertices. Characterize each of the resulting digraphs in terms of binary relations.

9-4. An irreflexive, asymmetric, transitive relation on a set is called a *strict partial order*. Give two examples of strict partial orders. Show that the digraph of a strict partial order is acyclic. Is the converse also true?

9-5. The combinations of reflexivity, symmetry, and transitivity define eight ($2^3 = 8$) types of binary relations. Two such relations are equivalence and partial order. List the other six and sketch a digraph for each.

9-6. Define and study the directed Hamiltonian circuit and semi-Hamiltonian circuit in a digraph.

9-7. Prove that every edge in a digraph belongs either to a directed circuit or a directed cut-set.

9-8. For an n-vertex digraph, define an n by n *accessibility* (or *reachability*) *matrix* $R = [r_{ij}]$ as follows:

$$r_{ij} = 1, \quad \text{if there is a directed path of length one or more from } i \text{ to } j,$$
$$= 0, \quad \text{otherwise.}$$

Devise a method of obtaining R from the powers of the adjacency matrix X. (Note that this reachability matrix is slightly different from that in [9-3], because we do not include paths of zero length; i.e., r_{ii} is not necessarily one.)

9-9. Is it possible for two nonisomorphic digraphs to have the same reachability matrix R? Explain.

9-10. Show that if R is the reachability matrix of a digraph G, the value of the ith entry in the principal diagonal of R^2 gives the number of vertices included in the strongly connected fragment containing the ith vertex.

9-11. Show that the following procedure applied to the adjacency matrix $X = [x_{ij}]$ of a digraph G will yield the reachability matrix R of G.
Step 1: Let $x_{1i}, x_{1j}, \ldots, x_{1m}$ be the nonzero elements in the first row. Add the ith, jth, $\ldots$, mth rows to the first row. Replace each nonzero element by a 1 (Boolean sum).
Step 2: Suppose that there are k additional nonzero elements $p, q, \ldots, r$ generated in the first row as a result of step 1. Add the pth, qth, $\ldots$, rth rows to the first row, and replace each nonzero element by a 1.
Step 3: Repeat step 2 until no additional 1's can be added to the first row by this process.
Step 4: Repeat the process on every row of X.

9-12. Prove that an n-vertex digraph is strongly connected if and only if the matrix M, defined by

$$M = X + X^2 + X^3 + \cdots + X^n,$$

has no zero entry. X is the adjacency matrix.

9-13. Prove that every Euler digraph (without isolated vertices) is strongly connected. Also show, by constructing a counterexample, that the converse is not true.

9-14. List all 16 distinct directed Euler lines in Fig. 9-10.

9-15. The Euler digraph in Fig. 9-10 is called the teleprinter diagram or the Good diagram for $r = 4$ [abbreviated as GD(4)]. Sketch and label GD(3) and GD(5). Find one directed Euler line and one directed Hamiltonian circuit in each. [*Hint:* GD(r) has 2^{r-1} vertices and 2^r edges. A vertex in GD($r + 1$) corresponds to an edge in GD(r).]

9-16. An *edge digraph* or a *line digraph* $L(G)$ of a digraph G is defined as follows:
1. There is exactly one vertex v_i in $L(G)$ for every edge e_i in G.
2. Whenever edges e_i and e_j (for a self-loop $e_j = e_i$) are such that e_i is incident into a vertex v and e_j is incident out of the same vertex v, an edge is drawn from the corresponding v_i to v_j in $L(G)$.

Show that $GD(r + 1)$ is a line digraph of $GD(r)$.

9-17. If $E|G|$ is the number of Euler lines in an n-vertex Euler digraph G, show that $2^{n-1} \cdot E|G|$ is the number of Euler lines in $L(G)$.

9-18. Prove that the number of directed Euler lines in $GD(r)$ is

$$2^{2^{r-1}-r}$$

(*Hint:* Use the results of Problem 9-16 or use Theorem 9-13.)

9-19. A drum rotates in discrete steps of θ degrees, and you are to determine its precise position as follows. Divide the surface of the drum into $k = 360°/\theta$ sectors, and paint each sector black or white (or conducting or nonconducting). Mount r consecutive reading heads—each capable of detecting the color of the sector.

Given some θ, express k and r in terms of θ. Sketch one such arrangement of colors on the drum for $k = 16$.

9-20. What is the longest circular sequence formed out of three symbols (letters) x, y, and z such that no subsequence (words) of four symbols is repeated. Give one such sequence. [*Hint:* Form a regular Euler digraph with $d^-(v_i) = d^+(v_i) = 3$, in the manner of Fig. 9-10.]

9-21. Prove that any acyclic digraph G is an arborescence if and only if there is a vertex v in G such that every vertex is accessible from v.

9-22. Prove that for every $n \geq 3$ there exists at least one acyclic complete tournament of n vertices. (*Hint:* Use induction.)

9-23. Let $R(G)$ be the reachability matrix of a digraph G, and let the vertices in G be ordered such that the sums of the rows in $R(G)$ are nonincreasing; that is,

$$\sum_{j=1}^{n} r_{ij} \geq \sum_{j=1}^{n} r_{kj} \quad \text{for every } i < k.$$

Show with this ordering of vertices in $R(G)$ that digraph G is acyclic if and only if $R(G)$ is an upper triangular matrix.

9-24. Prove that a digraph G is acyclic if and only if every element on the principal diagonal of its reachability (or accessibility) matrix $R(G)$ is zero.

9-25. Prove that an acyclic digraph G of n vertices has a unique directed Hamiltonian path if and only if the number of nonzero elements in $R(G)$ is $n(n-1)/2$.

9-26. There are 15 computer programs that must be processed according to the following set of orders:

$$1 > 2, 7, 13,$$
$$2 > 3, 8, 14,$$
$$3 > 9, 15,$$
$$4 > 3,$$
$$5 > 4, 11,$$
$$6 > 5, 12,$$
$$7 > 6,$$

$$8 > 7, 9, 14,$$
$$9 > 15,$$
$$10 > 4, 9,$$
$$11 > 10,$$
$$12 > 11,$$
$$13 > 7, 12,$$
$$14 > 13, 15,$$

where $1 > 2, 7, 13$ means that programs 2, 7, and 13 can be processed only after program 1 has been processed. Is it possible for the programs to be processed? If so, give a processing sequence. [*Hint:* Write $X(G)$; derive $R(G)$ from $X(G)$ using Problem 9-11. Use Problem 9-23 to check if G is acyclic.]

9-27. A digraph defined on the relation "is a parent of" is called a genetic digraph. (Genetic digraphs are useful in biology.) Investigate the properties of genetic digraphs.

9-28. Use digraphs to solve the classical problem of "three cannibals and three edible missionaries seeking to cross a river in a boat that can hold at most two people, and all the missionaries and one of the cannibals can row the boat. Also, at no time should the cannibals outnumber the missionaries on either shore." (*Hint:* Represent each state by a vertex and a possible transition by a directed edge.)

10 ENUMERATION OF GRAPHS

Arthur Cayley (1857), one of the founding fathers of graph theory, became interested in graph theory for the purpose of counting trees. The number of different trees of n vertices gave him the number of isomers of the saturated hydrocarbon with n carbon atoms, that is, C_nH_{2n+2}. Since Cayley's classic paper, a great deal of work has been done on counting (also called enumeration) of different types of graphs, and the results have been applied in solving some practical problems.

Some enumeration problems have already been introduced in earlier chapters. For example, in Chapter 2 the number of edge-disjoint Hamiltonian circuits in the complete graph of n vertices was discussed. Enumeration of trees in Section 3-6; finding all spanning trees in Section 3-9; the number of different edge sequences of length r between a specified pair of vertices (Theorem 7-8); Problems 7-20, 7-23, and 7-24; the number of different arborescences rooted at a given vertex in Chapter 9; and the number of different directed Euler lines in a digraph, also in Chapter 9, were all problems of counting graphs. In this chapter a more unified approach to enumerating graphs will be taken. Certain enumerative techniques will be developed and used for counting certain types of graphs. A thorough exposition of Pólya's counting theorem, the most powerful tool in graph enumeration, is the central feature of this chapter.

10-1. TYPES OF ENUMERATION

All graph-enumeration problems fall into two categories:

1. Counting the number of different graphs (or digraphs) of a particular

kind, for example, all connected, simple graphs with eight vertices and two circuits.

2. Counting the number of subgraphs of a particular type in a given graph G, such as the number of edge-disjoint paths of length k between vertices a and b in G.

The second type of problem usually involves a matrix representation of graph G and manipulations of this matrix. Such problems, although often encountered in practical applications, are not as varied and interesting as those in the first category. We shall not consider such problems in this chapter.

In problems of type 1 the word "different" is of utmost importance and must be clearly understood. If the graphs are labeled (i.e., each vertex is assigned a name distinct from all others), all graphs are counted. On the other hand, in the case of unlabeled graphs the word "different" means non-isomorphic, and each set of isomorphic graphs is counted as one.

As an example, let us consider the problem of constructing all simple graphs with n vertices and e edges. There are $n(n-1)/2$ unordered pairs of vertices. If we regard the vertices as distinguishable from one another (i.e., labeled graphs), there are

$$\binom{\frac{n(n-1)}{2}}{e} \tag{10-1}$$

ways of selecting e edges to form the graph. Thus expression (10-1) gives the number of simple *labeled* graphs with n vertices and e edges.

Many of these graphs, however, are isomorphic (that is, they are the same except for the labels of their vertices). Hence the number of simple, *unlabeled* graphs of n vertices and e edges is much smaller than that given by (10-1).

Among a collection of graphs, isomorphism is an equivalence relation (Problem 10-1). The number of different unlabeled graphs (of a certain type) equals the number of equivalence classes, under isomorphism, of the labeled graphs. For example, we have 16 different labeled trees of four vertices (Fig. 3-15), and these trees fall into two equivalence classes, under isomorphism. In Fig. 3-15 the 4 trees in the top row fall into one equivalence class, and the remaining 12 into another. Thus we have only two different unlabeled trees of four vertices (Fig. 3-16).

Let us now proceed with counting certain specific types of graphs.

THEOREM 10-1

The number of simple, labeled graphs of n vertices is

$$2^{n(n-1)/2}. \tag{10-2}$$

240 ENUMERATION OF GRAPHS CHAP. 10

Proof: The numbers of simple graphs of n vertices and $0, 1, 2, \ldots, n(n-1)/2$ edges are obtained by substituting $0, 1, 2, \ldots, n(n-1)/2$ for e in expression (10-1). The sum of all such numbers is the number of all simple graphs with n vertices. Then the use of the following identity proves the theorem:

$$\binom{k}{0} + \binom{k}{1} + \binom{k}{2} + \cdots + \binom{k}{k-1} + \binom{k}{k} = 2^k. \blacksquare$$

10-2. COUNTING LABELED TREES

Expression (10-1) can be used to obtain the number of simple labeled graphs of n vertices and $n - 1$ edges. Some of these are going to be trees and others will be unconnected graphs with circuits. Let us now prove Theorem 3-10, which gives the number of trees.

THEOREM 3-10

There are n^{n-2} labeled trees with n vertices ($n \geq 2$).

Proof of Theorem 3-10: Let the n vertices of a tree T be labeled $1, 2, 3, \ldots, n$. Remove the pendant vertex (and the edge incident on it) having the smallest label, which is, say, a_1. Suppose that b_1 was the vertex adjacent to a_1. Among the remaining $n - 1$ vertices let a_2 be the pendant vertex with the smallest label, and b_2 be the vertex adjacent to a_2. Remove the edge (a_2, b_2). This operation is repeated on the remaining $n - 2$ vertices, and then on $n - 3$ vertices, and so on. The process is terminated after $n - 2$ steps, when only two vertices are left. The tree T defines the sequence

$$(b_1, b_2, \ldots, b_{n-2}) \tag{10-3}$$

uniquely. For example, for the tree in Fig. 10-1 the sequence is $(1, 1, 3, 5, 5, 5, 9)$. Note that a vertex i appears in sequence (10-3) if and only if it is not pendant (see Problem 10-2).

Conversely, given a sequence (10-3) of $n - 2$ labels, an n-vertex tree can be

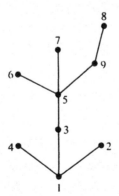

Fig. 10-1 Nine-vertex labeled tree, which yields sequence $(1, 1, 3, 5, 5, 5, 9)$.

constructed uniquely, as follows: Determine the first number in the sequence

$$1, 2, 3, \ldots, n \tag{10-4}$$

that does not appear in sequence (10-3). This number clearly is a_1. And thus the edge (a_1, b_1) is defined. Remove b_1 from sequence (10-3) and a_1 from (10-4). In the remaining sequence of (10-4) find the first number that does not appear in the remainder of (10-3). This would be a_2, and thus the edge (a_2, b_2) is defined. The construction is continued till the sequence (10-3) has no element left. Finally, the last two vertices remaining in (10-4) are joined. For example, given a sequence

$$(4, 4, 3, 1, 1),$$

we can construct a seven-vertex tree as follows: (2, 4) is the first edge. The second is (5, 4). Next, (4, 3). Then (3, 1), (6, 1), and finally (7, 1), as shown in Fig. 10-2.

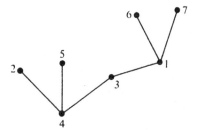

Fig. 10-2 Tree constructed from sequence (4, 4, 3, 1, 1).

For each of the $n - 2$ elements in sequence (10-3) we can choose any one of n numbers, thus forming

$$n^{n-2} \tag{10-5}$$

$(n - 2)$-tuples, each defining a distinct labeled tree of n vertices. And since each tree defines one of these sequences uniquely, there is a one-to-one correspondence between the trees and the n^{n-2} sequences. Hence the theorem. ■

Rooted Labeled Trees: In a rooted graph one vertex is marked as the root. For each of the n^{n-2} labeled trees we have n rooted labeled trees, because any of the n vertices can be made a root. Therefore,

THEOREM 10-2

The number of different rooted, labeled trees with n vertices is

$$n^{n-1}. \tag{10-6}$$

All rooted trees for $n = 1, 2$, and 3 are given in Fig. 10-3.

10-3. COUNTING UNLABELED TREES

The problem of enumeration of unlabeled trees is more involved and requires familiarity with the concepts of *generating functions* and *partitions*.

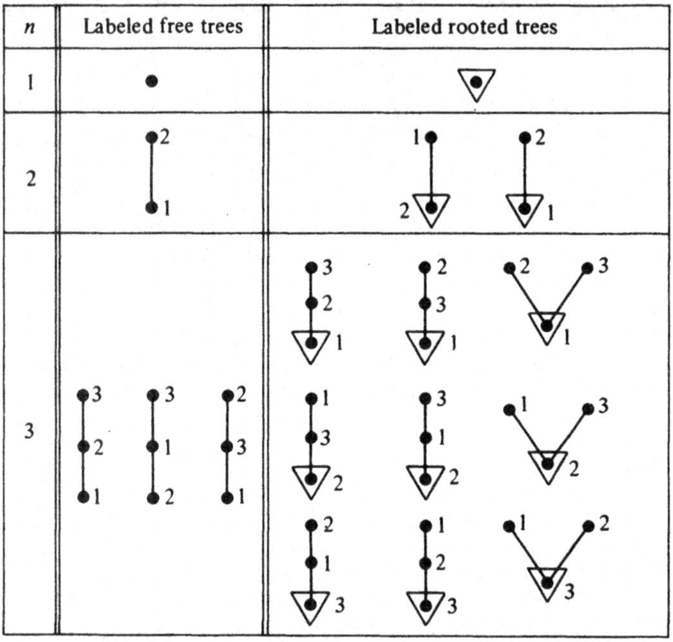

Fig. 10-3 Rooted labeled trees of one, two, and three vertices.

Generating Functions

One of the most useful tools in enumeration techniques is the generating function. A *generating function* $f(x)$ is a power series

$$f(x) = a_0 + a_1 x + a_2 x^2 + \cdots \qquad (10\text{-}7)$$

in some dummy variable x. The coefficient a_k of x^k is the desired number, which depends on a collection of k objects being enumerated. For example, in the generating function

$$(1 + x)^n = \binom{n}{0} + \binom{n}{1} x + \binom{n}{2} x^2 + \binom{n}{3} x^3 + \cdots + \binom{n}{n} x^n, \qquad (10\text{-}8)$$

the coefficient of x^k gives the number of distinct combinations of n different objects taken k at a time.

As another example, consider the following generating function:

$$(1 - x)^{-n} = (1 + x + x^2 + x^3 + \cdots)^n$$

$$= \sum_{k=0}^{\infty} \binom{n + k - 1}{k} x^k. \qquad (10\text{-}9)$$

The coefficient of x^k in (10-9) gives the ways of selecting k objects from n

(distinct) objects with unlimited repetitions.† Note that the variable x has no significance. We are interested only in the coefficients.

The generating function is used as a counting device and is therefore also called a *counting series* or an *enumerator*. An operation on a generating function is simpler than the corresponding operation on the sequence of coefficients $a_0, a_1, a_2, \ldots$. For a detailed treatment of generating functions, the reader is referred to Chapter 2 in [3-11] or Chapter 3 in [10-1].

Partitions

Another useful and important concept in enumerative combinatorics is that of a *partition of a positive integer*. When a positive integer p is expressed as a sum of positive integers

$$p = \lambda_1 + \lambda_2 + \lambda_3 + \cdots + \lambda_q,$$

such that
$$\lambda_1 \geq \lambda_2 \geq \lambda_3 \geq \cdots \geq \lambda_q \geq 1, \qquad (10\text{-}10)$$

the q-tuple is called a partition of integer p. For example, (5), (4 1), (3 2), (3 1 1), (2 2 1), (2 1 1 1), and (1 1 1 1 1) are the seven different partitions of the integer 5.

The integers, λ_i's, are called *parts* of the *partitioned number* p. It is convenient to represent the repeated parts by means of exponents; for example, partition (2 1 1 1) is written as (2 1^3).

The partitions of an integer p may be unrestricted or may have some restrictions on them, such as no repetition of any part [i.e., $\lambda_i \neq \lambda_j$ in (10-10)], or no part greater than k is allowed. The number of partitions of a given integer p is often obtained with the help of some generating function. For example, the coefficient of x^k in the polynomial

$$(1 + x)(1 + x^2)(1 + x^3) \ldots (1 + x^p) \qquad (10\text{-}11)$$

gives the number of partitions, without repetition, of an integer $k \leq p$ (see page 111, [3-11]).

Partitions are important to us because many graph-enumeration problems can be expressed in the form of partition problems.

Rooted Unlabeled Trees

Coming back to counting trees, let us recall that a rooted, unlabeled tree is one in which all vertices except the root are assumed alike. Let u_n be the

---

†The result can be proved as follows: Let the n objects be labeled $1, 2, 3, \ldots, n$, and let a specific selection be a list of k integers $a_1, a_2, \ldots, a_k$ arranged in nondecreasing order. The a_i's are not necessarily distinct. From this list we get a new list $a_1, a_2 + 1, a_3 + 2, \ldots, a_k + k - 1$ by adding 0 to a_1, 1 to a_2, and so on. Each term in the new list is distinct. Thus every selection with unlimited repetitions can be identified uniquely as a selection of k distinct integers from integers $1, 2, \ldots, n + k - 1$.

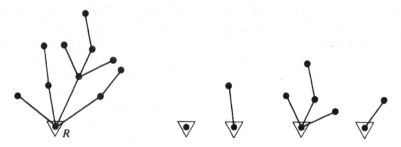

Fig. 10-4 Rooted tree decomposed into rooted subtrees.

number of unlabeled, rooted trees of n vertices, and let $u_n(m)$ be the number of those rooted trees of n vertices in which the degree of the root is exactly m. Then

$$u_n = \sum_{m=1}^{n-1} u_n(m).$$

Any rooted tree T of n vertices and with root R of degree m can be looked upon as composed of m rooted subtrees, each attached to R by means of an edge between its root and R. For example, in Fig. 10-4 an 11-vertex, rooted tree is composed of four rooted subtrees.

In an n-vertex tree T the $n-1$ vertices are distributed among the m subtrees, and thus T defines an m-part partition of the number $n-1$. Suppose that k_j is the number of such subtrees (in T) with j vertices. Then

$$k_1 + 2k_2 + 3k_3 + \cdots + (n-1)k_{n-1} = n-1 \qquad (10\text{-}12)$$

and
$$k_1 + k_2 + k_3 + \cdots + k_{n-1} = m. \qquad (10\text{-}13)$$

Note that Eqs. (10-12) and (10-13) represent an m-part partition of integer $n-1$, in which integer i appears k_i times ($0 \leq k_i \leq n-1$).

In Fig. 10-4, for example,

$$n = 11, \quad m = 4,$$
$$k_1 = 1, \quad k_2 = 2, \quad k_5 = 1,$$
$$k_3 = k_4 = k_6 = k_7 = \cdots = k_{10} = 0.$$

Thus
$$\sum k_j = 4$$
and
$$\sum j k_j = 10.$$

One can construct u_j distinct rooted trees with j unlabeled vertices. Out of these trees we select k_j trees to form subtrees of T. Since the same tree may appear more than once as a subtree of T, we have the problem of finding the number of ways of selecting k_j objects out of u_j objects with unlimited repetition. According to Eq. (10-9), this number is

$$\binom{u_j + k_j - 1}{k_j}. \tag{10-14}$$

Since each such selection can be made independently, the possible number of distinct trees for this specific partition is

$$u_n(k_1, k_2, \ldots, k_{n-1}) = \binom{u_1 + k_1 - 1}{k_1}\binom{u_2 + k_2 - 1}{k_2} \cdots \binom{u_{n-1} + k_{n-1} - 1}{k_{n-1}}, \tag{10-15}$$

where $u_n(k_1, k_2, \ldots, k_{n-1})$ stands for the number of n-vertex, rooted trees corresponding to the partition

$$1^{k_1} 2^{k_2} 3^{k_3} \ldots (n-1)^{k_{n-1}}.$$

Addition of $u_n(k_1, k_2, \ldots, k_{n-1})$ over all possible partitions of $n-1$ yields the total number of spanning trees. That is,

$$u_n = \sum_{\substack{\text{partitions} \\ \text{of } n-1}} \prod_{j=1}^{n-1} \binom{u_j + k_j - 1}{k_j}. \tag{10-16}$$

What we have obtained in (10-16) is a recurrence relation—a solution typical of many combinatorial problems. It gives u_n, the number of rooted, unlabeled trees of n vertices, in terms of $u_1, u_2, \ldots, u_{n-1}$. To use this relation, one builds up numerical tables in a step-by-step fashion. For example,

$$u_1 = 1$$

$$u_2 = \binom{u_1 + 1 - 1}{1} = 1,$$

$$u_3 = \binom{u_2 + 1 - 1}{1} + \binom{u_1 + 2 - 1}{2} = 2.$$

To evaluate u_4, we first have to find all partitions of integer 3. These are

$$(3), \quad (2, 1), \quad \text{and} \quad (1, 1, 1).$$

The sum of the respective terms contributed by these partitions is

$$u_4 = \binom{u_3 + 1 - 1}{1} + \binom{u_2 + 1 - 1}{1}\binom{u_1 + 1 - 1}{1} + \binom{u_1 + 3 - 1}{3} = 4.$$

Similarly, to evaluate u_5 we observe that the integer 4 has five different

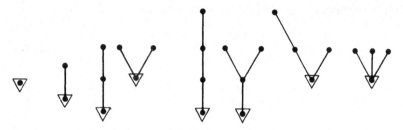

Fig. 10-5 Rooted, unlabeled trees of one, two, three, and four vertices.

partitions, and these are

$$(4), \quad (3, 1), \quad (2, 2), \quad (2, 1, 1), \quad \text{and} \quad (1, 1, 1, 1).$$

The number of rooted trees corresponding to each of these five partitions is obtained using (10-15). The sum yields u_5:

$$u_5 = \binom{u_1 + 3}{4} + \binom{u_1 + 1}{2} u_2 + \binom{u_2 + 1}{2} + u_3 + u_4$$
$$= 1 + 1 + 1 + 2 + 4 = 9.$$

And so on. In Fig. 10-5, all rooted, unlabeled trees of one, two, three, and four vertices are shown.

Clearly, computation of u_n for, say, $n = 20$, using (10-16) is extremely tedious and involved. It requires obtaining all possible partitions of integer 19 (there are 490 partitions of 19), computing $u_{19}, u_{18}, \ldots, u_2, u_1$, evaluation of the combinatorial product term

$$\prod_j \binom{u_j + k_j - 1}{k_j}$$

for each partition, and then taking the sum of all 490 such terms.

Counting Series for u_n: To circumvent some of these difficulties in computation of u_n, let us find its counting series (i.e., the generating function) $u(x)$, where

$$u(x) = u_1 x + u_2 x^2 + u_3 x^3 + \cdots$$
$$= \sum_{n=1}^{\infty} u_n x^n \qquad (10\text{-}17)$$
$$= x \sum_{n=1}^{\infty} u_n x^{n-1}.$$

Substitution of (10-16) in (10-17) and substitution of $n - 1$ by its partition as in (10-12) yields

$$u(x) = x \sum_{n=1}^{\infty} \sum_{\substack{\text{partitions} \\ \text{of } n-1}} \binom{u_1 + k_1 - 1}{k_1} x^{k_1} \binom{u_2 + k_2 - 1}{k_2} x^{2k_2}$$
$$\cdots \binom{u_{n-1} + k_{n-1} - 1}{k_{n-1}} x^{(n-1)k_{n-1}}. \tag{10-18}$$

Observing that every sequence of positive integers forms a partition of some integer, (10-18) can be rearranged as

$$u(x) = x \left[\sum_{k_1=0}^{\infty} \binom{u_1 + k_1 - 1}{k_1} x^{k_1} \right] \left[\sum_{k_2=0}^{\infty} \binom{u_2 + k_2 - 1}{k_2} x^{2k_2} \right]$$
$$\cdots \left[\sum_{k_{n-1}=0}^{\infty} \binom{u_{n-1} + k_{n-1} - 1}{k_{n-1}} x^{(n-1)k_{n-1}} \right] \cdots. \tag{10-19}$$

Substituting the identity

$$(1 - x^m)^{-p} = \sum_{j=0}^{\infty} \binom{p + j - 1}{j} x^{mj}$$

in (10-19) gives us the desired counting series. That is,

$$u(x) = x(1 - x)^{-u_1}(1 - x^2)^{-u_2}(1 - x^3)^{-u_3} \cdots$$
$$= x \prod_{r=1}^{\infty} (1 - x^r)^{-u_r}. \tag{10-20}$$

Calculation of u_n from (10-20) involves building up a table of u_i for $i = 1, 2, 3, \ldots, n - 1$, and substituting the values in (10-20). The first 10 terms in the series (10-20) are

$$u(x) = x + x^2 + 2x^3 + 4x^4 + 9x^5 + 20x^6 + 48x^7$$
$$+ 115x^8 + 286x^9 + 719x^{10} + \cdots. \tag{10-20a}$$

The reader should verify (10-20a) himself and extend the expansion through another 10 terms.

The generating function $u(x)$ can be expressed in an alternative form as follows:

Taking the natural logarithms of both sides of Eq. (10-20), we get

$$\ln u(x) = \ln x - \sum_{r=1}^{\infty} u_r \ln(1 - x^r)$$
$$= \ln x + \sum_{r=1}^{\infty} u_r \sum_{i=1}^{\infty} \frac{x^{ri}}{i}$$
$$= \ln x + \sum_{i=1}^{\infty} \frac{1}{i} \sum_{r=1}^{\infty} u_r x^{ri}$$
$$= \ln x + \sum_{i=1}^{\infty} \frac{1}{i} u(x^i).$$

Therefore,

$$u(x) = xe^{\sum_{i=1}^{\infty} (1/i)u(x^i)} \qquad (10\text{-}21)$$

Form (10-21) is due to George Pólya, whereas (10-20) is Arthur Cayley's.

To obtain the generating function for (free) unlabeled trees from rooted unlabeled trees, one can look at a (free) tree as composed of subtrees, rooted at some sort of central vertex distinct from all other vertices in the tree. For this, we shall use the concept of centroid in a tree.

Centroid

In a tree T, at any vertex v of degree d, there are d subtrees with only vertex v in common. The *weight of each subtree* at v is defined as the number of branches in the subtree. Then the *weight of the vertex* v is defined as the weight of the heaviest of the subtrees at v. A vertex with the smallest weight in the entire tree T is called a *centroid* of T.

Just as in the case of centers of a tree (Section 3-4), it can be shown that every tree has either one centroid or two centroids. It can also be shown that if a tree has two centroids, the centroids are adjacent. In Fig. 10-6 a tree with a centroid (called a *centroidal tree*) and a tree with two centroids (called a *bicentroidal tree*) are shown. The centroids are shown enclosed in circles, and the numbers next to the vertices are the weights.

Free Unlabeled Trees

Let $t'(x)$ be the counting series for centroidal trees, and $t''(x)$ be the counting series for bicentroidal trees. Then $t(x)$, the counting series for all (unlabeled, free) trees, is the sum of the two. That is,

$$t(x) = t'(x) + t''(x). \qquad (10\text{-}22)$$

To obtain $t''(x)$, observe that an n-vertex bicentroidal tree can be regarded as consisting of two rooted trees each with $n/2 = m$ vertices, and joined at their roots by an edge. (A bicentroidal tree will always have an even number of vertices; why?) Thus the number of bicentroidal trees with $n = 2m$ vertices is

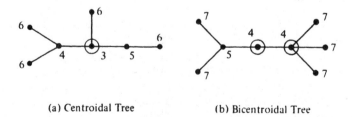

(a) Centroidal Tree   (b) Bicentroidal Tree

Fig. 10-6 Centroid and bicentroids.

given by

$$t_n'' = \binom{u_m + 1}{2} = \frac{u_m(u_m + 1)}{2},$$

and therefore

$$t''(x) = \sum_{m=1}^{\infty} \frac{u_m(u_m + 1)}{2} x^{2m}$$

$$= \frac{1}{2} \sum_{m=1}^{\infty} u_m x^{2m} + \frac{1}{2} \sum_{m=1}^{\infty} (u_m x^m)^2 \quad (10\text{-}23)$$

$$= \frac{1}{2} u(x^2) + \frac{1}{2} \sum_{m=1}^{\infty} (u_m x^m)^2.$$

The number of vertices, n, in a centroidal tree can be odd or even. If n is odd, the maximum weight the centroid could have is $\frac{1}{2}(n - 1)$. This maximum is achieved only when the tree consists of a path of $n - 1$ edges. On the other hand, if n is even and the tree is centroidal, the maximum weight the centroid could possibly have is $\frac{1}{2}(n - 2)$. This maximum is achieved when the degree of the centroid is three, and one of the subtrees consists of just one edge.

Thus, regardless whether n is odd or even, it is clear that an n-vertex (free) centroidal tree can be regarded as composed of several rooted trees, rooted at the centroid, and none of these rooted trees can have more than $\lfloor (n - 1)/2 \rfloor$ edges, where $\lfloor x \rfloor$ denotes the largest integer no greater than x. In view of this observation, an involved manipulation of Eq. (10-21) leads to the following (for missing steps see [10-3]):

$$t'(x) = u(x) - \frac{1}{2} u^2(x) - \frac{1}{2} \sum_{m=1}^{\infty} (u_m x^m)^2. \quad (10\text{-}24)$$

Adding (10-23) and (10-24), we get the desired counting series:

$$t(x) = u(x) - \frac{1}{2}\left(u^2(x) - u(x^2)\right). \quad (10\text{-}25)$$

This relation, which gives the tree-counting series in terms of the rooted-tree counting series, was first obtained by Richard Otter in 1948 and is known as Otter's formula. The first 10 terms of (10-25) are

$$t(x) = x + x^2 + x^3 + 2x^4 + 3x^5 + 6x^6 + 11x^7$$
$$+ 23x^8 + 47x^9 + 106x^{10} + \cdots.$$

The reader is encouraged to extend it by another 10 terms. The first 26 terms of both $u(x)$ and $t(x)$ are given in Riordan's book [3-11], page 138.

By now you must have the impression that enumeration of graphs is an involved subject. And indeed it is. So far we have enumerated only four types of graphs—rooted and free trees, both labeled and unlabeled varieties. It is difficult to proceed further without some additional enumerative tool. This is provided by a general counting theorem due to Pólya. We shall first state and discuss Pólya's theorem and then show how it can be applied for counting graphs.

10-4. PÓLYA'S COUNTING THEOREM

To understand Pólya's theorem, we need a few additional concepts in combinatorial theory. In this section we shall first define a permutation and see how it can be represented in different ways. Then we shall show how a set of permutations P can form a group (called a permutation group) under a binary operation called composition. Then we shall introduce a polynomial called the cycle index of a permutation group P. Finally, we shall show that all mappings f_i's from a domain D to a range R (both D and R being finite) are divided into equivalence classes by any permutation group P acting on the domain D.

After introducing these concepts we shall define figure-counting series and configuration-counting series. And this will be followed by the statement of the celebrated theorem of Pólya, which expresses the configuration-counting series in terms of the figure-counting series and the cycle index of the permutation group. The statement of the theorem will be followed by discussion and two illustrative examples.

If Pólya's theorem and the buildup to it do not appear very intuitive to you, don't worry; you are not alone. What is important is to understand the theorem and be able to use it for counting different types of graphs.

Permutation

On a finite set A of some objects, a permutation π is a one-to-one mapping from A onto itself. For example, consider a set $\{a, b, c, d\}$. A permutation

$$\pi_1 = \begin{pmatrix} a & b & c & d \\ b & d & c & a \end{pmatrix}$$

takes a into b, b into d, c into c, and d into a. Alternatively, we could write

$$\pi_1(a) = b,$$
$$\pi_1(b) = d,$$
$$\pi_1(c) = c,$$
$$\pi_1(d) = a.$$

The number of elements in the object set on which a permutation acts is called the *degree* of the permutation. The degree of π_1 in the above example is four.

A permutation can also be represented by a digraph, in which each vertex represents an element of the object set and the directed edges represent the mapping. For example, the permutation $\pi_1 = \begin{pmatrix} a\ b\ c\ d \\ b\ d\ c\ a \end{pmatrix}$ is represented diagrammatically by Fig. 10-7.

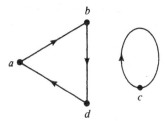

Fig. 10-7 Digraph of a permutation.

Observe that the in-degree and the out-degree of every vertex in the digraph of a permutation is one. Such a digraph must decompose into one or more vertex-disjoint directed circuits (why?). This suggests yet another way of representing a permutation—as a collection of the vertex-disjoint, directed circuits (called the *cycles of the permutation*). Permutation $\begin{pmatrix} a\ b\ c\ d \\ b\ d\ c\ a \end{pmatrix}$ can thus be written as $(a\ b\ d)(c)$. This compact and popular representation is called the *cyclic representation* of a permutation. The number of edges in a permutation cycle is called the *length of the cycle in the permutation*.

Often the only information of interest about a permutation is the number of cycles of various lengths. A permutation π of degree k is said to be of *type* $(\sigma_1, \sigma_2, \ldots, \sigma_k)$ if π has σ_i cycles of length i for $i = 1, 2, \ldots, k$. For example, permutation $(a\ b\ d)(c)$ is of type $(1, 0, 1, 0)$ and permutation $(a\ b\ f)(c)(d\ e\ h)(g)$ is of type $(2, 0, 2, 0, 0, 0, 0, 0)$. Clearly,

$$1\sigma_1 + 2\sigma_2 + 3\sigma_3 + \cdots + k\sigma_k = k. \qquad (10\text{-}26)$$

Another useful method for indicating the type of a permutation is to introduce k dummy variables, say, $y_1, y_2, \ldots, y_k$, and then show the type of permutation by the expression

$$y_1^{\sigma_1} y_2^{\sigma_2} \ldots y_k^{\sigma_k}. \qquad (10\text{-}27)$$

Expression (10-27) is called the *cycle structure* of π. For example, the cycle structure of the eight-degree permutation $(a\ b\ f)(c)(d\ e\ h)(g)$ is

$$y_1^2 y_2^0 y_3^2 y_4^0 y_5^0 y_6^0 y_7^0 y_8^0 = y_1^2 y_3^2.$$

Note that the dummy variable y_i has no significance except as a symbol to which subscripts (indicating the lengths) and exponents (indicating the number of cycles) are attached. Two distinct permutations (acting on the same object set) may have the same cycle structure (page 149 in [10-1]).

So far we have discussed only the representation and properties of a permutation individually. Let us now examine a set of permutations collectively.

On a set A with k objects, we have a total of $k!$ possible permutations—including the *identity permutation*, which takes every element into itself. For example, the following are the six permutations on a set of three elements $\{a, b, c\}$:

$$(a)(b)(c), \quad (a\,b)(c), \quad (a\,c)(b), \quad (a)(b\,c), \quad (a\,b\,c), \quad (a\,c\,b).$$

Their cycle structures, respectively, are

$$y_1^3, \quad y_1 y_2, \quad y_1 y_2, \quad y_1 y_2, \quad y_3, \quad y_3. \tag{10-28}$$

Composition of Permutations

Consider the two permutations π_1 and π_2 on an object set $\{1, 2, 3, 4, 5\}$:

$$\pi_1 = \begin{pmatrix} 1 & 2 & 3 & 4 & 5 \\ 2 & 1 & 4 & 5 & 3 \end{pmatrix} \quad \text{and} \quad \pi_2 = \begin{pmatrix} 1 & 2 & 3 & 4 & 5 \\ 3 & 4 & 1 & 2 & 5 \end{pmatrix}.$$

A *composition* of these two permutations $\pi_2 \pi_1$ is another permutation obtained by first applying π_1 and then applying π_2 on the resultant. That is,

$$\pi_2 \pi_1(1) = \pi_2(2) = 4,$$
$$\pi_2 \pi_1(2) = \pi_2(1) = 3,$$
$$\pi_2 \pi_1(3) = \pi_2(4) = 2,$$
$$\pi_2 \pi_1(4) = \pi_2(5) = 5,$$
$$\pi_2 \pi_1(5) = \pi_2(3) = 1.$$

Thus
$$\pi_2 \pi_1 = \begin{pmatrix} 1 & 2 & 3 & 4 & 5 \\ 4 & 3 & 2 & 5 & 1 \end{pmatrix}.$$

Thus among a collection of permutations on the same object set, composition is a binary operation.

Permutation Group

A collection of m permutations $P = \{\pi_1, \pi_2, \ldots, \pi_m\}$ acting on a set

$$A = \{a_1, a_2, \ldots, a_k\}$$

forms a group under composition, if the four postulates† of a group, that is, closure, associativity, identity, and inverse (see Section 6-1), are satisfied. Such a group is called a *permutation group*. For example, it can be easily verified that the set of four permutations

$$\{(a)(b)(c)(d), \ (a\ c)(b\ d), \ (a\ b\ c\ d), \ (a\ d\ c\ b)\} \qquad (10\text{-}29)$$

acting on the object set $\{a, b, c, d\}$ forms a permutation group.

The number of permutations m in a permutation group is called its *order*, and the number of elements in the object set on which the permutations are acting is called the *degree of the permutation group*. In the example just cited, both the degree and order of the permutation group is four. It can be shown that the set of all $k!$ permutations on a set A of k elements forms a permutation group. Such a group, of order $k!$ and degree k, is called the *full symmetric group*, S_k.

Cycle Index of a Permutation Group

For a permutation group P, of order m, if we add the cycle structures of all m permutations in P and divide the sum by m, we get an expression called the *cycle index* $Z(P)$ of P. For example, the cycle index of S_3, the full symmetric group of degree three, according to (10-28) comes out to be

$$Z(S_3) = \frac{1}{6}(y_1^3 + 3y_1 y_2 + 2y_3). \qquad (10\text{-}30)$$

Similarly, the cycle index of the permutation group (of degree four and order four) shown in (10-29) is

$$\frac{1}{4}(y_1^4 + y_2^2 + 2y_4).$$

Since the cycle index is the most important concept in this section, let us illustrate it with another example. Let us find $Z(S_4)$.

Table 10-1 gives the different types of permutations possible in S_4, the full symmetric group of degree four.

Table 10-1 is easy to understand and to construct. For example, we have six permutations of type (2, 1, 0, 0) on the object set $\{a, b, c, d\}$:

$$(a)(b)(c\ d), \quad (a)(c)(b\ d), \quad (a)(d)(b\ c),$$
$$(b)(c)(a\ d), \quad (b)(d)(a\ c), \quad (c)(d)(a\ b).$$

†In fact, it can be shown that if a collection of permutations is closed with respect to composition, the remaining three postulates are automatically satisfied (Problem 10-4).

Permutation Type	Number of Such Permutations	Cycle Structures
(4, 0, 0, 0)	1	y_1^4
(2, 1, 0, 0)	6	$y_1^2 y_2$
(1, 0, 1, 0)	8	$y_1 y_3$
(0, 2, 0, 0)	3	y_2^2
(0, 0, 0, 1)	6	y_4

Table 10-1

To get the cycle index of S_4 from Table 10-1, we multiply the corresponding entries in the second and third columns, add the products, and then divide the sum by 4!, the order of the group. Thus

$$Z(S_4) = \frac{1}{24}(y_1^4 + 6y_1^2 y_2 + 8y_1 y_3 + 3y_2^2 + 6y_4). \tag{10-31}$$

To display the variables involved, the cycle index of a permutation group P is often written as

$$Z(P) = Z(P; y_1, y_2, \ldots, y_k).$$

It is evident that computation of the cycle index of an arbitrary permutation group can become quite involved and laborious. There are certain groups, such as S_k, whose cycle indices have been derived in closed forms. These are related to the partitions of integer k satisfying Eq. (10-26). For more on methods of obtaining cycle indices, the reader should see [10-1] and [1-5].

Cycle Index of the Pair Group

When the n vertices of a graph G are subjected to permutation, the $n(n-1)/2$ unordered vertex pairs also get permuted. For example, let $V = \{a, b, c, d\}$ be the set of vertices of a four-vertex graph. The permutation

$$\beta = \begin{pmatrix} a & b & c & d \\ d & b & a & c \end{pmatrix}$$

on the vertices induces the following permutation on the six unordered vertex pairs:

$$\beta' = \begin{pmatrix} ab & ac & ad & bc & bd & cd \\ db & da & dc & ba & bc & ac \end{pmatrix}.$$

The diagrams of permutation β and the induced permutation are shown in Fig. 10-8.

Notice that a $y_1 y_3$ permutation on the vertex set induces a y_3^2 permutation on the vertex-pair set.

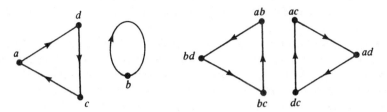

Fig. 10-8 Permutation on vertex set and the induced permutation on vertex-pair set.

Similarly, each of the $n!$ possible permutations on the n vertices of a graph results in some permutation of the $n(n-1)/2$, unordered, vertex pairs (or $n(n-1)$ ordered vertex pairs, in the case of digraphs). Furthermore, it can be shown that if a set of permutations on the vertices forms a group, the induced set of permutations on the pair of vertices will also form a group (Problem 10-5). For instance, the full symmetric group S_n on n vertices of a graph induces a group R_n of $n!$ permutations on the pairs of vertices.† Such an induced group is called the *pair group* R_n. Let us work out the pair group R_4 induced by S_4, the full symmetric group on the vertices of a four-vertex graph.

The identity permutation on the four vertices of a graph produces an identity permutation on the six pairs of vertices. A permutation with two cycles of length one and one cycle of length two produces two cycles of length one and two cycles of length two. And so on. The cycle structures of permutations in S_4 and the corresponding cycle structures of the induced permutations on the pairs of vertices are shown in Table 10-2.

Term in $Z(S_4)$	Induced Term in $Z(R_4)$	No. of Permutations
y_1^4	y_1^6	1
$y_1^2 y_2$	$y_1^2 y_2^2$	6
$y_1 y_3$	y_3^2	8
y_2^2	$y_1^2 y_2^2$	3
y_4	$y_2 y_4$	6

Table 10-2

Therefore, the cycle index of the pair group R_4 (induced on the pairs of vertices by S^4) is

$$Z(R_4) = \frac{1}{24}(y_1^6 + 9y_1^2 y_2^2 + 8y_3^2 + 6y_2 y_4). \quad (10\text{-}32)$$

For a general expression for $Z(R)$ see [10-2].

†Except for $n = 2$, in which case the number of possible permutations on the pair is 1, rather than 2.

Equivalence Classes of Functions

As a further preliminary to describing Pólya's theorem, let us introduce some additional concepts. Consider two sets D and R, with the number of elements $|D|$ and $|R|$, respectively. Let f be a mapping (or function) which maps each element d from domain D to a unique image $f(d)$ in range R. Since each of the $|D|$ elements can be mapped into any of the $|R|$ elements, the number of different functions from D to R is $|R|^{|D|}$.

Now let there be a permutation group P on the elements of set D. Then define two mappings f_1 and f_2 as *P-equivalent* if there is some permutation π in P such that for every d in D we have

$$f_1(d) = f_2[\pi(d)]. \qquad (10\text{-}33)$$

That the relationship defined by (10-33) is an equivalence relation can be shown as follows:

1. Since P is a permutation group, it contains the identity permutation, and thus (10-33) is reflexive.
2. If P contains permutation π, it also contains the inverse permutation π^{-1}. Therefore, the relation is symmetric also.
3. Furthermore, if P contains permutations π_1 and π_2, it must also contain the permutation $\pi_1 \pi_2$. This makes P-equivalence a transitive relation.

Since an equivalence relation divides a set into equivalence classes, all mappings from D to R are divided into equivalence classes by a permutation group P acting on set D. As an example, let $D = \{a, b, c\}$ and $R = \{s, t\}$. There are $2^3 = 8$ mappings $f_1, f_2, \ldots, f_8$ from D to R, as shown in Table 10-3.

	f_1	f_2	f_3	f_4	f_5	f_6	f_7	f_8
$f(a)$	s	s	s	t	s	t	t	t
$f(b)$	s	s	t	s	t	s	t	t
$f(c)$	s	t	s	s	t	t	s	t

Table 10-3

Now suppose a permutation group $P = \{(a)(b)(c), (a\ b\ c), (a\ c\ b)\}$ is acting on D. The reader can verify that the eight mappings in Table 10-3 will be divided into four equivalence classes. They are

$$\{f_1\}, \quad \{f_2, f_3, f_4\}, \quad \{f_5, f_6, f_7\}, \quad \{f_8\}.$$

Pólya's Counting Theorem

Let us consider two finite sets, domain D and range R, together with a permutation group P on D. To each element $\rho \in R$ let us assign a quantity $w[\rho]$ and call it the *content* (or *weight*) of the element ρ. The weight $w[\rho]$ can be a symbol or a real number. A mapping f from D to R can be described by a sequence of $|D|$ elements of set R such that the ith element in the sequence is the image of the ith element of set D under f. Therefore the content $W(f)$ of a mapping f can be defined as the product of the contents of all its images. That is,

$$W(f) = \prod_{d \in D} w[f(d)].$$

Clearly, all functions belonging to the same equivalence class defined by (10-33) have identical weights. Therefore, we define the weight of an entire equivalence class (of functions from domain D to range R) to be the (common) weight of the functions in this class. Our problem is to count the number of equivalence classes with various weights, given D, R, permutation group P on D, and weights $w[\rho]$ for each $\rho \in R$. This is exactly what Pólya's counting theorem gives.

In Pólya's terminology, elements ρ of set R are called *figures*, and functions f from D to R are called *configurations*. Often the weights of the elements of R can be expressed as powers of some common quantity x. In that case the weight assignment to elements of set R can be neatly described by means of a counting series $A(x)$

$$A(x) = \sum_{q=0}^{\infty} a_q x^q, \tag{10-34}$$

where a_q is the number of elements in set R with weight x^q.† Likewise, the number of configurations can be expressed in terms of another series, called *configuration counting series* $B(x)$, such that

$$B(x) = \sum_{m=0}^{\infty} b_m x^m, \tag{10-35}$$

where b_m is the number of different configurations having weight x^m. Now we can state the following powerful result known as Pólya's counting theorem.

Theorem 10-3

The configuration-counting series $B(x)$ is obtained by substituting the figure-counting series $A(x^i)$ for each y_i in the cycle index $Z(P; y_1, y_2, \ldots, y_k)$ of the

---

†If the content assigned to figures cannot be expressed as powers of a single quantity x, then the figure-counting series will be a multinomial in different variables, rather than in just one variable x.

permutation group P. That is,

$$B(x) = Z(P; \sum a_q x^q, \sum a_q x^{2q}, \sum a_q x^{3q}, \ldots, \sum a_q x^{kq}). \qquad (10\text{-}36)$$

The proof of Pólya's theorem, although not complicated, is not particularly illuminating and is therefore left out. The reader can find it in [10-1], page 157. Our interest is mainly in the application of the theorem; let us illustrate it with some examples.

Example 1: Suppose that we are given a cube and four (identical) balls. In how many ways can the balls be arranged on the corners of the cube? Two arrangements are considered the same if by any rotation of the cube they can be transformed into each other.

The answer is seven, as can be seen by inspection in Fig. 10-9. In Pólya's terms the domain D is the set of the eight corners of the cube, and the range

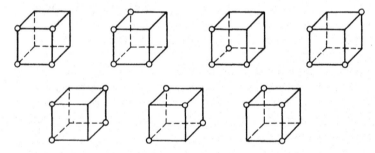

Fig. 10-9 Attaching four balls to corners of a cube.

R consists of two elements (i.e., figures), "presence of a ball" or "absence of a ball," with contents x^1 and x^0, respectively. The figure-counting series is

$$A(x) = \sum_{q=0}^{\infty} a_q x^q = a_0 x^0 + a_1 x^1 = 1 + x, \qquad (10\text{-}37)$$

since a_0, the number of figures with content 0, is one, and a_1, the number of figures with content 1, is also one. The configurations are $2^8 = 256$ different mappings that assign balls to the corners of the cube. The permutation group P on D is the set of all those permutations that can be produced by rotations of the cube. These permutations with their cycle structures are

1. One identity permutation. Its cycle structure is y_1^8.

2. Three 180° rotations around lines connecting the centers of opposite faces. Its cycle structure is y_2^4.

3. Six 90° rotations (clockwise and counterclockwise) around lines connecting the centers of opposite faces. The cycle structure is y_4^2.

4. Six 180° rotations around lines connecting the midpoints of opposite edges. The corresponding cycle structure is y_2^4.

5. Eight 120° rotations around lines connecting opposite corners in the cube. The cycle structure of the corresponding permutation is $y_1^2 y_3^2$.

The cycle index of this group consisting of these 24 permutations is, therefore,

$$Z(P) = \frac{1}{24}(y_1^8 + 9y_2^4 + 6y_4^2 + 8y_1^2 y_3^2). \tag{10-38}$$

Using Polya's theorem, we now substitute the figure-counting series, that is $1 + x$ for y_1, $1 + x^2$ for y_2, $1 + x^3$ for y_3, and $1 + x^4$ for y_4. This yields the configuration-counting series.

$$B(x) = 1 + x + 3x^2 + 3x^3 + 7x^4 + 3x^5 + 3x^6 + x^7 + x^8. \tag{10-39}$$

The coefficient of x^4 in $B(x)$ gives the number of P-inequivalent configurations of content x^4 (i.e., with four balls). This verifies the answer obtained by exhaustive inspection in Fig. 10-9.

The total number of P-inequivalent configurations (with contents $x^0, x^1, x^2, \ldots, x^8$) is obtained by adding all coefficients in (10-39), which is 23. It may be observed that this is the number of distinct ways of painting the eight vertices of a cube with two colors (one color corresponds to the "presence of a ball" and the other with the "absence of a ball").

Example 2: In example 1 we were given four identical balls. Now suppose that we are given two red balls and two blue balls, and are again asked to find the number of distinct arrangements on the corners of the cube. Clearly, D, P, and $Z(P)$ will remain the same as they were in example 1. Only the range R and the figure-counting series $A(x)$ will change. The range will contain three elements: (1) presence of no ball, (2) presence of a red ball, and (3) presence of a blue ball. Choosing x to indicate the presence of a red ball and x' to indicate the presence of a blue ball, the three elements in the range mentioned above will have the contents $x^0 x'^0$, $x^1 x'^0$, and $x^0 x'^1$, respectively. Therefore the figure-counting series is

$$A(x, x') = x^0 x'^0 + x^1 x'^0 + x^0 x'^1 = 1 + x + x'.$$

Substituting this figure-counting series in (10-38), we get the configuration-counting series

$$B(x, x') = \frac{1}{24}[(1 + x + x')^8 + 9(1 + x^2 + x'^2)^4 + 6(1 + x^4 + x'^4)^2$$
$$+ 8(1 + x + x')^2(1 + x^3 + x'^3)^2]$$
$$= 1 + x + x' + 3x^2 + 3x'^2 + 3xx' + 3x^3 + 3x'^3$$
$$+ 7x^2x' + 7xx'^2 + 7x^4 + 7x'^4 + 13x^3x' + 13xx'^3$$
$$+ 22x^2x'^2 + 3x^5 + 3x'^5 + 13x^4x' + 13xx'^4$$
$$+ 24x^3x'^2 + 24x^2x'^3 + 3x^6 + 3x'^6 + 7x^5x' \qquad (10\text{-}40)$$
$$+ 7xx'^5 + 22x^4x'^2 + 22x^2x'^4 + 24x^3x'^3 + x^7 + x'^7$$
$$+ 3x^6x' + 3xx'^6 + 7x^5x'^2 + 7x^2x'^5 + 13x^3x'^4 + 13x^4x'^3$$
$$+ x^8 + x'^8 + x^7x' + xx'^7 + 3x^6x'^2 + 3x^2x'^6 + 3x^5x'^3$$
$$+ 3x^3x'^5 + 7x^4x'^4.$$

The coefficient of $x^r x'^b$ in (10-40) is the number of distinct arrangements with r red balls, b blue balls and $8 - r - b$ corners with no balls. The number of arrangements with two red and two blue balls is, therefore, 22.

For some other non-graph-theoretic examples of the applications of Pólya's theorem, the reader should work out Problems 10-10, 10-11, 10-14, and 10-15. Let us now return to the counting of graphs.

10-5. GRAPH ENUMERATION WITH POLYA'S THEOREM

Enumeration of Simple Graphs: Let us consider the problem of counting all unlabeled, simple graphs of n vertices. Any such graph G can be regarded as a mapping (i.e., configuration) of the set D of all $\frac{1}{2}n(n-1)$ unordered pairs of vertices (for digraphs $n(n-1)$ pairs of vertices). Range R consists of two elements s and t, with contents x^1 and x^0, respectively. If a vertex pair is joined by an edge in G, the vertex pair maps into s, an element with content x^1; otherwise, into t, an element with content $x^0 = 1$. Thus the figure-counting series is

$$A(x) = \sum a_q x^q = 1 + x.$$

The relevant permutation group in this case is R_n, the group of permutations on the pairs of vertices induced by S_n (the full symmetric group on the n vertices of the graph).† Therefore, the configuration-counting series is

†Because in an unlabeled graph, all n vertices are indistinguishable. Were we to count labeled graphs the permutation group would have consisted of only the identity permutation. Substitution of $1 + x$ in its cycle index would have yielded the simple result of expression (10-1).

obtained by substituting $1 + x$ for y_1, $1 + x^2$ for y_2, $1 + x^3$ for y_3, and so on in $Z(R_n)$. Some specific cases are

(1) For $n = 3$,

$$Z(R_3) = \frac{1}{6}(y_1^3 + 3y_1 y_2 + 2y_3).$$

Therefore, the configuration-counting series is

$$B(x) = \frac{1}{6}[(1 + x)^3 + 3(1 + x)(1 + x^2) + 2(1 + x^3)]$$
$$= 1 + x + x^2 + x^3.$$

The coefficient of x^i in $B(x)$ is the number of configurations with content x^i. The content of a configuration here is the number of edges in the corresponding graph. Thus the number of nonisomorphic simple graphs of three vertices with 0, 1, 2, and 3 edges is each one. This is how it should be, as shown in Fig. 10-10.

(2) For $n = 4$, the cycle index $Z(R_4)$ is given in (10-32). Substituting $1 + x^i$ for y_i in (10-32), we get

$$B(x) = \frac{1}{24}[(1 + x)^6 + 9(1 + x)^2(1 + x^2)^2 + 8(1 + x^3)^2$$
$$+ 6(1 + x^2)(1 + x^4)] \tag{10-41}$$
$$= 1 + x + 2x^2 + 3x^3 + 2x^4 + x^5 + x^6.$$

In (10-41) the coefficient of x^r gives the number of simple graphs with four vertices and r edges. The validity of series (10-41) is verified in Fig. 10-11.

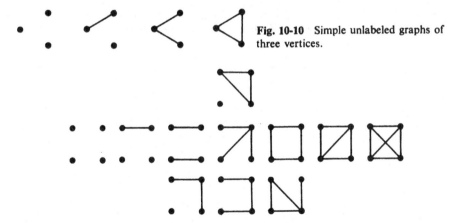

Fig. 10-10 Simple unlabeled graphs of three vertices.

Fig. 10-11 Simple unlabeled graphs of four vertices.

(3) For $n = 5$, the cycle index $Z(R_5)$ is given in Problem 10-9. Substituting $1 + x^i$ for y_i in $Z(R_5)$, we get the counting series $B(x)$ for simple graphs of five vertices, as follows:

$$B(x) = \frac{1}{120}[(1 + x)^{10} + 10(1 + x)^4(1 + x^2)^3 + 20(1 + x)(1 + x^3)^3$$
$$+ 15(1 + x)^2(1 + x^2)^4 + 30(1 + x^2)(1 + x^4)^2$$
$$+ 20(1 + x)(1 + x^3)(1 + x^6) + 24(1 + x^5)^2] \quad (10\text{-}42)$$
$$= 1 + x + 2x^2 + 4x^3 + 6x^4 + 6x^5 + 6x^6 + 4x^7 + 2x^8$$
$$+ x^9 + x^{10}.$$

Again, for each r the coefficient of x^r in (10-42) gives the number of simple graphs of five vertices and r edges.

The number of simple, unlabeled graphs with n vertices for any n can be counted similarly.

Enumeration of Multigraphs: Suppose that we are interested in counting multigraphs of n vertices, in which at most two edges are allowed between a pair of vertices.

In this case the domain and the permutation group are the same as they were for simple graphs. The range, however, is different. A pair of vertices may be joined by (1) no edge, (2) one edge, or (3) two edges. Thus range R contains three elements, say, s, t, u, with contents x^0, x^1, and x^2, respectively; that is, x^i indicates the presence of i edges between a vertex pair, for $i = 0, 1, 2$. Therefore, the figure-counting series becomes

$$1 + x + x^2. \quad (10\text{-}43)$$

Substitution of $1 + x^r + x^{2r}$ for y_r in $Z(R_n)$ will yield the desired configuration-counting series. For $n = 4$, using the cycle index from (10-32), we get

$$\frac{1}{24}[(1 + x + x^2)^6 + 9(1 + x + x^2)^2(1 + x^2 + x^4)^2 + 8(1 + x^3 + x^6)^2$$
$$+ 6(1 + x^2 + x^4)(1 + x^4 + x^8)] \quad (10\text{-}44)$$
$$= 1 + x + 3x^2 + 5x^3 + 8x^4 + 9x^5 + 12x^6 + 9x^7 + 8x^8$$
$$+ 5x^9 + 3x^{10} + x^{11} + x^{12}.$$

The coefficient of x^i in (10-44) is the number of distinct, unlabeled, multigraphs of four vertices and i edges (such that there are at most two parallel edges between any vertex pair). For example, the coefficient of x^3 is 5, and these five multigraphs are shown in Fig. 10-12.

Instead of allowing at most two parallel edges between a pair of vertices,

SEC. 10-5   GRAPH ENUMERATION WITH PÓLYA'S THEOREM   263

Fig. 10-12 Unlabeled multigraphs of four vertices, three edges, and at most two parallel edges.

had we allowed any number of parallel edges the figure-counting series would be the infinite series

$$A(x) = 1 + x + x^2 + x^3 + \cdots = \frac{1}{1-x}. \quad (10\text{-}45)$$

Enumeration of Digraphs: For enumerating digraphs we have to consider all $n(n-1)$ *ordered* pairs of vertices as constituting the domain. The relevant permutation group will consist of permutations induced on all ordered pairs of vertices by S_n. The cycle index of this permutation group, M_n, can be obtained in the same fashion as was done in the case of R_n. For example, for $n = 4$, Table 10-4 gives the terms in $Z(M_n)$ induced by each term in $Z(S_n)$.

Term in $Z(S_4)$	Induced Term in $Z(M_4)$
y_1^4	y_1^{12}
$y_1^2 y_2$	$y_1^2 y_2^5$
$y_1 y_3$	y_3^4
y_2^2	y_2^6
y_4	y_4^3

Table 10-4

Therefore, the cycle index is

$$Z(M_4) = \frac{1}{24}(y_1^{12} + 6y_1^2 y_2^5 + 8y_3^4 + 3y_2^6 + 6y_4^3). \quad (10\text{-}46)$$

For a simple digraph the figure-counting series $A(x) = 1 + x$ is applicable, because a given ordered pair of vertices (a, b) either does or does not have an edge (directed) from a to b. On substituting $1 + x^i$ for every y_i in (10-46), we get the following configuration-counting series for four-vertex, simple digraphs.

$$\begin{aligned}B(x) &= \frac{1}{24}[(1+x)^{12} + 6(1+x)^2(1+x^2)^5 + 8(1+x^3)^4 \\ &\quad + 3(1+x^2)^6 + 6(1+x^4)^3] \\ &= 1 + x + 5x^2 + 13x^3 + 27x^4 + 38x^5 + 48x^6 \\ &\quad + 38x^7 + 27x^8 + 13x^9 + 5x^{10} + x^{11} + x^{12}.\end{aligned} \quad (10\text{-}47)$$

Fig. 10-13 Simple unlabeled digraphs of four vertices and two edges.

The coefficient of x^j in (10-47) is the number of simple digraphs with four vertices and j edges. For example, the five digraphs of two edges are shown in Fig. 10-13.

The general expression for the cycle index, $Z(M_n)$, of the permutation group on $n(n-1)$ ordered pairs induced by S_n is given in [1-5], page 180. Digraphs with parallel edges can be enumerated by substituting the appropriate figure-counting series, say (10-43), in $Z(M_n)$.

SUMMARY

Enumeration of graphs is one of the most involved areas in graph theory and deserves an entire volume to itself. In this chapter, we have briefly presented some enumerative techniques—the most important of them being Pólya's counting theorem. The major problem in using Pólya's theorem is finding the appropriate permutation group and then obtaining its cycle index.

One could think of a hundred different types of graphs to be counted—each presenting a special problem. We have, in this chapter, counted the following five types of unlabeled graphs (enumeration of labeled graphs is much easier): (1) rooted trees, (2) free trees, (3) simple graphs, (4) multigraphs, and (5) simple digraphs. Important as these types of graphs are, they were enumerated mainly as illustrations. One could, for example, be interested in counting all unlabeled, simple graphs with n vertices that are (1) connected, or (2) planar, or (3) nonseparable, or (4) self-dual, and so on.

Many such types of graphs have been enumerated and reported as research papers in the literature, but there are many types of graphs that have yet to be counted.

REFERENCES

Chapters 1, 3, 4, 5, and 6 of [10-1] are strongly recommended to supplement the material presented in this chapter.

For an exhaustive survey of the literature in graph enumeration, see the paper by Harary [10-4], as well as Chapter 15 of [1-5], which contains a list of 66 solved problems with the appropriate references where the solutions are to be found. The latest list of 27 unsolved problems in graph enumeration is discussed in another article by Harary [10-5]. For a lucid exposition of Pólya's counting theorem, see the paper by deBruijn, which appears as Chapter 5 in [10-1], or see Chapter 5 of the book by Liu [8-3]. For some excellent illustrations of applications of Pólya's theorem to graph enumeration, see [10-2], [10-6], [10-7], [10-8], and Chapter 6 of [3-11].

Many counting problems in chemistry, physics, biology, information theory, and so on, can be regarded as graph-enumeration problems. A survey of such applications is given in Chapter 6 of [10-1] and in [10-4]. A detailed treatment of an application to a problem in statistical mechanics is given in [10-9]. An application to counting of distinct automata is given in [12-5].

The pioneering papers of Cayley, Redfield, and Pólya are not included in the following list. They have been referred to in most of the following:

10-1. BECKENBACH, E. F. (ed.), *Applied Combinatorial Mathematics*, John Wiley & Sons, Inc., New York, 1964.

10-2. HARARY, F., "The Number of Linear, Directed, Rooted and Connected Graphs," *Trans. Am. Math. Soc.*, Vol. 78, 1955, 445–463.

10-3. HARARY, F., "Note on the Pólya and Otter Formulas for Enumerating Trees," *Michigan Math Journal*, Vol. 3, 1956, 109–112.

10-4. HARARY, F., "Graphical Enumeration Problems," in *Graph Theory and Theoretical Physics* (F. Harary, ed.), Academic Press, Inc., New York, 1967, 1–41.

10-5. HARARY, F., "Enumeration Under Group Action: Unsolved Problems in Graphical Enumeration IV," *J. Combinatorial Theory*, Vol. 8, 1970, 1–11.

10-6. PALMER, E. M., "Methods for the Enumeration of Multigraphs," in *The Many Facets of Graph Theory* (G. Chartrand and S. F. Kapoor, eds.), Springer-Verlag New York, Inc., New York, 1969, 251–261.

10-7. READ, R. C., "On the Number of Self-Complementary Graphs and Digraphs," *J. London Math. Soc.*, Vol. 38, 1963, 99–104.

10-8. ROBINSON, R. W., "Enumeration of Non-separable Graphs," *J. Combinatorial Theory*, Vol. 9, No. 4, Dec. 1970, 327–356.

10-9. UHLENBECK, G. E., and G. W. FORD, "Theory of Linear Graphs with Applications to the Theory of the Virial Development of the Properties of Gases," in *Studies in Statistical Mechanics*, Vol. 1 (J. de Boer and G. E. Uhlenbeck, eds.), North-Holland Publishing Company, Amsterdam, 1962, 123–211.

PROBLEMS

10-1. Satisfy yourself that for a set of graphs isomorphism (as defined in Section 2-1) is indeed an equivalence relation. That is, the relation is reflexive, symmetric, and transitive.

10-2. Prove that a vertex v appears in sequence (10-3) m times if and only if degree of $v = m - 1$.

10-3. Prove that a digraph in which the in-degree as well as the out-degree of every vertex is one can be decomposed into one or more vertex-disjoint directed circuits. (*Hint:* In such a digraph every component is a directed circuit.)

10-4. Prove that a subset A of a finite group forms a subgroup if the subset satisfies the closure postulate. (*Hint:* Show the existence of the inverse of an element $a \in A$ as follows: Elements $a, a^2, a^3, \ldots$ cannot all be distinct because the group is finite, but they must all be in A because of the closure property. Suppose that $a^p = a^q$, where $p > q$. Therefore, $a^{p-q} = 1$ or $a^{-1} = a^{p-q-1}$.)

10-5. Prove that if a set of permutations P on an object set S forms a group, the set R of all permutations induced by P on set $S \times S$ also forms a group. [*Hint:* Prove closure by showing that the composition of two permutations on $S \times S$ induced by any two permutations π_1, π_2 (in P) is the permutation induced by the composition $(\pi_2 \cdot \pi_1)$. Then use Problem 10-4.]

10-6. Show that the cycle index of a group consisting of the identity permutation only is y_1^k, k being the number of elements in the object set.

10-7. Show that the cycle index of the induced pair group R_3 is the same as that of S_3. That is,
$$Z(R_3) = \frac{1}{6}(y_1^3 + 3y_1y_2 + 2y_3).$$

10-8. Show that the cycle index of S_5, the full symmetric group of degree five, is
$$Z(S_5) = \frac{1}{5!}(y_1^5 + 10y_1^3 y_2 + 20y_1^2 y_3 + 15y_1 y_2^2$$
$$+ 30y_1 y_4 + 20y_2 y_3 + 24y_5). \qquad (10\text{-}48)$$

10-9. Show that the cycle index of the unordered pair group R_5 (on the set of 10 unordered pairs induced by S_5) is
$$Z(R_5) = \frac{1}{5!}(y_1^{10} + 10y_1^4 y_2^3 + 20y_1 y_3^3 + 15y_1^2 y_2^4$$
$$+ 30y_2 y_4^2 + 20y_1 y_3 y_6 + 24y_5^2).$$

(*Hint:* Use the result of Problem 10-8.)

10-10. Find the different ways of painting the six vertices of an octahedron with three colors. Two octahedrons are colored distinctly if they cannot be made to coincide by any rotation. [*Hint:* First show that the cycle index of the permutation group is
$$\frac{1}{24}(y_1^6 + 6y_1^2 y_4 + 3y_1^2 y_2^2 + 6y_2^3 + 8y_3^2).$$
Then substitute the figure-counting series $1 + x + x'$.]

10-11. List all partitions of 5, and use them to find u_6, the number of unlabeled trees of six vertices. (You may use the values of $u_1, u_2, \ldots, u_5$ given in this chapter.)

10-12. Given a square, show that there are exactly eight distinct motions (combinations of rotations and reflections) which bring the square into coincidence with itself. Show that these motions form a group (called dihedral group D_4). Furthermore, show that the cycle index of this group is
$$Z(D_4) = \frac{1}{8}(y_1^4 + 2y_1^2 y_2 + 3y_2^2 + 2y_4).$$

10-13. Show that the order of D_n, the group of symmetries of a regular n-sided polygon, is $2n$. Find the cycle index of D_n.

10-14. Suppose that we are to make necklaces with four beads—some blue and some green. How many distinct necklaces are possible? Two necklaces are considered indistinguishable if one can be made identical to the other by any combination of rotation and flipping. [*Hint:* Use $Z(D_4)$ and follow the procedure of example 1.]

10-15. Find the number of different ways of painting the four faces of a pyramid with two colors.

10-16. Find the counting series for unlabeled, simple, connected graphs with exactly one circuit. [*Hint:* Use $Z(D_n)$ and consider the graph as consisting of a single circuit with one or more trees attached to its vertices.]

10-17. Find the counting series for the structural isomers of saturated alcohols $C_nH_{2n+1}OH$. (*Hint:* Consider the compound as an n-vertex rooted tree in which each vertex is a carbon atom. The carbon atom carrying the OH radical corresponds to the root. Then find the counting series for unlabeled, rooted trees in which the root is at most of degree three and the nonroot vertices are at most of degree four.)

10-18. A permutation π applied on the vertex set V of a graph G is called an *automorphism* of G, if π preserves the adjacency. That is, an automorphism of G is an isomorphism with itself. Prove that the set of all automorphisms $\Omega(G)$ on G forms a group. (*Hint:* This group will obviously be a subgroup of S_n. Use the result of Problem 10-4, after observing that an automorphism followed by another is also an automorphism.)

10-19. Find the automorphism group $\Omega(G)$ of a graph G if G is (a) a complete graph of n vertices, and (b) a circuit with n vertices. Find a graph with minimum number of vertices $n > 1$ in which $\Omega(G)$ consists of only the identity permutation.

10-20. Prove that the number of ways an unlabeled n-vertex graph can be labeled is $n!/|\Omega(G)|$, where $|\Omega(G)|$ is the order of the automorphism group $\Omega(G)$ of G. (*Hint:* The problem requires some additional knowledge of group theory. The proof can be found on page 180 in [1-5].)

11 GRAPH-THEORETIC ALGORITHMS AND COMPUTER PROGRAMS

To be able to use a digital computer in solving graph-theoretic problems is undoubtedly an important part of learning graph theory, especially for those interested in applications. Most of the practical problems which call for graph theory involve large graphs—graphs that are virtually impossible for hand computation. In fact, one of the reasons for the recent growth of interest in graph theory has been the arrival of the high-speed electronic computer. Problems that hitherto were of academic interest only are suddenly being solved by the computer, and their solutions are applied to practical situations. Computer programs have been written to handle successfully large graphs encountered in PERT, flow problems, transportation networks, electrical networks, circuit layouts, and the like.

We must hasten to add, however, that although our computers are very fast and operate at nanosecond (10^{-9} second) speeds, they quickly reach their limit if used as a brute force to solve graph-theory problems (in fact, any combinatorial problem). Consider, for example, the problem of finding a lowest-weight Hamiltonian circuit in a weighted complete graph of n vertices, that is, the traveling salesman problem. There are $\frac{1}{2} \cdot (n-1)!$ different Hamiltonian circuits. One may be tempted to use brute force and generate all Hamiltonian circuits and compare their weights. For a graph with 10 vertices, the number of Hamiltonian circuits is $\frac{1}{2} \cdot 9! = 181{,}440$, and this method may be all right. But for a graph of 20 vertices, we have

$$\tfrac{1}{2} \cdot (n-1)! = \tfrac{1}{2} \cdot 19! \simeq 6 \times 10^{16},$$

and to perform $\frac{1}{2} \cdot 19!$ operations at the rate of even one operation per nanosecond would require about

$$\frac{6 \times 10^{16}}{10^9 \times 3 \times 10^7} = 2 \text{ years.}$$

Thus it is amply clear that without the aid of mathematical tools one cannot hope to get the desired numerical answer, regardless of the speed of the electronic computer. The power of the computer must be combined with the ingenuity of mathematical techniques.

As is the case with all combinatorial problems, the manipulation and analysis of graphs and subgraphs is essentially nonnumerical. That is, in graph-theoretic programs it is primarily the decision-making ability of the computer that is used rather than its ability to perform arithmetic operations.

In this chapter it is assumed that the reader has some familiarity with computer programming.

11-1. ALGORITHMS

An algorithm is, in essence, a recipe for solving a certain mathematical problem. It consists of a set of instructions that when followed step by step will lead to the solutions of the problem. Every step in an algorithm must be precisely and unambiguously defined, and an algorithm must terminate after having solved the given problem in a finite number of steps. As pointed out by Knuth [11-39], page 4, every algorithm must have five important features: finiteness, definiteness, input, output, and effectiveness.

An algorithm can be expressed in different forms: (1) the steps may be written in English; (2) it may be in the form of a computer program written in complete detail in the language understandable by the machine in use; or (3) the algorithm may be expressed in a form between these two extremes, such as a flow chart. Each form has certain advantages and shortcomings. Usually, an algorithm is first expressed in ordinary language, then converted into a flow chart, and finally written in the detailed and precise language so that a machine can execute it.

For our purpose the flow chart is the best. It is the most popular form of expressing an algorithm. It is independent of the programming language and of the computer the student may have at his disposal. As examples of actual programs, listings of several tested programs are provided at the end of this chapter. One of the programs is in APL (*A Programming Language*), and the others are in FORTRAN.

Efficiency of Algorithms: An algorithm must not only do what it is supposed to do, but must do it efficiently. The two main criteria for efficiency of an algorithm are the memory and computation-time requirements as a function of the size of the input. In our case the input is a graph, and its size is the number of vertices, n, and the number of edges, e. For most graph problems the memory requirement is generally not the bottleneck, but the computation time can be (as we saw in the opening remarks of this chapter).

In evaluating the figure of merit of an algorithm one may seek the "worst-case" execution time (i.e., the time taken for the worst possible choice of a graph of the given size), or the "best-case" execution time or the "average-case."

Often, more than one algorithm is available for one graph-theoretic problem. Sometimes one algorithm can easily be seen to be more efficient than others, for all nontrivial graphs. In many cases, however, the relative efficiencies can be compared only in the context of the size and structure of the graph, detailed implementation of the algorithms, and the computer used.

A detailed analysis of the performance of a graph-theoretic algorithm is extremely involved. We will not indulge in such analyses, as that would require a chapter in itself. We will, however, make some gross observations on complexities of the algorithms; namely how the computation time grows as a function of n or e, as n and e become very large, assuming that the worst-case graph is provided as the input. Such an index of performance, too unrealistic to be useful in estimating the expected computation time of a program, is often valuable in classifying algorithms and in their theoretical studies.

11-2. INPUT: COMPUTER REPRESENTATION OF A GRAPH

An algorithm has some inputs—the data with which the algorithm begins (just as a recipe for a dish calls for raw ingredients). Naturally, the input for our algorithms here will be one or more graphs (or digraphs). A graph is generally presented to and is stored in a digital computer in one of the following five forms. Each has advantages and disadvantages. The choice depends on the graph, the problem, the language, the type of machine, and whether or not the graph is modified during the course of the computation.

(a) *Adjacency Matrix:* The most popular form in which a graph or digraph is fed to a computer is its adjacency matrix. For example, algorithms described in [11-25] and [11-47] use the adjacency matrix. After assigning a distinct number to each of the n vertices of the given graph (or digraph) G, the n by n binary matrix $X(G)$ is used for representing G during input, storage, and output. Since each of the n^2 entries is either a 0 or a 1, the adjacency matrix requires n^2 bits of computer memory. Bits can be packed into words. Let w be the word length (i.e., the number of bits in a computer word) and n be the number of vertices in the graph. Then each row of the adjacency matrix may be written as a sequence of n bits in $\lceil n/w \rceil$ machine words. ($\lceil x \rceil$ denotes the smallest integer not less than x.) The number of words required to store the adjacency matrix is, therefore, $n\lceil n/w \rceil$.

The adjacency matrix of an undirected graph is symmetric, and therefore storing only the upper triangle is sufficient. This requires only $n(n-1)/2$ bits of storage. This saving in storage, however, often costs in increased complexity and computation time. In some problems it is worth it.

It must be kept in mind that the adjacency matrix is defined for graphs without parallel edges. As discussed in Chapter 7, it is not possible to represent parallel edges in an adjacency matrix.

(b) *Incidence Matrix:* Occasionally, an incidence matrix is also used for storing and manipulation of a graph. The algorithm in [11-68], for example, uses the incidence matrix $A(G)$. An incidence matrix requires $n \cdot e$ bits of storage, which might be more than the n^2 bits needed for an adjacency matrix, because the number of edges e is usually greater than the number of vertices n. On rare occasions it may be advantageous to use the incidence matrix rather than the adjacency matrix, in spite of the increased requirements in storage. Incidence matrices are particularly favored for electrical networks and switching networks.

(c) *Edge Listing:* Another representation often used is to list all edges of the graph as vertex pairs, having numbered the n vertices in some arbitrary order. For example, the digraph in Fig. 11-1 would appear as a set of the following ordered pairs: (1, 2), (2, 1), (2, 4), (3, 2), (3, 3), (3, 4), (4, 1), (4, 1), (5, 2). Had this graph been undirected, we would simply ignore the ordering in each vertex pair.

Clearly, parallel edges and self-loops *can* be included in this representation of a graph or digraph.

The number of bits required to label (1 through n) each vertex is b, where

$$2^{b-1} < n \leq 2^b.$$

And since each of the e edges requires storing two such numbers, the total storage required is

$$2e \cdot b \text{ bits.}$$

Comparing this with n^2, we see that this representation is more economical than the adjacency matrix if

$$2e \cdot b < n^2.$$

In other words, for a graph whose adjacency matrix is sparse†, edge listing is a more efficient method of storing the graph.

Edge listing is a very convenient form for inputting a graph into the computer, but the storage, retrieval, and manipulation of the graph within the computer become quite difficult. For example, extensive search techniques would be required for finding out whether or not a graph is connected (Algorithm 1 in Section 11-4).

(d) *Two Linear Arrays:* A slight variation of edge listing is to represent the graph by two linear arrays, say $F = (f_1, f_2, \ldots, f_e)$ and $H = (h_1, h_2, \ldots,$

†A matrix that contains many zero elements is called a sparse matrix. A sparse adjacency matrix implies a small e/n ratio.

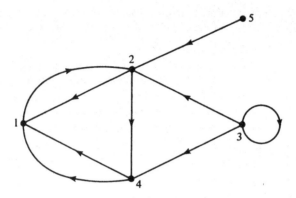

Fig. 11-1 A digraph.

h_e). Each entry in these arrays is a vertex label. The *i*th edge e_i is from vertex f_i to vertex h_i if G is a digraph. (If G is undirected, just consider e_i as between f_i and h_i.) For example, the digraph in Fig. 11-1 would be represented by the two arrays

$$F = (5, 2, 1, 3, 2, 4, 4, 3, 3),$$
$$H = (2, 1, 2, 2, 4, 1, 1, 4, 3).$$

This representation, which was used in the algorithm in [11-58], lends itself to convenient sorting in weighted graphs. The storage requirements are the same as in (c).

(e) *Successor Listing:* Another efficient method used frequently for graphs in which the ratio e/n is not large is by means of n linear arrays. After assigning the vertices, in any order, the numbers $1, 2, \ldots, n$, we represent each vertex k by a linear array, whose first element is k and whose remaining elements are the vertices that are *immediate successors* of k, that is, the vertices which have a directed path of length one from k. (In an undirected graph these are simply vertices adjacent to k.) The five-vertex digraph in Fig. 11-1 will appear as follows in this representation.

1 : 2
2 : 1, 4
3 : 2, 3, 4
4 : 1, 1
5 : 2

For an undirected graph the neighbors (rather than the successors) of every vertex are listed. Therefore, each edge appears twice—an obvious redundancy.

To compare its storage efficiency with that of the adjacency matrix, let d_{av} be the average degree (out-degrees in the case of a digraph) of the vertices in the graph. Assuming that one computer word is needed for the label of each vertex, the total storage requirement for an n vertex graph is $n(1 + d_{av})$ words. Thus the successor listing is more efficient than the adjacency matrix if

$$d_{av} < \left\lceil \frac{n}{w} \right\rceil - 1,$$

w being the word length.

The successor or neighbor listing form is extremely convenient for path-finding algorithms, and for a depth-first search on the graph [11-61].

You must have observed that the foregoing methods of graph representation are not entirely different. In fact, they are necessarily related in that they convey the same information. Simple programs can be written to convert one form into another (Problem 11-2). Additional variations in these representations can be made to suit the requirements at hand. For instance, a weighted graph can be represented by an n by n weight matrix (also called cost matrix or distance matrix), which is like the adjacency matrix except that instead of 1's the weights of the edges appear as the entries in the matrix. It should, however, be kept in mind that in many problems the efficiency of the algorithm may depend on the form in which the graph is presented. Thus the proper choice of the data structure is important.

11-3. OUTPUT

Every algorithm has an output—the cooked dish from the recipe. Unlike the input, which is one or more graphs, the output will vary from problem to problem. If the output consists of subgraphs, we may make the program print the appropriate adjacency matrices. On the other hand, if an output is, for instance, a yes or no to the question of planarity of a given graph, we may ask the program to simply print YES or NO. In addition, if the answer is YES, we may choose to get the planar representation of the graph; or if the answer is NO, we may ask for the thickness of the graph. For a shortest-path algorithm, we may simply wish to print the distance (shortest) between a pair of specified vertices x and y. Or one may desire to output a sequence of edges (or vertices) which describes a shortest path between x and y. And so forth. The outputs are as varied as the algorithms.

Let us now proceed with some specific algorithms.

11-4. SOME BASIC ALGORITHMS

Algorithm 1: Connectedness and Components

The first questions one is most likely to ask when encountering a new graph G will be: Is G connected? If G is not connected, what are the components of G? Therefore, our first algorithm will be one that determines the connectedness and components of a given graph.

In addition to being an important question in its own right, the question of connectedness and components arises in many other algorithms. For example, before testing a graph G for separability, planarity, or isomorphism with another graph, it may be better for the sake of efficiency to determine the components of G and then subject each component to the desired scrutiny. The connectedness algorithm is very basic and may serve as a subroutine in more involved graph-theoretic algorithms. (The reader may be reminded here that although in drawing a graph one might see whether a graph is connected or not, the connectedness is by no means obvious to a computer or human being if the graph is presented in other forms, such as those discussed in Section 11-2.)

Given the adjacency matrix X of a graph, it is possible to determine whether or not the graph is connected by trying various permutations of rows and the corresponding columns of X, and then checking if it is in a block-diagonal form. (See observation 5 in Section 7-9.) This, however, is an inefficient method, because it may involve $n!$ permutations. A more efficient method would be to use Corollary B of Theorem 7-8, and check for zeros in the matrix

$$Y = X + X^2 + \cdots + X^{n-1}.$$

This too is not very efficient, as it involves a large number of matrix multiplications. The following is an efficient algorithm:

Description of the Algorithm: The basic step in this algorithm is the fusion of adjacent vertices (recall Section 2-7). We start with some vertex in the graph and fuse all vertices that are adjacent to it. Then we take the fused vertex and again fuse with it all those vertices that are adjacent to it now. This process of fusion is repeated until no more vertices can be fused. This indicates that a connected component has been "fused" to a single vertex. If this exhausts every vertex in the graph, the graph is connected. Otherwise, we start with a new vertex (in a different component) and continue the fusing operation.

In the adjacency matrix the fusion of the jth vertex to the ith vertex is accomplished by OR-ing, that is, logically adding the jth row to the ith row as well as the jth column to the ith column. (Remember that in logical adding $1 + 0 = 0 + 1 = 1 + 1 = 1$ and $0 + 0 = 0$.) Then the jth row and the jth column are discarded from the matrix. (If it is difficult or time consuming to discard the specified rows and columns, one may leave these rows and columns in the matrix, taking care that they are not considered again in any fusion.)

Note that a self-loop resulting from a fusion appears as a 1 in the main diagonal, but parallel edges are automatically replaced by a single edge because of the logical addition (or OR-ing) operation. These, of course, have no effect on the connectedness of a graph.

The maximum number of fusions that may have to be performed in this algorithm is $n - 1$, n being the number of vertices. And since in each fusion one performs at most n logical additions, the upper bound on the execution time is proportional to $n(n - 1)$.

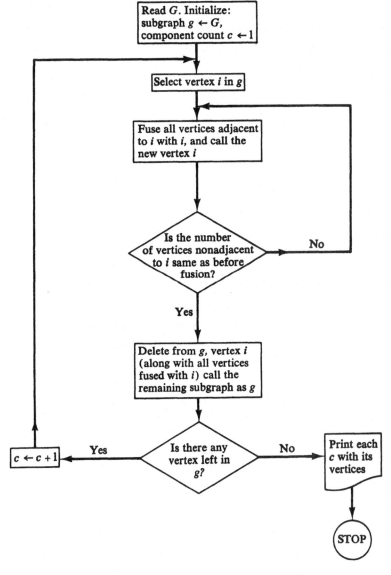

Fig. 11-2 Algorithm 1: Components of G.

A proper choice of the initial vertex (to which adjacent vertices are fused) in each component would improve the efficiency, provided one did not pay too much of a price for selecting the vertex itself (see Problem 11-6).

A flow chart of the "Connectedness and Components Algorithm" is shown in Fig. 11-2.

A complete computer program, ready to be executed, written in APL\360, together with the legend identifying the variables used in the program, is given at the end of the chapter. Note that the program selects a vertex with maximum degree as the initial vertex in each component.

To illustrate the program, an input and the resulting output are shown as follows: The input is a 20 by 20 adjacency matrix representing a 20-vertex graph, and the output is a list of components (COMP) followed by the names of vertices (VERT) included in each component. Vertex i corresponds to the ith row and column in X.

A slightly modified program (without selecting vertices of maximum

INPUT

N = 20

$$X = \begin{bmatrix} 0 & 1 & 1 & 0 & 1 & 0 & 0 & 0 & 0 & 0 & 1 & 0 & 0 & 0 & 0 & 0 & 0 & 0 & 0 & 0 \\ 1 & 0 & 1 & 0 & 1 & 0 & 0 & 0 & 0 & 0 & 1 & 0 & 0 & 0 & 0 & 0 & 0 & 0 & 0 & 0 \\ 1 & 1 & 0 & 0 & 1 & 0 & 0 & 0 & 0 & 0 & 0 & 0 & 0 & 0 & 0 & 0 & 0 & 0 & 0 & 0 \\ 0 & 0 & 0 & 0 & 0 & 0 & 0 & 0 & 0 & 0 & 0 & 0 & 0 & 0 & 0 & 0 & 0 & 0 & 0 & 0 \\ 1 & 1 & 1 & 0 & 0 & 0 & 0 & 0 & 0 & 0 & 1 & 0 & 0 & 0 & 0 & 0 & 0 & 0 & 0 & 0 \\ 0 & 0 & 0 & 0 & 0 & 0 & 1 & 1 & 0 & 1 & 0 & 0 & 0 & 0 & 0 & 1 & 0 & 0 & 1 & 1 \\ 0 & 0 & 0 & 0 & 0 & 1 & 0 & 1 & 0 & 1 & 0 & 0 & 0 & 0 & 0 & 0 & 0 & 0 & 1 & 1 \\ 0 & 0 & 0 & 0 & 0 & 1 & 1 & 0 & 0 & 1 & 0 & 0 & 0 & 0 & 0 & 0 & 0 & 0 & 1 & 0 \\ 0 & 0 & 0 & 0 & 0 & 0 & 0 & 0 & 0 & 0 & 1 & 1 & 0 & 0 & 0 & 1 & 0 & 0 & 0 & 0 \\ 0 & 0 & 0 & 0 & 0 & 1 & 1 & 1 & 0 & 0 & 0 & 0 & 0 & 0 & 0 & 1 & 0 & 0 & 1 & 1 \\ 1 & 1 & 0 & 0 & 1 & 0 & 0 & 0 & 0 & 0 & 0 & 0 & 0 & 0 & 0 & 0 & 0 & 0 & 0 & 0 \\ 0 & 0 & 0 & 0 & 0 & 0 & 0 & 0 & 1 & 0 & 0 & 0 & 0 & 0 & 0 & 0 & 1 & 0 & 0 & 0 \\ 0 & 0 & 0 & 0 & 0 & 0 & 0 & 0 & 1 & 0 & 0 & 0 & 0 & 0 & 0 & 0 & 0 & 0 & 0 & 0 \\ 0 & 0 & 0 & 0 & 0 & 0 & 0 & 0 & 0 & 0 & 0 & 0 & 0 & 0 & 1 & 0 & 0 & 0 & 0 & 0 \\ 0 & 0 & 0 & 0 & 0 & 0 & 0 & 0 & 0 & 0 & 0 & 0 & 0 & 1 & 0 & 0 & 0 & 0 & 0 & 0 \\ 0 & 0 & 0 & 0 & 0 & 1 & 0 & 0 & 1 & 0 & 0 & 0 & 0 & 0 & 0 & 0 & 0 & 0 & 1 & 1 \\ 0 & 0 & 0 & 0 & 0 & 0 & 0 & 0 & 1 & 0 & 0 & 1 & 0 & 0 & 0 & 0 & 0 & 0 & 0 & 0 \\ 0 & 0 & 0 & 0 & 0 & 0 & 0 & 0 & 0 & 0 & 0 & 0 & 0 & 0 & 0 & 0 & 0 & 0 & 0 & 0 \\ 0 & 0 & 0 & 0 & 0 & 1 & 1 & 1 & 0 & 1 & 0 & 0 & 0 & 0 & 0 & 1 & 0 & 0 & 0 & 1 \\ 0 & 0 & 0 & 0 & 0 & 1 & 1 & 0 & 0 & 1 & 0 & 0 & 0 & 0 & 0 & 1 & 0 & 0 & 1 & 0 \end{bmatrix}$$

OUTPUT

```
COMP 1; VERT:  6  7  8  10  16  19  20
COMP 2; VERT:  1  2  3  5  11
COMP 3; VERT:  9  12  13  17
COMP 4; VERT:  14  15
COMP 5; VERT:  4
COMP 6; VERT:  18
```

degree as the initial vertices and without discarding the rows and columns after they are OR-ed with the initial row) took 35 FORTRAN statements to write. The execution time of this FORTRAN program for a typical 50-vertex graph (with varying number of edges and components) on the IBM 7044 was $\frac{9}{60}$ second.

Algorithm 2: A Spanning Tree

Perhaps the best known and most frequently used algorithms in graph theory are the spanning-tree algorithms. In its simplest form a spanning-tree algorithm yields one spanning tree in a given connected graph. If the graph is disconnected, the algorithm should produce a spanning forest containing $n - p$ edges, where $p > 1$ is the number of components in the disconnected graph. Clearly then, as a by-product of such an algorithm, we can find out whether or not the graph is connected, and if the graph is disconnected, its components can be identified. In fact, sometimes a spanning-tree algorithm is used for testing connectedness of a graph. If, on the other hand, the given graph has a weight or distance associated with each edge (weighted graph), we may wish to find a spanning tree with smallest possible weight. The significance of an algorithm for such a tree (called minimal or shortest spanning tree) was discussed in Section 3-10. A spanning tree is also needed for obtaining a fundamental set of circuits. As we saw in Section 3-9, some algorithms for generation of *all* spanning trees (a much more difficult task) in a given connected graph G also start by first obtaining one spanning tree.

Description of the Algorithm: Let the given undirected self-loop-free (if the graph has any self-loops, they may be discarded) graph G contain n vertices and e edges. Let the vertices be labeled $1, 2, \ldots, n$, and the graph be described by two linear arrays F and H [i.e., in the form (d) of Section 11-2] such that $f_i \in F$ and $h_i \in H$ are the end vertices of the ith edge in G.

At each stage in the algorithm a new edge is tested to see if either or both of its end vertices appear in any tree formed so far.† At the kth stage, $1 \leq k \leq e$, in examining the edge (f_k, h_k) five different conditions may arise:

1. If neither vertex f_k nor h_k is included in any of the trees constructed so far in G, the kth edge is named as a new tree and its end vertices f_k, h_k are given the component number c, after incrementing the value of c by 1.

2. If vertex f_k is in some tree T_i ($i = 1, 2, \ldots, c$) and h_k in tree T_j ($j = 1, 2, \ldots, c$, and $i \neq j$), the kth edge is used to join these two trees; therefore, every vertex in T_j is now given the component number of T_i. The value of c is decremented by 1.

3. If both vertices are in the same tree, the edge (f_k, h_k) forms a fundamental circuit and is not considered any further.

†Initially there is no tree formed. The very first edge (f_1, h_1) considered will always occur in a spanning tree (or forest). Thus the spanning tree (or forest) generated by this algorithm is very much dependent on the ordering of the edges.

4. If vertex f_k is in a tree T_i and h_k is in no tree, the edge (f_k, h_k) is added to T_i by assigning the component number of T_i to h_k also.

5. If vertex f_k is in no tree and h_k is in a tree T_j, the edge (f_k, h_k) is added to T_j by assigning the component number of T_j to f_k also.

These five cases are marked by circled numbers in the flow chart of the algorithm shown in Fig. 11-3.

The efficiency of a computer program based on this algorithm depends mainly on the speed with which we can test whether or not the end vertices of the edge under consideration have occurred in any tree formed so far. For this testing, we maintain a linear array (called VERTEX in the program listing) of size n. When an edge (i, j) is included in the cth tree, the ith and jth entries in this array are set to c. Subsequently, when another edge (f_k, h_k) is examined, it is only necessary to check if the f_kth and the h_kth entries in array VERTEX are nonzero. A zero in the qth position in the array indicates that the vertex q has not so far been included in any tree. At the end of the execution, this array VERTEX identifies the components of the graph.

Unlike a component, a tree cannot be described by a set of vertices alone. Therefore, we must have an array of edges as the output. Let this linear array be called EDGE. If the kth edge (in the original order in which the edges were placed) is in the cth tree, EDGE(k) = c; otherwise, it is zero. All zero entries in array EDGE correspond to the chords (i.e., the edges not included in the spanning tree or forest). This array, together with arrays F and H, uniquely identifies the spanning tree (or forest) generated by this algorithm.

In this algorithm the main loop is executed e times (e being the number of edges). The time required to test whether or not the end vertices have appeared in any tree is constant—independent of both e and n. Thus the time bound for the execution of the algorithm is proportional to e.† In case the ratio e/n is high, execution time can be reduced by introducing a new variable to keep count of the edges included in the tree. When this variable reaches the value of $n - 1$, the program would terminate (only if the graph is connected; otherwise, we must examine every edge).

A ready-to-be-executed program in FORTRAN language, based on this algorithm, is given at the end of this chapter. For an ALGOL listing of the same program, see [11-58]. A randomly generated graph of 50 vertices took $\frac{2}{60}$ second on the IBM 7044, using FORTRAN IV.

Minimal-Spanning-Tree Algorithms: As discussed in Section 3-10, we can use the algorithm suggested by Kruskal to find a shortest spanning tree in a graph G in which every edge has a distance (or weight) associated with it. This can be accomplished with the algorithm just described. The only addi-

---

†The time required for merging two partial trees (T_i, T_j) as implemented in the FORTRAN program is not independent of n. There are, however, very efficient set-merging algorithms available which almost accomplish this.

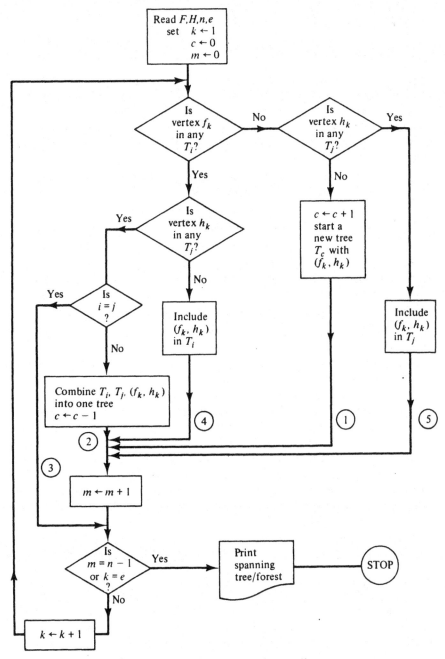

Fig. 11-3 Algorithm 2: Spanning tree/forest.

tional work required is to first sort the edges in a nondecreasing order of their weights, before representing them by F and H arrays. That is, the following set of inequalities must be satisfied:

$$\text{wt of edge } (f_i, h_i) \leq \text{wt of } (f_{i+1}, h_{i+1}), \quad \text{for all } 1 \leq i \leq e-1.$$

Because of the sorting involved, Kruskal's algorithm is not as efficient as the one due to Prim [3-10] (which was also discovered independently by Dijkstra [11-16]). The latter algorithm, as outlined in Sec. 3-10, requires no sorting of edges, but builds up a minimal spanning tree by successively connecting the partially formed tree to its nearest neighbor. For a FORTRAN listing of an efficient implementation of Prim's minimal-spanning-tree algorithm, see [11-70]. The computation time of this algorithm is proportional to n^2, n being the number of vertices, [11-22].

Minimal spanning tree has been found quite useful in providing a lower bound on the length of the traveling salesman's route [11-28].

Spanning Trees with Desired Properties: Instead of a shortest spanning tree, one may wish to find a longest spanning tree. Or one may be interested in trees with other desired properties and constraints, such as a spanning tree with a specified maximum degree or diameter. Algorithm 2 with appropriate additional sorting or testing can be used for such purposes also.

Generating All Spanning Trees: As we shall see in Chapter 13, analysis of electrical networks basically reduces to finding all spanning trees in graphs. Because of this important application, more than a dozen different algorithms for generation of all spanning trees have been proposed. In Section 3-9 we discussed one of these methods, the method of cyclic interchange. Since the number of spanning trees even in a small graph is very large, the efficiency of these algorithms is of paramount importance. A survey of these methods was done by Chase in his Ph.D. thesis [11-8]. He concludes that the most efficient algorithm is of the type suggested by Minty [11-43], which essentially consists of successively reducing a graph by operations of deletion of an edge and fusion of its end vertices. From the spanning trees of reduced graphs (which are much smaller) the spanning trees of the original graph are obtained. To ensure that the algorithm terminates, graphs below a certain size are not reduced any further; instead their spanning trees are obtained directly. A compact ALGOL program based on this method is given in [11-42].

Algorithm 3: A Set of Fundamental Circuits

Sometimes we are required to find a set of fundamental circuits in a given graph. The spanning-tree algorithm just described can be used for generating fundamental circuits if the following additional work is performed:

While examining the kth edge (f_k, h_k) in Algorithm 2, if condition 3 arises (i.e., both vertices f_k and h_k occur in the same tree T_i), then instead of simply rejecting this edge we must find those edges in T_i that form the path between f_k and h_k. This path and the edge (f_k, h_k) constitute a fundamental circuit. Finding this path is the main problem here. In [11-52] a tree-felling procedure has been suggested, where the edge (f_k, h_k) is added to T_t, and all pendant vertices of the resulting graph are deleted iteratively. This method, however, turns out to be inefficient. More efficient methods have been proposed by Welch [11-68], Gottlieb and Corniel [11-25], and Paton [11-47]. Among these three, Paton's algorithm appears to be the most efficient, and is as follows:

Description of the Algorithm: Here also each edge is tested to see if it forms a circuit with the tree constructed so far; but instead of taking the edges themselves in an arbitrary order (as was done in Algorithm 2), we select a vertex z and examine this vertex by looking at every edge incident on z. (Vertex z, as we shall shortly see, is the vertex added most recently to the partially formed tree.) Let the vertices of the given connected graph $G = (V, E)$ be labeled $1, 2, \ldots, n$, and the graph be given by its adjacency matrix X. Let T be the current set of vertices in the partially formed tree, and let W be the set of vertices that are yet to be examined (i.e., those vertices, in T as well as not in T, which have one or more unexamined edges incident on them). Initially, $T = \emptyset$ and $W = V$, the entire set of vertices.

We start the algorithm by setting $T = 1$, the first vertex, and $W = V$. Vertex 1 will be regarded as the root of the tree to be formed. After initialization, the following procedure is used:

1. If $T \cap W = \emptyset$, then the algorithm is terminated.
2. If $T \cap W \neq \emptyset$, choose a vertex z in $T \cap W$.
3. Examine z by considering every edge incident on z. If there is no such edge left, remove z from W, and go to step 1.
4. If there is such an edge (z, p), test if vertex p is in T.
5. If $p \in T$, find the fundamental circuit consisting of edge (z, p) together with the unique path from z to p in the tree (formed so far). Delete edge (z, p) from the graph, and go to step 3.
6. If $p \notin T$, add edge (z, p) to the tree and vertex p to set T. Delete edge (z, p) from the graph, and go to step 3.

As mentioned earlier, the only tricky part in this algorithm is in step 5. How do we find the unique path from z to p in the tree? The following procedure provides an answer:

We maintain a pushdown list (a stack) $TW = T \cap W$, which stores those vertices in the tree that have not yet been examined. The most recently added vertex is at the top of the stack. Each time a vertex is taken for examination it is taken from the top of this stack, and is removed from the stack. Two linear arrays of length n are employed: LEVEL(i) being the distance of vertex i from the root of the spanning tree (i.e., vertex 1), and PRED(i) being a vertex v such that (i, v) is an edge in the tree with v nearer the root. In other words, PRED(i) is the predecessor of i in the path from the root to i. LEVEL(i) $= -1$ if and only if vertex i is not in set T, the current set of tree vertices. Initially, LEVEL(1) is set to 0 and LEVEL(i) to -1 for $i = 2, 3, \ldots, n$.

In step 5 vertex z is under examination and an edge (z, p) has been found such that vertex $p \in T$. To find the fundamental circuit formed by (z, p) with the tree, we trace the unique path from z to p in the tree by successively finding the predecessors PRED(z), PRED(PRED(z)), ..., till we encounter PRED(p), the predecessor of p. In other words, as shown in Fig. 11-4, the

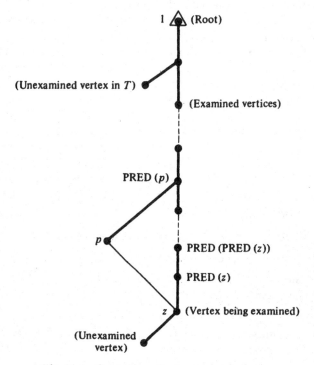

Fig. 11-4 Generation of a fundamental circuit.

fundamental circuit generated is

$$z, \text{PRED}(z), \text{PRED}(\text{PRED}(z)), \ldots, \text{PRED}(p), p, z.$$

The most important thing to note is that the predecessor PRED(k) of every vertex k in T is a vertex which is either already examined or is being examined. That is, if $k \in T \cap W$, then

$$\text{PRED}(k) \notin W \quad \text{but} \quad \text{PRED}(k) \in T.$$

A flow chart of the algorithm is given in Fig. 11-5, and a ready-to-be-executed FORTRAN program is provided at the end of the chapter.

The execution time is bounded by n^v, $2 \leq v \leq 3$, and the value of v depends on the structure of the graph and also on the labeling of vertices [11-47].

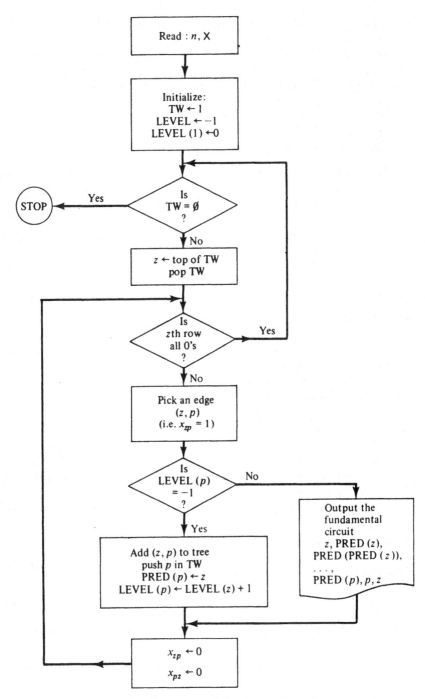

Fig. 11-5 Algorithm 3: Fundamental circuits.

Although for simplicity we assumed that the graph is connected, the algorithm will work for disconnected graphs also. First, it will produce all fundamental circuits in the component containing the starting vertex 1. After having exhausted the first component, we select a vertex y such that LEVEL$(y) = -1$, and start with y as the root of a spanning tree in the second component. This procedure continues till there is no vertex left with -1 as its LEVEL.

Typically for a graph of 50 vertices and 132 edges the IBM 7044 (using FORTRAN IV) took $\frac{14}{60}$ second to generate a set of fundamental circuits.

All Circuits in an Undirected Graph: All circuits of an undirected graph might be found by first forming the set of all linear combinations (i.e., ring sums) of the fundamental circuits and then discarding from this set all those circuit sets that contain other circuits. (This is because a linear combination of circuits can be either a circuit or a union of edge-disjoint circuits. And a union of edge-disjoint circuits contains other circuits.) Such an algorithm, however, would be very inefficient. From μ fundamental circuits $2^\mu - \mu - 1$ linear combinations must be made, and then each of these must be compared pairwise with every other to test for containment. Therefore, a different approach has to be taken.

To reduce the storage requirement and the number of comparisons, one may be tempted to suggest discarding every edge-disjoint union of circuits as soon as it is generated. This approach, however, is faulty. For we might find that a genuine circuit was a combination of some discarded circuit set and another circuit generated later.

Welch [11-68] proposed a scheme of ordering the fundamental circuits so that one could discard a union of edge-disjoint circuits as it is produced before generating all $2^\mu - \mu - 1$ combinations and then make pairwise comparisons. In a more recent paper [11-24], Gibbs has pointed out an error in Welch's algorithm and has proposed a modification. Gibbs's algorithm for generation of all circuits from a set of fundamental circuits is essentially an exhaustive method and requires storage proportional to 2^μ. Finding an efficient algorithm for identifying all circuits in a graph is an open problem. For a survey of circuit-generation algorithms see Prabhaker [11-50].

Algorithm 4: Cut-Vertices and Separability

Having found out that a graph G is connected, the next question one is most likely to ask is: Is the graph G separable? That is, is there one or more cut-vertices in G? If the answer is yes, one would like to find the cut-vertices and the blocks (maximal nonseparable subgraphs) of G.

As pointed out in Section 4-5, cut-vertices are important in the study of vulnerability of a communication network. Moreover, this algorithm may serve as a subroutine for other algorithms, such as for planarity and isomorphism.

Preliminary Simplification: In this algorithm, as in most others, it pays to perform some preliminary simplification. If the given graph has the possibility of being disconnected, we could apply Algorithm 1 and consider each component as a connected graph. It will be a waste of computer memory and execution time to drag along all the components of a disconnected graph. Similarly, if the graph is not simple, we can immediately discard all the self-loops and parallel edges, since their presence or absence has no effect on separability. Third, if the graph has any pendant vertices, we can prune the graph by repeatedly deleting pendant vertices. (In the pruning process we must keep in mind that every vertex adjacent to a pendant vertex is a cut-vertex, except in the trivial case where the graph consists of just one edge.) Usually these simplifications will have substantially reduced the size of the original graph.†

A straightforward method (which was used in [5-8], for example) for testing separability of a graph would be to remove each vertex in turn (by deleting the corresponding row and column from its adjacency matrix X) and then to test the resulting graph for connectedness, using Algorithm 1. But this is an inefficient method, and we can do better using a different approach, suggested by Read [11-53].

Let us recall a result from Chapter 4: Two edges are in the same block if and only if there exists at least one circuit that contains both these edges (Problem 4-10). At first sight it may appear that to use this characterization of a block we would have to generate all circuits—an obviously time- and storage-consuming affair. The following two results, however, reveal that it would suffice to generate only a set of fundamental circuits:

LEMMA 1

A nonempty intersection of two fundamental circuits in a graph is always a path.

Proof: With respect to some specified spanning tree T, let e_1 and e_2 be two chords forming fundamental circuits f_1 and f_2, respectively. Then if $f_1 \cap f_2$ contains two edges, x and y, not connected in $f_1 \cap f_2$, there is a path P_1 in f_1 between x and y (that is, a path between one of the end vertices of edge x and one of the end-vertices of edge y); and this path does not contain chord e_1. Similarly, there is a path P_2 in f_2 between x and y that does contain chord e_2. Then the subgraph

$$P_1 \cup P_2 \cup \{x, y\}$$

contains a circuit without containing any chord, which is impossible. ∎

†A preliminary simplification is to be performed only if it produces a net saving in running time.

Lemma 2

In a graph G if edges a and b belong to a fundamental circuit f_i, and if edges b and c belong to another fundamental circuit f_j such that $a \notin f_j$ and $c \notin f_i$, then there exists some circuit Γ in G such that a and c both are in Γ.

Proof: The proof follows from Lemma 1, because $f_i \cap f_j$ is a path containing b but not a or c. Therefore, $f_i \oplus f_j$ is a circuit (not an edge-disjoint union of circuits) containing a and c. ■

Description of the Algorithm: If we generate fundamental circuits one by one, and as each fundamental circuit is generated we label† (or relabel) all its edges identically, using the following procedure, we will have identified the blocks in the graph:

Each edge in the first fundamental circuit is labeled with 2's. When the second fundamental circuit is found, it will have either all its edges unlabeled, or some of its edges would be labeled 2. In the former case, label every edge of the second fundamental circuit with 3's, and in the latter case with 2's. When this process reaches the mth ($1 \leq m \leq e - n + 1$) fundamental circuit, we may find any one of three conditions:

1. If every edge in the mth fundamental circuit is unmarked, label all of them with a new integer $q + 1$.

2. If some edges in the mth fundamental circuit are marked u and all others are unmarked, label each of the unlabeled ones as u also.

3. Suppose that some edges in the mth fundamental circuit are marked u, others v, and others $w, \ldots$, and some are unmarked. Let $u < v < w < \ldots$. Then relabel all edges marked $v, w, \ldots$, in G as u, and label all unmarked edges in the mth fundamental circuit as u also.

When this process terminates, after having generated fundamental circuits and labeled the edges in each of them, the following has been accomplished:

Every edge that belongs to a circuit has been labeled. Moreover, any two edges have the same label if and only if they are together in some circuit (not necessarily a fundamental circuit). In other words, each set of edges carrying identical labels constitutes a block. If there is more than one block in the graph, the graph is separable. Any vertex incident on edges with different labels is a cut-vertex. An edge that has not acquired any label is a *bridge*. (A bridge is an edge whose removal disconnects the graph.)

In this algorithm an edge belonging to a circuit gets relabeled many

---

†This labeling can be conveniently performed using the adjacency matrix X and by writing over it. The edge between the ith and jth vertices is labeled q ($q = 2, 3, \ldots, e - n + 1$), simply by replacing x_{ij} and x_{ji} with q (x_{ij} only, if the upper triangle is used). Entries that are still 1's correspond to unlabeled edges. Others will have labels $2, 3, \ldots,$ and so on.

times—an obvious source of inefficiency. An improvement suggested by Paton [11-48] reduces the relabeling of edges by the following devices: Instead of relabeling the edges in a fundamental circuit as soon as it is generated, we wait till a vertex z in Algorithm 3 has been completely examined, and then assign identical labels to the fundamental circuits (passing through z) thus generated. Therefore, labeling has to be performed only n times and not μ times. Moreover, we need not label every edge in the graph (Problem 11-11). It is left for the reader to construct a flow chart for the block-identification algorithm in which Algorithm 3 is completely embedded. Remaining details can be found in [11-48].

Using an entirely different approach, Hopcroft and Tarjan [11-31] and Tarjan [11-61] have given an algorithm which is faster than the algorithm described here for certain types of graphs. Their algorithm uses depth-first search on the graph (to be discussed later in this chapter), and the graph is to be input in the successor-listing form. Its execution time is proportional to e, whereas the time bound for the Read-Paton algorithm described here is proportional to n^γ, where $1 \leq \gamma \leq 2$, depending on the structure of the graph. Analysis and extensive tests show that for a typical graph of n vertices and e edges, Hopcroft and Tarjan's algorithm outperforms (on IBM 7044) Paton's algorithm as long as $e \leq 5n$. For graphs of much higher densities (Problem 11-1) Paton's algorithm performed better. Thus for planar graphs (since $e \leq 3n - 6$), Hopcroft and Tarjan's algorithm will in general be faster. Typically, for an 80-vertex 400-edge graph, the IBM 7044 took about 7 seconds for block identification with either of the two algorithms.

Algorithm 5: Directed Circuits

One of the most important things about a digraph is its directed circuits (also called cycles). The significance of directed circuits in many applications was discussed in Chapter 9. Unlike the case of undirected graphs, no technique is known by which we can obtain a basic set of directed circuits such that every directed circuit in the digraph is obtained as a linear combination of this basic set. Therefore, Algorithm 3 is of little help in obtaining all directed circuits of a digraph. We must generate every directed circuit individually. For this we must examine each edge (unless the edge is known a priori to belong to no directed circuit) many times.

Preliminary Simplification: Although it is not necessary, in most cases the prior application of the following two steps will simplify a given digraph. First, if the digraph is likely to be disconnected, use Algorithm 1 [with slight modification for a digraph (Problem 11-7)] to identify the connected components, and then consider one component at a time. Second, successively delete all vertices (and the edges incident on them) that have zero in-degree or zero out-degree. Clearly, such a vertex cannot lie in any directed circuit. These vertices are easy to identify because they correspond to entire rows [for $d^+(v) = 0$] or columns [for $d^-(v) = 0$] of zeros in the adjacency matrix X. For example, if the digraph given was the one in Fig. 9-16, edges a, b, and h would be eliminated. Then in the next go-round edges e and c would be deleted, leaving us a digraph of only three edges, d, f, and g. On the other hand, this method of simplification will not reduce the digraph shown in Fig. 9-21.

Description of the Algorithm: This algorithm, first proposed by Roberts and Flores [11-56] and subsequently systematized by Tiernan [11-63], uses an exhaustive search to find all directed circuits in a given digraph G. As usual, the vertices of G are assigned integers $1, 2, \ldots, n$ as their names. The algorithm depends on starting from a vertex p_1 and building a directed path $P = (p_1, p_2, \ldots, p_k)$ until no further vertices (satisfying certain conditions) are "available" at vertex p_k. At p_k, when it is not possible to extend the directed path any further, the algorithm checks to see if there is a directed edge from p_k to p_1. If there is such an edge, a directed circuit $(p_1, p_2, \ldots, p_k, p_1)$ has been found and is duly recorded. If there is no such edge in the digraphs, we move back one vertex to p_{k-1} and try extending the path again from p_{k-1} along a different edge (if there is one). Whether a directed circuit is found or not, the algorithm makes vertex p_k forbidden for the next extension from p_{k-1} (thus avoiding going over the same path).

This process of looking for directed circuits and then moving back a vertex is continued till we finally backtrack to the vertex p_1 itself. Thus all directed paths starting from p_1 have been examined and directed circuits recorded. Starting with the next vertex, the entire process is repeated. The iteration starts with vertex $p_1 = 1$ and ends with $p_1 = n$.

In this exhaustive search for directed paths we must take the following precautions:

1. In the process of extending each directed path, going round and round a directed circuit must be avoided. This is achieved by insisting that any vertex that has already been included in the directed path is "not available" for extending the path.

2. Generating a directed circuit of q vertices q times—once at each vertex in the circuit—must be avoided. This is accomplished by insisting that no vertex $i \leq p_1$ is available for path extension, if the path begins with vertex p_1. This rule assures that the search for a particular directed circuit commences only when its lowest-numbered vertex is at the path initiation.

3. The same path must not be considered more than once during the path extension. This is accomplished by keeping an updated list of forbidden vertices in a binary n by n matrix $H = [h_{ij}]$. The 1 entries in the ith row correspond to the vertices that are forbidden from vertex i(i.e., if vertex j is forbidden from vertex i, set $h_{ij} \leftarrow 1$). A 0 entry indicates that the vertex is not forbidden (i.e., if $h_{ij} = 0$, then vertex j is not forbidden

from vertex i). Matrix H is reset to zero each time a new vertex is chosen as the starting vertex.

The digraph is inputted as its adjacency matrix [see Section 11-2(a)]. The vertices are labeled as usual with integers $1, 2, \ldots, n$. The directed path under consideration is represented by a linear array

$$P = (p_1, p_2, \ldots, p_{k-1}, p_k, 0, 0, \ldots, 0, 0)$$

of order n. The first vertex of every path is p_1 and the last one is p_k.

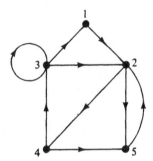

Fig. 11-6 Digraph.

The algorithm can be best explained with an example. When applied to the digraph of Fig. 11-6, the following steps will be performed:

Path P

p_1	p_2	p_3	p_4	p_5	Action on P
1	0	0	0	0	
1	2	0	0	0	
1	2	4	0	0	
1	2	4	3	0	Report circuit: set $h_{4,3} \leftarrow 1$
1	2	4	0	0	
1	2	4	5	0	No circuit: set $h_{4,5} \leftarrow 1$
1	2	4	0	0	No circuit: $h_{4,1} \leftarrow h_{4,2} \leftarrow h_{4,3} \leftarrow h_{4,4} \leftarrow h_{4,5} \leftarrow 0$; $h_{2,4} \leftarrow 1$
1	2	0	0	0	
1	2	5	0	0	No circuit: $h_{2,5} \leftarrow 1$
1	2	0	0	0	No circuit: $h_{2,1} \leftarrow h_{2,2} \leftarrow h_{2,3} \leftarrow h_{2,4} \leftarrow h_{2,5} \leftarrow 0$; $h_{1,2} \leftarrow 1$
1	0	0	0	0	No circuit: $p_1 \leftarrow p_1 + 1$; H $\leftarrow 0$
2	0	0	0	0	
2	4	0	0	0	
2	4	3	0	0	Report circuit: $h_{4,3} \leftarrow 1$
2	4	0	0	0	
2	4	5	0	0	Report circuit: $h_{4,5} \leftarrow 1$
2	4	0	0	0	No circuit: $h_{4,1} \leftarrow h_{4,2} \leftarrow h_{4,3} \leftarrow h_{4,4} \leftarrow h_{4,5} \leftarrow 0$; $h_{2,4} \leftarrow 1$
2	0	0	0	0	
2	5	0	0	0	Report circuit: $h_{2,5} \leftarrow 1$
2	0	0	0	0	No circuit: $p_1 \leftarrow p_1 + 1$; H $\leftarrow 0$

Path P (Cont.)

$p_1\ p_2\ p_3\ p_4\ p_5$	Action on P
3 0 0 0 0	Report circuit: $p_1 \leftarrow p_1 + 1$; $H \leftarrow 0$
4 0 0 0 0	
4 5 0 0 0	No circuit: $h_{4,5} \leftarrow 1$
4 0 0 0 0	No circuit: $p_1 \leftarrow p_1 + 1$; $H \leftarrow 0$
5 0 0 0 0	No circuit: stop

The flow chart of this algorithm, which is a modified version of the algorithm given in [11-63], is shown in Fig. 11-7.

You must have observed that this algorithm is nothing more than a systematic and exhaustive search for directed circuits. As shown in the example, the same directed path is traversed many times. Even a directed circuit is usually examined and rejected several times before its turn to be accepted arrives. Consequently, the algorithm is very slow, and there is room for considerable improvement. To quote Tiernan [11-63], this algorithm "would be costly to utilize on a graph containing more than 50 arcs or 7 vertices"—a small graph indeed.

The algorithm could be easily modified to generate all directed Hamiltonian circuits. This, in fact, was the original purpose of the algorithm as reported by Roberts and Flores [11-56].

A random directed graph of 20 vertices, 55 edges, and 434 directed circuits took about 17 seconds on the IBM 7044. This indicates that this method, involving a systematic but exhaustive search, is quite inefficient in terms of execution time.

A similar algorithm but somewhat more involved, considerably faster, but requiring more storage was proposed by Weinblatt [11-67]. See also [11-50].

11-5. SHORTEST-PATH ALGORITHMS

A large number of optimization problems are mathematically equivalent to finding shortest paths in a graph. Consequently, shortest-path algorithms have been worked over more thoroughly than any other algorithm in graph theory. More than 100 papers have been published and dozens of algorithms have been proposed. Some of these algorithms are better than others, some are more suited for a particular structure than others, and some are only minor variations of earlier algorithms. For a good comparative study of various shortest-path algorithms through the year 1968, a survey paper by Dreyfus [11-17] is highly recommended.

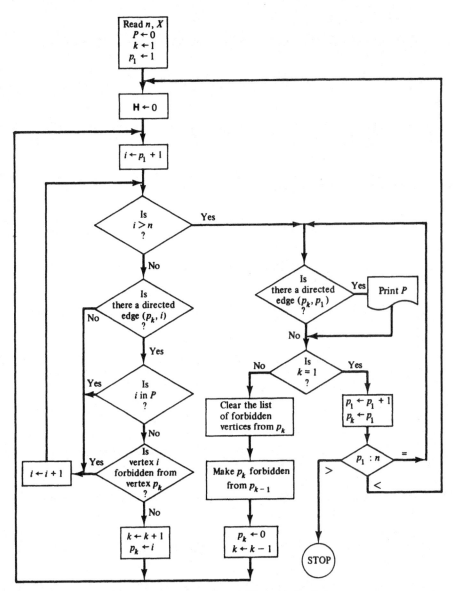

Fig. 11-7 Algorithm 5: Directed circuits.

There are different types of shortest-path problems. Most frequently encountered among these are the following five, of which we shall solve the first three:

1. Shortest path between two specified vertices.
2. Shortest paths between all pairs of vertices.

3. Shortest paths from a specified vertex to all others.
4. Shortest path between specified vertices that passes through specified vertices.
5. The second, third, and so on, shortest paths.

In a worst-case situation, type 1 becomes identical to 3, because (as we shall see shortly) in the process of finding the shortest path from a specified vertex to another specified vertex, we may have to determine the shortest paths to all other vertices. Let us deal with type 1 first.

Algorithm 6: Shortest Path from a Specified Vertex to Another Specified Vertex

The problem of finding the shortest path from a specified vertex s to another specified vertex t, can be stated as follows:

A simple weighted digraph† G of n vertices is described by an n by n matrix $\mathbf{D} = [d_{ij}]$, where

d_{ij} = length (or distance or weight) of the directed edge from vertex i to vertex j, $d_{ij} \geqslant 0$,

$d_{ii} = 0$,

$d_{ij} = \infty$, if there is no edge from i to j (in carrying out a program ∞ is replaced by a large number, say 9999999).

In general, $d_{ij} \neq d_{ji}$, and the triangle inequality need not be satisfied. That is, $d_{ij} + d_{jk}$ may be less than d_{ik}. [In fact, if the triangle inequality is satisfied, for every i, j, and k, the problem would be trivial because the direct edge (x, y) would be the shortest path from vertex x to vertex y.] The distance of a directed path P is defined to be the sum of the lengths of the edges in P. The problem is to find the shortest possible path and its length from a starting vertex s to a terminal vertex t.

Among several algorithms that have been proposed for the shortest path between a specified vertex pair, perhaps the most efficient one is an algorithm due to Dijkstra [11-16].

Description of the Algorithm: Dijkstra's algorithm labels the vertices of the given digraph. At each stage in the algorithm some vertices have permanent labels and others temporary labels. The algorithm begins by assigning

---

†If the given digraph is not simple, it can be simplified by discarding all self-loops, and replacing every set of parallel edges by the shortest (least-weight) edge among them. Also, the graph need not be directed. For an undirected graph $d_{ij} = d_{ji}$, and effectively each undirected edge is replaced by two oppositely directed edges of the same weight. If the graph is not weighted, assume $d_{ij} = 1$, and the adjacency matrix becomes the distance matrix.

a permanent label 0 to the starting vertex s, and a temporary label ∞ to the remaining $n - 1$ vertices. From then on, in each iteration another vertex gets a permanent label, according to the following rules:

1. Every vertex j that is not yet permanently labeled gets a new temporary label whose value is given by

$$\min [\text{old label of } j, (\text{old label of } i + d_{ij})], \qquad (11\text{-}1)$$

where i is the latest vertex permanently labeled, in the previous iteration, and d_{ij} is the direct distance between vertices i and j. If i and j are not joined by an edge, then $d_{ij} = \infty$.

2. The smallest value among all the temporary labels is found, and this becomes the permanent label of the corresponding vertex. In case of a tie, select any one of the candidates for permanent labeling.

Steps 1 and 2 are repeated alternately until the destination vertex t gets a permanent label.

The first vertex to be permanently labeled is at a distance of zero from s. The second vertex to get a permanent label (out of the remaining $n - 1$ vertices) is the vertex closest to s. From the remaining $n - 2$ vertices, the next one to be permanently labeled is the second closest vertex to s. And so on. The permanent label of each vertex is the shortest distance of that vertex from s. This statement can be proved by induction (Problem 11-13). As an illustration of Dijkstra's procedure, let us find the distance from vertex B to G in the digraph shown in Fig. 11-8. We shall use a vector of length seven to show the temporary and permanent labels of the vertices as we go through the solution. The permanent labels will be shown enclosed in a square, and the most recently assigned permanent label in the vector is indicated by a tick □✓ The labeling proceeds as follows:

A	B	C	D	E	F	G	
∞	[0]	∞	∞	∞	∞	∞	: Starting Vertex B is labeled 0.
7	[0]	1✓	∞	∞	∞	∞	: All successors of B get labeled.
7	[0]	[1]	∞	∞	∞	∞	: Smallest label becomes permanent.
4	[0]	[1]	∞	5	4✓	∞	: Successors of C get labeled.
4	[0]	[1]	∞	5	[4]	∞	
4✓	[0]	[1]	14	5	[4]	11	
[4]	[0]	[1]	14	5	[4]	11	
[4]	[0]	[1]	12	5✓	[4]	11	
[4]	[0]	[1]	12	[5]	[4]	11	
[4]	[0]	[1]	12	[5]	[4]	7✓	
[4]	[0]	[1]	12	[5]	[4]	[7]	: Destination vertex gets permanently labeled.

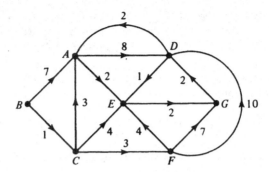

Fig. 11-8 Simple weighted digraph.

All steps are easily programmed except for the job of distinguishing the permanently labeled vertices from the temporarily labeled ones, which is slightly tricky. An efficient method of accomplishing this is to associate indices $1, 2, \ldots, n$ with the vertices, and keep a binary vector VECT of order n. When the ith vertex becomes permanently labeled, the ith element in this binary vector changes from 0 to 1.

A flow chart of this algorithm is given in Fig. 11-9, and a FORTRAN listing of the program is provided at the end of the chapter.

The algorithm described does not actually list the shortest path from the starting vertex to the terminal vertex; it only gives the shortest distance. The shortest path can be easily constructed by working backward from the terminal vertex such that we go to that predecessor whose label differs exactly by the length of the connecting edge. (A tie indicates more than one shortest path.) Alternatively, the shortest path can be determined by keeping a record of the vertices from which each vertex was labeled permanently. This record can be maintained by another linear array of length n, such that whenever a new permanent label is assigned to vertex j, the vertex from which j is directly reached is recorded in the jth position of this array.

Remarks

1. In this algorithm, had we continued the labeling until every vertex got a permanent label (rather than stopping at the permanent labeling of the destination vertex t), we would have gotten an *algorithm for the shortest paths from starting vertex s to all other vertices*. A computer program for this purpose, written in ALGOL, is given in [11-6].

2. If we take a shortest path from the starting vertex s to each of the other vertices (which are accessible from s), then the union of these paths will be an arborescence T rooted at vertex s. Every path in T from s is the (unique) shortest path in the digraph (or graph, as the case may be). Such a tree is called the *shortest-distance arborescence* (and *shortest-distance tree* in an

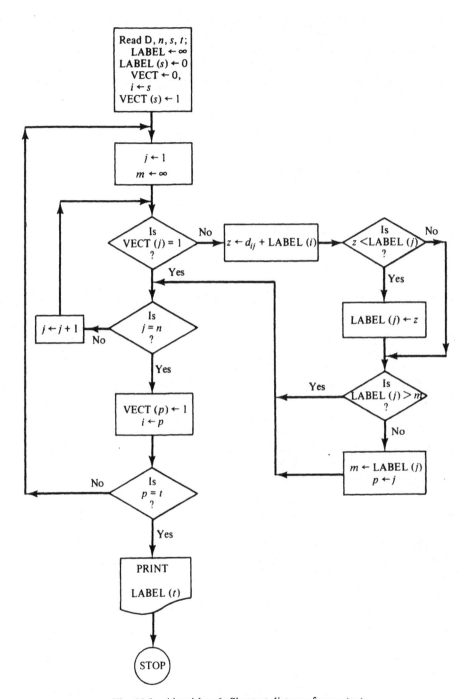

Fig. 11-9 Algorithm 6: Shortest distance from s to t.

undirected graph—not to be confused with the shortest spanning tree introduced in Chapter 3). This arborescence may be constructed as a by-product in Algorithm 6, if the labeling is continued till every vertex gets a permanent label, and if each time a vertex is labeled permanently, the corresponding edge is added to the arborescence. For example, the shortest-distance arborescence of Fig. 11-8 is given in Fig. 11-10.

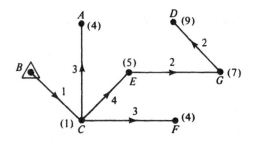

Fig. 11-10 Shortest-distance arborescence of Fig. 11-8.

3. In this algorithm, as more vertices acquire permanent labels the number of additions and comparisons needed to modify the temporary labels continues to decrease. In the case where every vertex gets permanently labeled, we need $n(n-1)/2$ additions and $2n(n-1)$ comparisons. Thus the computation time is proportional to n^2.

4. Notice that for a given n the computation time is independent of the number of edges the digraph may have. This is because it is tacitly assumed that the digraph is complete—each missing edge is simply given a very large weight. This observation is also borne out by the following typical data: On the IBM 7044, for a random digraph of 80 vertices and 3200 edges, it took $\frac{36}{60}$ second to find the shortest distance from a given vertex to all others. Another random graph with 80 vertices but only 1000 edges also took $\frac{36}{60}$ second for the same computation.

5. If the digraph is sparse [i.e., the number of edges e is much smaller than $n(n-1)$], it is possible to reduce the time of computation. This can be achieved by incorporating another test which alters the temporary labels of only those vertices that are successors of the most recent permanently labeled vertex. There is, of course, a trade off here between the time taken for testing and the time that is saved as a result of this test. An ALGOL program of Dijkstra's algorithm, which takes advantage of the sparseness of the graph, is given in [11-38], pages 43–44. For another technique that reduces computation time in sparse graphs, see remark 3 in Algorithm 7.

6. If the given digraph G is not weighted, every edge in G has a weight of one, and matrix D is the same as the adjacency matrix. Then the problem is simpler. We perform logical operations rather than real arithmetic.

7. We have assumed the distances d_{ij} are all nonnegative numbers. If some

of the distances are negative, Algorithm 6 will not work. (Negative distances in a network may represent costs and the positive ones profits.) The reason for the failure of Algorithm 6 is that once a vertex is permanently labeled its label cannot be altered. Shortest-path algorithms have, however, been proposed (see [11-17]) that will solve this problem, provided the sum of all d_{ij} around every directed circuit is positive. (The problem has no solution if a negative-weight circuit having a vertex on a directed path from s to t exists, because then one could continue minimizing the distance to $-\infty$ by going round and round this circuit.) The computation time of the existing algorithms that can handle negative d_{ij} is n^3 and not n^2.

8. It was suggested by T. A. J. Nicholson that carrying the shortest-path algorithm simultaneously from both ends s and t would improve the speed. Dreyfus [11-17] has, however, shown that the double-ended procedure would improve the efficiency only in certain types of digraphs. In the case where nearly all n vertices must be permanently labeled from either one end or the other, the double-ended procedure is actually less efficient than Dijkstra's one-ended procedure. For an ALGOL listing of Nicholson's double-ended program see Algorithm 22 in [11-6].

Algorithm 7: Shortest Path Between All Pairs of Vertices

Sometimes one is interested in finding the shortest paths between all $n(n-1)$ ordered pairs of vertices in a digraph (or $n(n-1)/2$ unordered pairs of vertices in an undirected graph). If we were to use Algorithm 6 for this purpose, the computation time would be proportional to n^4. There are several algorithms available that can do better. Among these, two are considered best, both being equally efficient. One is due to Dantzig [11-15] as improved by Tabourier [11-60]; the other one is due to Floyd [11-21], based on a procedure by Warshall [11-65]. Both algorithms require computation time proportional to n^3. We shall describe the Warshall–Floyd algorithm.

Description of the Algorithm: The algorithm works by inserting one or more vertices into paths, whenever it is advantageous to do so.

Starting with the n by n matrix $D = [d_{ij}]$ of direct distances, n different matrices $D_1, D_2, \ldots, D_n$ are constructed sequentially. Matrix D_k, $1 \leqslant k \leqslant n$, may be thought of as the matrix whose (i,j)th entry gives the length of the shortest directed path among all directed paths from i to j, with vertices 1, 2, ..., k allowed as the intermediate vertices. Matrix $D_k = [d_{ij}^{(k)}]$ is constructed from D_{k-1} according to the following rule:

$$d_{ij}^{(k)} = \min\,[d_{ij}^{(k-1)}, (d_{ik}^{(k-1)} + d_{kj}^{(k-1)})], \tag{11-2}$$

for $\quad k = 1, 2, \ldots, n,$

and $\quad d_{ij}^{(0)} = d_{ij}.$

That is, in iteration 1, vertex 1 is inserted in the path from vertex i to j if $d_{ij} > d_{i1} + d_{1j}$. In iteration 2, vertex 2 is inserted, and so on.

Suppose, for example, that the shortest directed path from vertex 7 to 3 is 7 4 1 9 5 3. The following replacements occur:

Iteration 1: $d_{49}^{(0)}$ is replaced by $(d_{41}^{(0)} + d_{19}^{(0)})$.

Iteration 4: $d_{79}^{(3)}$ is replaced by $(d_{74}^{(3)} + d_{49}^{(3)})$.

Iteration 5: $d_{93}^{(4)}$ is replaced by $(d_{95}^{(4)} + d_{53}^{(4)})$.

Iteration 9: $d_{73}^{(8)}$ is replaced by $(d_{79}^{(8)} + d_{93}^{(8)})$.

Once the shortest distance is obtained in $d_{73}^{(9)}$, the value of this entry will not be altered in subsequent operations.

The flow chart of the algorithm is given in Fig. 11-11 and a FORTRAN listing is given at the end of the chapter. Its ALGOL listing can be found in [11-21].

The algorithm described so far does not actually list the path; it only gives the shortest distances. Obtaining the path is slightly more involved than in Algorithm 6, because now there are $n(n-1)$ paths required, not just one. An efficient method of obtaining the intermediate vertices in each of the shortest paths is by constructing a matrix $Z = [z_{ij}]$ (referred to as the *optimal-policy matrix*), such that entry z_{ij} is the first vertex from i along the shortest path from i to j. The optimal-policy matrix Z can be constructed as follows:

Initially we set

$$z_{ij} = j, \quad \text{if } d_{ij} \neq \infty,$$
$$= 0, \quad \text{if } d_{ij} = \infty.$$

In the kth iteration if vertex k is inserted between i and j, element z_{ij} is replaced by the current value of z_{ik}, for all i and j. This updating of the Z matrix is done during each iteration k, where $k = 1, 2, \ldots, n$. At the end, the shortest path $(i, v_1, v_2, \ldots, v_q, j)$ from i to j is derived as a sequence of vertex numbers from matrix Z as follows (see Problem 11-15):

$$v_1 = z_{ij},$$
$$v_2 = z_{v_1,j},$$
$$v_3 = z_{v_2,j},$$
$$\cdot$$
$$\cdot$$
$$\cdot$$
$$j = z_{v_q,j}.$$

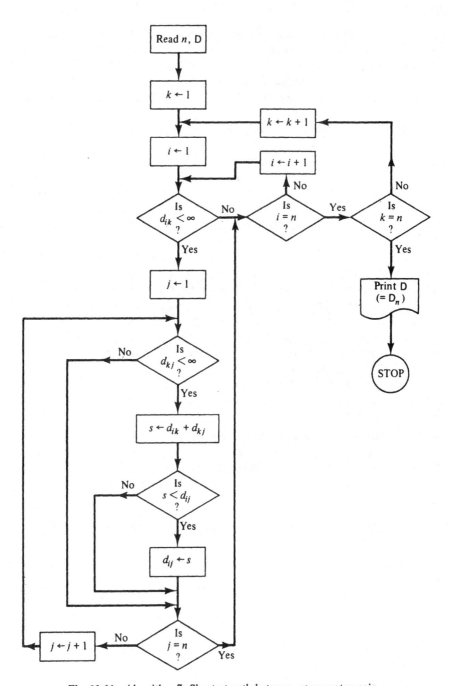

Fig. 11-11 Algorithm 7: Shortest path between every vertex-pair.

Remarks

1. Notice that for computational purposes we need memory space for only one n by n matrix. Other constructed matrices can be overwritten on this matrix.

2. To estimate the execution time, note that we have to construct n matrices $D_1, D_2, \ldots, D_n$, sequentially. For each matrix D_k the number of elements to be computed is $(n - 1)(n - 2)$, because in Eq. (11-2) $i \neq j$, $i \neq k$, $j \neq k$ (although, for simplicity, in the flow chart we have not taken advantage of this slight saving). Thus the execution time is proportional to $n(n - 1)(n - 2) \simeq n^3$.

3. Whenever $d_{ik}^{(k-1)} = \infty$ in Eq. (11-2), it is possible to circumvent $n - 1$ additions and comparisons in exchange for an additional test. This is a trade off just as in Algorithm 6, but since the execution time for Algorithm 7 is proportional to n^3, it pays to include this extra test for almost every digraph. (Note that the test is not included in the flow chart.)

4. If the graph is sparse, that is, the number of edges are far fewer than $n(n - 1)$, it is possible to take advantage of the special structure and reduce the labor by decomposing the graph. Shortest paths are obtained in each subgraph and these are put together to obtain the shortest paths in the entire graph. As an extreme example, consider the case in which a digraph of n vertices consists of the two digraphs of $n/2$ vertices each. The number of computations reduces from n^3 to $2(n/2)^3$, a reduction of 75 per cent. See Hu and Torres [11-33] for decomposition algorithms.

5. As in Algorithm 6, if the digraph is unweighted, that is, $D = X$, the computational time can be reduced by replacing the arithmetic operations with logical operations.

Transitive Closure of a Digraph: Let G be a simple, n-vertex digraph. Let us construct another simple n-vertex digraph by adding edges to G as follows: Add an edge (i, j) directed from vertex i to j if and only if there is a directed path (of any length 2, 3, $\ldots$, $n - 1$) from i to j in G. Digraph H is called the *transitive closure* of G. In other words,

$$X(H) = R(G),$$

where $X(H)$ is the adjacency matrix of H and $R(G)$ is the reachability (or accessibility) matrix of G.

It is easy to see that the transitive closure of a given digraph G can be obtained by applying ALGORITHM 7 to the adjacency matrix $X(G)$, i.e. by setting $D \leftarrow X$. The time taken by this method of obtaining transitive closure is proportional to n^3. We can, however, do better. For discussions of more efficient algorithms for transitive closure see [11-45].

Longest-Path Analysis: Sometimes, notably in critical path analysis of activity networks (see Chapter 14), one needs the longest paths (rather than the shortest) from a specified vertex to all others. One would expect that maximization procedures analogous to the minimization procedures in Algorithms 6 and 7 would yield the desired paths. But for an arbitrary digraph this will not work. For in the process of maximization, one could go round and round a directed circuit and the length would be made arbitrarily large. Another difficulty is the following: In the shortest-path problem, if $(s, t, u, \ldots, f)$ is a shortest directed path from s to t, then the subpath $(t, u, \ldots, f)$ is a shortest path from t to f. It is this property on which the shortest-path algorithms are based. On the other hand, if $(s, t, u, \ldots, f)$ is a longest directed path from s to f, there may well exist a directed path from t to f via s that is longer than the subpath $(t, u, \ldots, f)$.

Both these difficulties disappear if the given digraph is acyclic (which is the case for activity networks). Dijkstra's algorithm can then be used to find longest paths from a given vertex to all others in an acyclic digraph. The details are left as an exercise (Problem 11-16).

In addition to the three shortest-path problems dealt with in Algorithms 6 and 7, there are several other shortest-path problems. For example, one may be interested in finding the *second-shortest path* from s to f. Or one may be interested in finding a shortest path from s to f that passes through certain specified vertices. For these and more, [11-17] and [11-3] are recommended, while we move on to an altogether different problem.

11-6. DEPTH-FIRST SEARCH ON A GRAPH

In this section we shall discuss a powerful technique of systematically traversing the edges of a given graph such that every edge is traversed exactly once and each vertex is visited at least once. This technique, called the *depth-first search* (DFS) or *backtracking* on a graph was first formalized and used by Hopcroft and Tarjan [11-31] and was subsequently studied in some depth by Tarjan [11-61].

It is evident that for answering almost any nontrivial question about a given graph G we must examine every edge (and in the process every vertex) of G at least once. For example, before declaring a graph G to be disconnected we must have looked at every edge in G; for otherwise, it might happen that the one edge we had decided to ignore could have made the graph connected. The same can be said for questions of separability, planarity, and the like.

There are two natural ways of scanning or searching the edges of a graph as we move from vertex to vertex: (i) once at a vertex v we scan all edges incident on v and then move to an adjacent vertex w. At w we scan all edges incident on *w*. This process is continued till all edges in the graph are scanned. This method of fanning out at each vertex is referred to as the *breadth-first search* of the graph. This was the method used in Algorithm 3. It was also employed implicitly in Algorithms 1 and 6. (ii) An opposite approach is instead of scanning every edge incident on vertex v, we move to an adjacent vertex w (a vertex not visited before) as soon as possible, leaving v with possibly unexplored edges for the time being. In other words, we trace a walk through the graph going on to a new vertex whenever possible. This method of traversing the graph, called the *depth-first search* (DFS), has been found to be very useful in simplifying many graph-theoretic algorithms, because of the resulting numbering of the vertices and orientations imposed on the edges.

Numbering Vertices and Orienting Edges in DFS: During a DFS on a graph, whenever a vertex v is visited for the first time we assign it a distinct serial number NUM(v), so that NUM(v) = i if v was the ith vertex to be visited during the traversal. Also an orientation is imposed on each edge along the route of the traversal. When the search terminates the undirected graph G on which the DFS was being performed, becomes a digraph $\vec{G}$ with its vertices numbered $1, 2, \ldots, n$. The details of the DFS algorithm can be best described by the following steps:

Description of the DFS Algorithm: Let G be the given undirected graph, inputted as neighbor listings (i.e., representation (e) in Section 11-2). Let x be the specified vertex from which the search is to begin. PALM and FRON are two disjoint subsets into which the edges of G are to be partitioned.

Step 1: Set $v \leftarrow x$, $i \leftarrow 0$, PALM $\leftarrow \emptyset$, FRON $\leftarrow \emptyset$

Step 2: Set $i \leftarrow i + 1$, NUM(v) $\leftarrow i$

Step 3: Look for an untraversed edge incident on v.
    (a) If there is no such edge (i.e., every edge incident on v has already been traversed), go to Step 5; otherwise,
    (b) Pick the first untraversed edge at v, say (v, w), and traverse this edge. Orient the edge (v, w) from v to w. Now you are at vertex w.

Step 4: (a) If w is a vertex which has not been visited before during this search (that is, if NUM(w) is undefined), add edge (v, w) to the set PALM. Set $v \leftarrow w$ and go to Step 2.
    (b) If w is a vertex which has been visited earlier (that is, NUM(w) < NUM(v)), add edge (v, w) to the set FRON. Go to Step 3. You are back at vertex v.

Step 5: Check to see if there exists some traversed edge (u, v) in set PALM oriented toward v.
    (a) If there is such an edge move back to vertex u. (Note that u is the vertex from which v was visited for the first time.) Set $v \leftarrow u$ and go to Step 2.

(b) If there is no such edge (u, v), stop (we are back at root x, having traversed every edge and visited every vertex connected to x).

The DFS algorithm just described is illustrated by an example in Fig. 11-12. In the given graph G of five vertices and eight edges, the starting vertex x is specified. The order in which the edges are explored is given in Fig. 11-12 (b), and for this order of traversal $\vec{G}$ is given in Fig. 11-12(c).

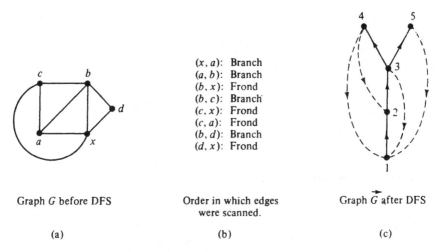

Graph G before DFS

(a)

Order in which edges were scanned.

(b)

Graph $\vec{G}$ after DFS

(c)

Fig. 11-12 Depth-first search on a graph.

Palm Tree and Fronds:

It is not difficult to see that if this DFS procedure is applied to any connected, undirected graph G (with n vertices and e edges), it will terminate after numbering the vertices $1, 2, \ldots, n$ and orienting every edge in G. Let $\vec{G}$ be the resulting digraph. Consider PALM, the set of $n - 1$ oriented edges, each of which led to a new vertex during the DFS. This subdigraph (defined by the edge set PALM) is a spanning arborescence in $\vec{G}$, because every vertex in this subdigraph, except the root x, has an in-degree equal to one, and the indegree of x is zero. (Review Sec. 9-6, to recall arborescence and its properties.) This spanning arborescence is referred to as a *palm tree*. Edges not in the palm tree (i.e., edges belonging to the set FRON) are called *fronds*. Since for every frond (a, b) vertex b was visited before a, NUM(b) < NUM(a). In other words, every frond is oriented from a higher-numbered vertex to a lower-numbered vertex.

The DFS by itself does not reveal properties of a given graph G (except whether or not G is connected). What it does, however, is to number the vertices in a systematic manner and partition the edges into two sets PALM and FRON, with properties just discussed. It is this which makes DFS a powerful tool in the construction of efficient algorithms for solving a surprisingly large number of graph-theory problems. The following are some of the problems for which algorithms have been constructed employing DFS.

(1) Identification of components. (2) Identification of blocks and cut-vertices, [11-61]. (3) Identification of maximal subgraphs of connectivity three or more [11-32]. (4) Planarity [11-31], and [11-62]. (5) Isomorphism of planar graphs [11-32]. (6) Identification of fragments (i.e., maximal strongly connected subgraphs) in a digraph [11-61].

It has been shown (see the appropriate reference) that the computation time of all these six algorithms is proportional to e, the number of edges, if the graph is given in the neighbor-listing form of Sec. 11-2(e). And since every edge must be examined at least once, this is also the lower bound on these algorithms, disregarding multiplicative constants, of course.

We shall now sketch the planarity algorithm, a problem to which DFS has been applied with spectacular success, resulting in drastic improvements in computation time over earlier methods.

Algorithm 8: Planarity Testing

The problem of determining whether or not a given graph is planar is an important one. As pointed out in Chapter 5, the planarity characterizations of Kuratowski, Whitney, or MacLane (although theoretically elegant) are unsuitable for testing by a computer. They are difficult to implement; and, besides, if a graph is planar, these methods do not yield a plane representation, which is often what is needed. It has been shown, for example, that if Kuratowski's characterization is used to test planarity of an n-vertex graph ($n > 5$), the computation time is at least proportional to n^6 (see [5-8]).

In recent years many algorithms for planarity testing have been proposed and programmed on computers (see [5-8] for a survey). Most of these methods employ the map-construction approach, which works as follows: A planar subgraph g (in most algorithms g is a circuit) of the given graph G is first selected and mapped on a plane. Then gradually the remaining edges are added to g, such that no crossings occur. If we succeed in the reconstruction, graph G is obviously planar, and we have obtained a plane representation of G. If we fail, G is nonplanar.

The only difficult part of such an algorithm is that in the early stages of adding edges to g we have choices available (i.e., ambiguity) in placement of edges. A wrong choice made earlier may later prevent us from adding an edge, even if the graph is planar. For example, in Fig. 11-13(a), suppose that the starting subgraph g was the circuit $\{e_1, e_2, e_3, e_4, e_5\}$. Then we add edges e_6, e_7, e_8, and e_9, without any crossover. Now we find that the last edge e_{10} cannot be added without a crossing. From this we might erroneously con-

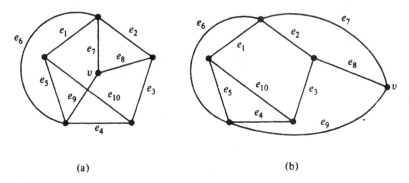

Fig. 11-13 Two mappings of a graph.

clude that G is nonplanar. On the other hand, had we selected a different face for placing vertex v, we would have obtained a planar representation, as shown in Fig. 11-13(b).

This essentially is *the* problem in the map-construction method of planarity testing, and different procedures have been devised to solve it.

Preliminary Simplification: As pointed out in Section 5-5, an arbitrary graph can, in general, be reduced to a much smaller graph if subjected to certain simplifying steps. These steps do not affect the planarity (or the nonplanarity) of a graph.

1. Apply Algorithm 1 to check for connectedness. If the graph is disconnected, consider only one component at a time.
2. Remove all self-loops, and replace each set of parallel edges by a single edge.
3. Eliminate every vertex of degree two by merging the two edges incident on the vertex. Apply steps 2 and 3 alternately and repeatedly, till the graph cannot be reduced any further.
4. Apply Algorithm 4 to partition the graph into its blocks (i.e., maximal nonseparable subgraphs).
5. Subject each block to reduction steps 3 and 2 alternately till no further reduction is possible.
6. Each simplified block thus obtained, with e edges and n vertices, is tested for

$$n \geqslant 5,$$
$$e \geqslant 9, \tag{11-3}$$
$$e \leqslant 3n - 6.$$

If any of these three inequalities is not satisfied, our job is finished, and we move on to the next block. Every graph with $n < 5$ or $e < 9$ is planar, and every simple graph with $e > 3n - 6$ is nonplanar.

One planarity testing algorithm due to Bruno, Steiglitz, and Weinberg [5-2] goes a step further in simplification. Each nonseparable graph is further broken down into its maximum 3-connected subgraphs (called 3-connected components). Then the following result, due to W. T. Tutte, is used: A graph is planar if and only if all of its 3-connected components are planar. This algorithm, however, is not as efficient as the one due to Hopcroft and Tarjan [11-31], which will be described next.

Description of the Algorithm: The planarity-testing algorithm is quite involved. We shall sketch only its essential features. To understand the main algorithm, let us consider the following decomposition procedure applied to a given simple, nonseparable graph G with n vertices and e edges:

Circuit–Path Decomposition:
Step 1: Find some circuit K in G. Set $g \leftarrow K$. Label the vertices and edges of g as $v_1, v_2, \ldots$, and $e_1, e_2, \ldots$, respectively. Set $i \leftarrow 1$.

Step 2: If there is an unlabeled edge in G, find a path p_i that begins and ends at labeled vertices but consists only of unlabeled edges. Store p_i. If there is no unlabeled edge left in G, go to step 4.

Step 3: Set $g \leftarrow g \cup p_i$. Set $i \leftarrow i + 1$. Label the unlabeled edges and vertices in g. Return to step 2.

Step 4: Stop. Print $g, p_1, p_2, \ldots, p_m$.

It can be shown ([5-8]) that the procedure just outlined decomposes the simple, nonseparable graph G into one circuit and $m = e - n$ paths. Since the circuit may be looked upon as two edge-disjoint paths, G is thus decomposed into $e - n + 2$ paths. It may be noted that although such a decomposition of a graph G may not be unique, the number of paths into which G is decomposed is constant and equals one circuit and $e - n$ paths. For example, in Fig. 11-13 consider two distinct decompositions, each with one circuit and four $(10 - 6 = 4)$ paths:

$$\{e_1, e_2, e_3, e_4, e_5\}, \quad \{e_6\}, \quad \{e_{10}\}, \quad \{e_7, e_9\}, \quad \{e_8\},$$
and
$$\{e_1, e_2, e_3, e_4, e_5\}, \quad \{e_7, e_8\}, \quad \{e_9\}, \quad \{e_6\}, \quad \{e_{10}\}.$$

In this circuit-path decomposition, we can map the circuit K on a plane, and continue to add new paths $p_1, p_2, \ldots$, as they are generated. A new path p_i will either divide an existing face into two new faces, or will make $g \cup p_i$ nonplanar, when added to the planar map g. This method of arbitrarily adding paths as they are generated may lead to a situation shown in Fig. 11-13, which has already been discussed. To solve this problem, one can either

SEC. 11-6   DEPTH-FIRST SEARCH ON A GRAPH   307

1. Continue adding paths till no path can be added. Then backtrack to explore the alternative choices he could have made earlier, or
2. Continue to look at different paths but not add them to K, till it is found which face a path must be placed in, or it is ascertained that it does not matter which face the path is placed in.

Some algorithms (see [11-53]) use approach 1, but Hopcroft and Tarjan [11-31] have used approach 2 and have shown that their algorithm is more efficient because of it. They use list processing and have an elaborate program (985 lines of ALGOL). The gist of their technique of resolving ambiguity in adding paths, is as follows:

Suppose that at any stage we have a path p_i (on top of a pushdown list of paths) whose ambiguity we are trying to resolve. Let a and b be the end vertices of p_i. The flow chart in Fig. 11-14 shows the different cases that may arise and what action is taken for each. These steps are explained by means of Fig. 11-15.

In Fig. 11-15(a) path p_i can be swiveled at vertices a and b, and therefore can divide either the face "above" or "below" the path p_j. This ambiguity of path p_i must be resolved. To resolve this ambiguity, starting from some vertex x on path p_i a new path p_k is constructed. (The path p_k consists of unlabeled edges and it terminates as soon as it touches a labeled vertex.)

If both end vertices x and y of p_k are on path p_i, as shown in Fig. 11-15(b), p_k can be swiveled at vertices x and y and thus divides either of the two faces—one "above" p_i and the other "below." Thus not only did we not resolve the ambiguity in placement of path p_i, but we have a new path p_k whose ambiguity must be resolved first. Path p_k is put above p_i in the stack, and we begin resolving its ambiguity just as we were doing for p_i.

Another possibility is that the end vertex y is neither on path p_i nor on p_j but on a different path, as shown in Fig. 11-15(c). In this case p_i cannot be swiveled, and therefore there is no ambiguity as to which face p_i divides.

As in Fig. 11-15(d), if path p_i ends on a vertex in path p_j between a and b, then p_i (together with p_k) can still be swiveled about a and b. Therefore, the ambiguity in p_i remains. The path p_k, however, divides the face $a\,x\,b\,y\,a$ unambiguously. Therefore, we shall have to generate another path from a vertex on p_i for resolving ambiguity in the placement of p_i.

Finally in the case where path p_k ends on a vertex on p_j but not between a and b, we have the situation shown in Fig. 11-15(e). The path p_i can still be swiveled. But, unlike Fig. 11-15(b), there is no ambiguity in path p_k with respect to p_i. Therefore, a new path must be generated to resolve ambiguity in p_i, and that path can be generated starting from any vertex in path $y\,a\,x\,b$, which is path p_i extended up to vertex y on p_j.

In summary, the planarity algorithm consists of routines (a) for finding blocks (i.e., maximal nonseparable subgraphs), (b) for partitioning each block

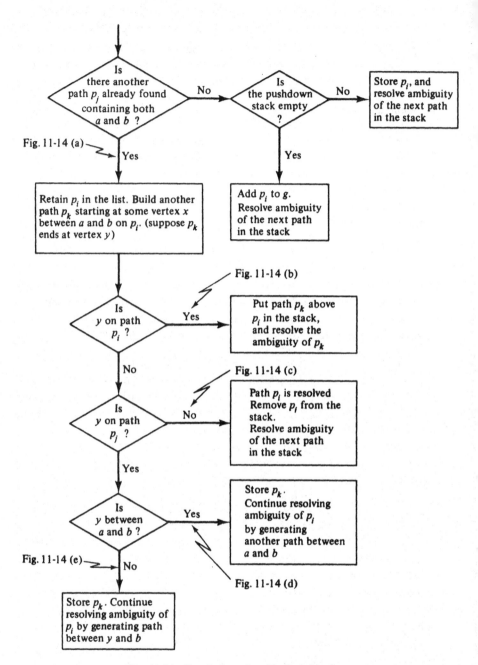

Fig. 11-14 Resolution of ambiguity of path p_i.

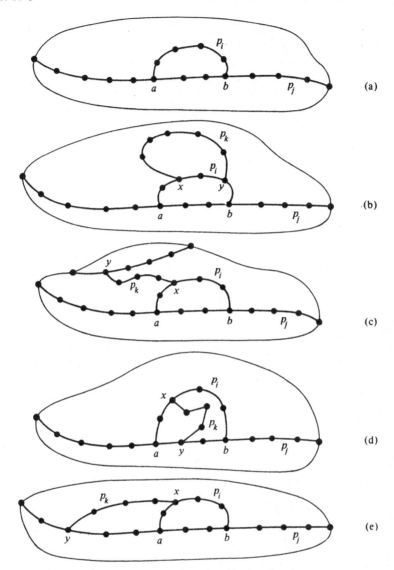

Fig. 11-15 Resolution of ambiguity of path p_l.

into paths, (c) for deciding where to add these paths, and (d) for building a planar representation. The DFS is used in both (a) and (b). The paths generated from the resulting palm tree and fronds have several important properties which are made use of in (c) and (d). The time required for each, except (d), is proportional to the number of edges, e. The building of planar representation requires time proportional to $n \log n$, n being the number of vertices.

Thus the theoretical time bound for the entire algorithm is proportional to $n \log n$. (Because of Euler's equation e is proportional to n, in graphs subjected to planarity algorithm.)

This algorithm has subsequently been improved and was reported in Tarjan's Ph.D. Thesis [11-61]. The time taken by the improved version is proportional to just n. A somewhat simplified form of the improved version appeared as a Cornell University technical report [11-61].

11-7. ALGORITHM 9: ISOMORPHISM

The graph isomorphism problem is to determine if there exists a one-to-one correspondence between the vertices of two graphs G_1 and G_2 that preserves the adjacency of vertices. The problem of graph isomorphism arises in many fields, such as chemistry, switching theory, information retrieval, and linguistics. Consequently, the isomorphism problem has been studied extensively. For a survey and references, see Corneil's Ph.D. thesis [11-9], and [11-12].

Theoretically, it is always possible to determine whether or not two graphs G_1 and G_2 are isomorphic by keeping G_1 fixed and reordering vertices of G_2 to check if their adjacency matrices become identical. This process may require all $n!$ reordering and comparisons, n being the number of vertices. Such an inefficient procedure, in which the running time grows factorially with n, is of limited use for practical problems. An algorithm guaranteeing a solution in running time proportional to a constant power of n is desirable, but no such algorithm has been discovered for determining if two arbitrary graphs are isomorphic.[†]

There are, however, efficient isomorphism algorithms available for certain types of graphs. Some of these are

1. Isomorphism in trees: [11-9], pages V-44–V-52.

2. Isomorphism in planar graphs: [11-32].

3. Isomorphism in graphs containing no k-strongly regular subgraphs for large k: see [11-9] for definition of a k-strongly regular graph.

4. Isomorphism in partially labeled graphs with special structure: [11-59] and [11-64]. (See also Chapter 15 in this book.)

Heuristic Procedure

If two given graphs G_1 and G_2 are arbitrary, the usual approach is first to try to show if G_1 and G_2 are *not* isomorphic. This is done by asking ques-

---

†There are problems for which only factorially or exponentially growing algorithms exist, and better ones may never be found. In such cases, one does have to live with an inefficient algorithm and use it for small graphs.

tions of the following type:

1. Do G_1 and G_2 have the same number of vertices?
2. Do G_1 and G_2 have the same number of edges?
3. Is the number of vertices n_i with degree i the same in both, for $i = 1, 2, \ldots$?
4. Do both graphs have the same number of components?
5. For each component are questions 1, 2, and 3 answered in the affirmative?
6. Are the characteristic polynomials† of their adjacency matrices $X(G_1)$ and $X(G_2)$ the same (taken in the field of real numbers)?

And so on.

Clearly, if the answer to any of these questions is no, G_1 and G_2 are nonisomorphic. A yes answer, however, does not guarantee an isomorphism.

An *invariant* of a graph G is a number that is the same for all graphs isomorphic to G. Some examples of invariants are number of vertices n, number of edges e, rank R, nullity μ, number of components p, connectivity, and coefficients in the characteristic polynomial of the adjacency matrix. A complete set of invariants is a set of invariants that completely describes a graph within isomorphism.

The problem of graph isomorphism is solved if we can find a complete set of invariants for G_1 and G_2 and then check to see if they are identical.

The problem of finding a complete set of invariants can also be thought of as the *coding of a graph*. If we could find a complete set of invariants, we would place them in a sequence. This sequence would contain all the essential information about a graph. Two graphs would be isomorphic if and only if their codes were the same. The problem of coding trees and some other types of graphs has been solved, [11-54], [11-66]. But there exists no method of coding an arbitrary graph with a large number of vertices.

Numerous heuristic procedures have been proposed (and programmed) based on the idea that if you compute many invariants, and if they are the same for both G_1 and G_2, it is *likely* that G_1 and G_2 are isomorphic. Heuristic approaches work well for graphs of small orders. For example, it can be shown that two simple, connected graphs with $n \leqslant 7$ which have affirmative answers for questions 3 and 6 are isomorphic. But these heuristic algorithms fail for large arbitrary graphs.

Sometimes heuristic procedures are also used for simplifying the last-resort method involving vertex-by-vertex correspondence between G_1 and

---

†It is not difficult to show that the characteristic polynomial, $\det(X - \lambda I)$, is independent of the order in which the vertices appear in the adjacency matrix X.

G_2. For example, the number of comparisons can always be reduced from $n!$ to

$$\sum_{i=1}^{k} n_i!,$$

where

$$\sum_{i=1}^{k} n_i = n.$$

This is because in reordering the vertices in G_2 it suffices to permute only vertices with the same degree. For a graph that does not have a large percentage of its vertices with the same degree, the number $\sum n_i!$ is much smaller than $n!$.

11-8. OTHER GRAPH-THEORETIC ALGORITHMS

The algorithms described so far, although very important and basic, are only a few samples out of scores of graph-theoretic algorithms available in the literature. Obviously, it is not possible to include them all. Some of the other commonly used algorithms for graphs are

1. Find all fragments (i.e., maximal strongly connected subgraphs) in a given digraph G [11-61].

2. Find a Hamiltonian path (if there is one) in a given undirected graph. See [11-49].

3. Find all directed Hamiltonian circuits in a given digraph. See [11-13], pages 35–37.

4. Find a maximal complete subgraph (clique) in a given graph G. See [11-1] and [11-44].

5. Find a maximal matching in a bipartite graph. An algorithm, known as the Hungarian method, is often used for solving this assignment problem. The Hungarian method is available as a standard subroutine in most operations research computer program packages. See [9-4], pages 265–269. The computation time for the Hungarian method, for a graph of n vertices and e edges, is proportional to $e \cdot n$. Recently a more efficient algorithm has been proposed by Hopcroft and Karp [11-30]; it takes time proportional to $(e + n) \cdot \sqrt{n}$.

6. Given a connected weighted digraph G in which the weight of an edge represents the maximum rate of flow through that edge, find the maximum possible flow from a vertex x to another vertex y in G. This is the well-known maximum-flow problem and is solved by the Ford–Fulkerson algorithm (discussed in Sections 4-6 and 14-1). The Ford–Fulkerson algorithm is also a standard program in operations research. See [9-4]. For a more recent and improved algorithm see Edmonds and Karp [11-19].

7. Find the chromatic number of a given graph. See [11-26].

8. Given an acyclic digraph, perform a topological sorting of its vertices. See Section 14-8 for definition, an algorithm, and references.

9. Given a connected weighted graph G, partition the vertices of G into subsets no larger than a given size so as to minimize the total weight of edges cut in the process. For a heuristic algorithm to solve this important but difficult problem, see [11-36]. Also, see [11-35].

10. Given a connected weighted graph G, find a Hamiltonian circuit with smallest weight. This is the traveling-salesman problem, for which no satisfactory solution has been found so far. For heuristic algorithms see [2-1].

11. In a given graph $G = (V, E)$, find a smallest subset (i.e., subset of minimum cardinality) of edges $E' \subseteq E$ such that every vertex of G is incident on at least one of the edges in E'. This problem of finding a smallest covering was discussed in Section 8-5. See [11-18] for an efficient algorithm.

12. In a given graph $G = (V, E)$, find a smallest subset of vertices V' such that every vertex of G not included in V' is adjacent to at least one vertex in V'. This problem of finding a smallest dominating set was discussed in Section 8-2. Superficially similar to 11, this problem is much more difficult.

13. In a given digraph G, find a smallest set of edges which when deleted from G would destroy all directed circuits. A method of finding this minimum decyclization edge set (or feedback arc set) was outlined in Section 9-11, but the computation time grew exponentially with n, the number of vertices.

14. In the previous problem instead of finding a smallest set of edges we now wish to find a smallest set of vertices in G whose removal destroys all directed circuits in G. Just as for 13 no efficient algorithm is known for this problem.

15. Given a connected weighted graph $G = (V, E)$, and a subset V' of V. Find a minimal tree in G which spans the vertices in V' (and possibly some more). Such a tree, known as a *Steiner tree*, is much more difficult to obtain than a minimal spanning tree for G. The difficulty arises from the fact that inclusion of extra vertices may lead to a different tree with a smaller total weight. Although no efficient algorithm has been found for an exact solution of this problem, an approximate solution can be obtained by an efficient algorithm given by S. K. Chang (in "The Generation of Minimal Trees with a Steiner Topology," *J. ACM*, Vol. 19, No. 4, October 1972, 699–711).

11-9. PERFORMANCE OF GRAPH-THEORETIC ALGORITHMS

As observed in Section 11-7, for a given graph-theory problem it would be desirable to have an algorithm which guarantees a solution in an execution time proportional to some constant power of n or e (as usual, n and e are the number of vertices and edges, respectively, in the given graph). In other words, the execution time t (for the worst possible graph) can be expressed as

$$t \leq \alpha n^k \quad \text{or} \quad t \leq \beta e^q,$$

and the lower the value of k and q the better. Such an algorithm (whose computation time is bounded by a polynomial in n or e) is called a *polynomial bounded* algorithm. For example, Algorithm 1 (connectedness and components) is polynomial-bounded, since $k = 2$. Algorithms 2, 3, 4, 6, 7, and 8 are also polynomial-bounded, but Algorithms 5 and 9 are not. Some important polynomial-bounded graph-theoretic algorithms along with their bounds

Problems	Run-Time Bounds	Relevant References
Connectedness and components	n^2 or e	[11-29], [11-61]
Spanning tree	e	[11-58]
Minimal spanning tree	n^2	[11-22], [11-70]
Fundamental circuit-set	n^v, $2 \leq v \leq 3$	[11-47]
Cut-vertices and blocks	n^2 or e	[11-48], [11-61]
Bridges	n^2 or e	[11-10], [11-61]
Shortest path between two vertices	n^2	[11-16], [11-73]
Shortest paths between all vertex-pairs	n^3	[11-21], [11-73]
Transitive closure	n^α, $2 < \alpha < 3$	[11-45]
Strong connectedness and fragments	n^2 or e	[11-29], [11-61]
Planarity	e	[11-62]
Topological sorting	e	[11-39]
Maximal matching in a bipartite graph	$n^{5/2}$	[11-30]
Minimal cut	n^β, $2 < \beta < 3$	[11-19]
Minimal edge cover	n^3	[11-18], [11-46]

Table 11-1 Polynomial-Bounded Algorithms

and relevant references are shown in Table 11-1. Note that since for a simple graph $e \leq n(n-1)/2 \leq n^2/2$, bounds in terms of e and n are convertible into each other. It should also be kept in mind that different algorithms bounded by the same power of n may take very different amounts of actual computer time (for the same graph) because of their different multiplicative constants.

On the other hand, there are graph-theoretic problems for which it is simply not possible to have a polynomial-bounded algorithm. Take, for example, the problem of generating all spanning trees of a given graph, as discussed in Section 11-4. The number of spanning trees in an n-vertex simple, labeled graph can be as high as n^{n-2}. Therefore, if each spanning tree were generated in c units of time, the algorithm to generate all spanning trees would consume $c \cdot n^{n-2}$ units of time. Thus no polynomial-bounded algorithm can be found for this problem. Similar arguments hold for problems of generating all cliques, all circuits/directed circuits, all paths, all cut-sets, and so forth, for a given graph.

There is a third category of graph-theoretic problems, for which so far no polynomial-bounded algorithms have been discovered, nor has it been possible to show that polynomial-bounded algorithms do not exist for these problems. Detection of isomorphism (Algorithm 9) is one such problem. A list of important problems of this type is given in Table 11-2. The computation time for solving these problems (using the best available algorithm at present, and the worst possible graph) grows exponentially or factorially (but not polynomially) with n. Such inefficient algorithms are obviously of very limited use for practical problems. Heuristic techniques are the mainstay of their solutions.

Based on a remarkable result of Stephen Cook (1971), Richard Karp [11-34] showed the following surprising result: Except for the isomorphism problem, all other problems in Table 11-2 are polynomially equivalent, that is, if a polynomial-bounded algorithm exists for one, polynomial-bounded

Problems	Relevant References
Chromatic number	[11-26], [11-41]
Smallest dominating set	[11-44]
Maximal clique	[11-20]
Hamiltonian circuit	[11-56]
Directed hamiltonian circuit	[11-56]
Traveling salesman problem	[11-28]
Minimal feedback edge-set	[9-6]
Minimal feedback vertex set	[9-6]
Steiner tree	Chang in Section 11-8
Isomorphism	[11-9], [11-12], [11-64]

Table 11-2 Nonpolynomial Algorithms

algorithms can be found for the others. The proof of this equivalence is involved and is not very relevant for us here. (The Cook-Karp class of problems includes a number of other combinatorial and graph-theoretic problems in addition to the top nine in Table 11-2.)

11-10. GRAPH-THEORETIC COMPUTER LANGUAGES

The increasing interest in graph-theoretic computations has led to the development of several programming languages for the sole purpose of handling graphs. The major goal of such a language is to enable the user to formulate operations on graphs in a compact and natural manner, as if he were communicating with another graph theorist. For example, in one such language [11-37], the statement SPANTREE @ G would call the subroutine for a spanning tree and would find a spanning tree of the graph G. Another statement IF (G, PLANAR) 17, 3 would transfer control to statement 17 if G is planar, otherwise to statement 3.

Once such a language is developed and implemented, its advantages are enormous. It makes the writing of graph-theory programs easy and compact. It frees the user from having to concern himself with many unnecessary details and allows him to concentrate on the essential features of his program. The disadvantage of such a language, as of all special-purpose languages, is that it takes a great deal of time, trouble, and expense to develop such a language, which can be used only for the purpose of writing programs in graph theory.

For such a graph-theoretic computer language to be useful to many users, with different problems in graph theory, the language must have a large number of primitives (i.e., basic graph-theoretic statements) such as "remove vertex k of G", "add an edge between vertices x and y", or "find the shortest distance from vertex u to vertex v in G". Moreover, the graph-theoretic language must have all the computing facilities of an existing symbolic language such as FORTRAN, so that the programmer can perform functions which are not covered by primitives. Since there is little to be gained by developing an entirely new language from scratch, all graph-theoretic languages available and being developed are extensions of some well-known programming language.

Some graph-theoretic computer languages available at present are

1. Graph-Theoretic Language (GTPL) at the University of West Indies, Jamaica [11-52]. It is an extension of FORTRAN.

2. Graph Algorithm Software Package (GASP) at the University of Illinois [11-7]. It is an extension of PL/1.

3. HINT at Michigan State University [11-27] is an extension of the list-processing language LISP 1.5.
4. GRASPE at the University of Houston, [11-23] and [11-51], is also an extension of LISP 1.5.
5. Directed Graph Processor (DIP) at Carnegie-Mellon University is an extension of PL/1. (See a report by Terry C. Gleason, 1969.)
6. An Interactive Graph Theory System at the University of Pennsylvania is an extension of FORTRAN [11-71].
7. Graphic Extended ALGOL (GEA) at Instituto di Elettrotecnica ed Elettronica del Politecnico di Milano, Italy is an extension of ALGOL: [11-13] and [11-14].
8. AMBIT/G, developed by C. Christensen for manipulation of digraphs, is an extension of AMBIT [11-57].
9. GIRL—Graph Information Retrieval Language [11-4].
10. FORTRAN Extended Graph Algorithmic Language (FGRAAL) at the University of Maryland [11-2].

The interested reader should consult the cited references for the details of these languages.

SUMMARY

Computational aspects of graph theory were presented in this chapter. For anyone interested in applying graph theory to solve physical problems (such as flow problems, assignment problems, identification of a chemical compound, topological analysis of an electrical network, layout of a printed-circuit board, etc.), it is essential to be able to enlist the help of the digital computer. Without the help of high-speed electronic computers, he cannot hope to handle a graph of a size generally encountered in solving a nontrivial practical problem.

A computer that has been "taught" elementary graph theory (such as, find out if graph G is separable, or pick out a spanning tree in G) can be of immense aid even to a "pure" graph theorist. It can, for instance, relieve him of the drudgery of finding graphs with special properties to serve as examples and counterexamples.

To be able to teach graph theory to a computer, one must obviously know both computer programming and graph theory. In addition, one must find efficient algorithms. The most important prerequisite of any useful graph-theoretic algorithm is that the running time of its program on the computer must not rise factorially or even exponentially with n. It should be proportional to n^k, where k is some fixed number—preferably a low number.

REFERENCES

As yet, there is no one reference where all or most of the graph-theoretic algorithms are available. Each algorithm has to be looked up individually in a published paper, thesis, or report. Consequently, this chapter contains a long list of references. It is unlikely that any one student would be interested in all of them.

For a general introduction to the nature of graph-theoretic computer algorithms, two papers of Read [11-52] and [11-53] are recommended. Lehmer's paper [11-40] is an excellent treatise on the use of computers for solving combinatorial problems. Among the few books that do discuss some aspects of using computers to solve graph-theoretic problems are Even [11-20], Berztiss [11-5], Chapters 3, 6, and 8, Knuth [11-39], which is at an advanced level, Knoedel [11-38], which is in German, and Bellman, Cooke, and Lockett [11-3]. A related book on combinatorial computing by Wells [11-69] is also recommended. It develops and uses a special programming language developed for combinatorial applications.

11-1. AUGUSTON, J. G., and J. MINKER, "An Analysis of Some Graph-Theoretical Cluster Techniques," *J. ACM*, Vol. 17, No. 4, 1970, 571–588. (Correction: *J. ACM*, Vol. 19, April 1972, 244–247).

11-2. BASILI, V. R., C. K. MESZTENYI, and W. C. REINBOLDT, "FGRAAL: Fortran Extended Graph Algorithmic Language," University of Maryland, Computer Science Center Report TR-179, March 1972. See also Report TR-225, January 1973 on implementation of FGRAAL.

11-3. BELLMAN, R., K. L. COOKE and J. A. LOCKETT, *Algorithms, Graphs and Computers*, Academic Press, Inc., New York, 1970.

11-4. BERKOWITZ, S., "GIRL—Graph Information Retrieval Language—Design of Syntax," in *Software Engineering*, Vol. 2 (J. T. Tou, ed.), Academic Press, Inc., New York, 1971, 119–139.

11-5. BERZTISS, A. T., *Data Structures; Theory and Practice*, Academic Press, Inc., New York, 1971.

11-6. BOOTHROYD, J., "Algorithm 23: Shortest Path Between Start Node and All Other Nodes of a Network," *The Computer Journal*, Vol. 10, No. 3, Nov. 1967, 307–308. Also see Algorithm 22.

11-7. CHASE, S. M., "GASP—Graph Algorithm Software Package," Quarterly Technical Progress Report, Oct., Nov., Dec. 1969, Department of Computer Science, University of Illinois, Urbana.

11-8. CHASE, S. M., "Analysis of Algorithms for Finding All Spanning Trees of a Graph," Report No. 401, Department of Computer Science, University of Illinois, Urbana, Oct. 1970.

11-9. CORNEIL, D. G., "Graph Isomorphism," Ph.D. Thesis, Department of Computer Science, University of Toronto, Toronto, Canada, 1968. Updated version Tech. Rep. No. 18, April 1970.

11-10. CORNEIL, D. G., "An n^2 Algorithm for Determining the Bridges of a Graph," *Information Processing Letters*, Vol. 1, No. 2, 1971, 51–55.

11-11. CORNEIL, D. G., "An Algorithm for Determining the Automorphism Partitioning of an Undirected Graph," *BIT*, Vol. 12, No. 2, 1972, 161–171.

11-12. CORNEIL, D. G., and C. C. GOTLIEB, "An Efficient Algorithm for Graph Isomorphism," *J. ACM*, Vol. 17, No. 1, Jan. 1970, 51–64.

11-13. CRESPI-REGHIZZI, S., and R. MORPURGO, "A Graph-Theory Oriented Extension to Algol," *Calcolo*, Vol. 5, No. 4, 1968, 1–43.

11-14. CRESPI-REGHIZZI, S., and R. MORPURGO, "A Language for Treating Graphs," *Comm. ACM*, Vol. 13, No. 5, May 1970, 319–323.

11-15. DANTZIG, G. B., "All Shortest Routes in a Graph," *Proceedings of the International Symposium on Graph Theory*, Rome, Italy, July 1966, 91–92. Published by Dunod Editeur, Paris.

11-16. DIJKSTRA, E. W., "A Note on Two Problems in Connection with Graphs," *Numerische Math*, Vol. 1, 1959, 269–271.

11-17. DREYFUS, S. E., "An Appraisal of Some Shortest-Path Algorithms," *J. Operations Research*, Vol. 17, No. 3, 1969, 395–412.

11-18. EDMONDS, J., "Paths, Trees, and Flowers," *Canadian J. of Math.*, Vol. 17, 1965, 449–467.

11-19. EDMONDS, J., and R. M. KARP, "Theoretical Improvements in Algorithmic Efficiency for Network Flow Problems," *J. ACM*, Vol. 9, No. 2, April 1972, 248–264.

11-20. EVEN, S., *Algorithmic Combinatorics*, The Macmillan Company, New York, 1973.

11-21. FLOYD, R. W., "Algorithm 97: Shortest Path," *Comm. ACM*, Vol. 5, 1962, 345.

11-22. FRAZER, W. D., "Analysis of Combinatory Algorithms—A Sample of Current Methodology," *Proc. AFIPS* 1972 S.J.C.C., 483–491.

11-23. FRIEDMAN, D. P., D. C. DICKSON, J. J. FRASER, and T. W. PRATT, "GRASPE 1.5 —A Graph Processor and Its Application," University of Houston, Report RS 1-69, August, 1969.

11-24. GIBBS, N. E., "A Cyclic Generation Algorithm for Finite Undirected Linear Graphs," *J. ACM*, Vol. 16, No. 4, Oct. 1969, 564–568.

11-25. GOTLIEB, C. C., and D. G. CORNEIL, "Algorithms for Finding a Fundamental Set of Cycles for an Undirected Linear Graph," *Comm. ACM*, Vol. 10, No. 12, Dec. 1967, 780–783.

11-26. GRAHAM, G. D., "An Algorithm to Determine the Chromatic Number of a Graph," Technical Report No. 47, Department of Computer Science, University of Toronto, Canada, Nov. 1972.

11-27. HART, R., "HINT: A Graph Processing Language," Research Report, Computer Institute for Social Science Research, Michigan State University, East Lansing, Feb. 1969.

11-28. HELD, M. and R. M. KARP, "The Traveling-Salesman Problem and Minimum Spanning Trees: Part II," *Mathematical Programming*, Vol. 1, 1971, North-Holland Publishing Company, 6–25.

11-29. HOLT, R. C., and E. M. REINGOLD, "On the Time Required to Detect Cycles and Connectivity in Graphs," *Mathematical Systems Theory*, Vol. 6, No. 2, 1972, 103–106.

11-30. HOPCROFT, J. E., and R. M. Karp, "A $n^{5/2}$. Algorithm for Maximum Matchings in Bipartite Graphs," IEEE Conf. Record of the Twelfth Annual Symp. on Switching and Automata Theory, 1971, 122–125.

11-31. HOPCROFT, J. E., and R. TARJAN, "Planarity Testing in Vlog V Steps," Computer Science Technical Report No. 201, Stanford University, Stanford, Calif., March 1971. Also in *Proc. IFIP Congress*, Ljubljana, Booklet Ta-2, August 1971, 18–23.

11-32. HOPCROFT, J. E. and R. E. TARJAN, "Isomorphism of Planar Graphs," in *Complexity of Computer Computation*, (R. E. Miller and J. W. Thatcher, eds.) Plenum Press, New York, 1972, 131–152.

11-33. HU, T. C., and W. T. TORRES, "Shortcut in Decomposition Algorithm for Shortest Paths in a Network," *IBM J. of Res. Develop.*, Vol. 13, No. 4, July 1969, 387–390.

11-34. KARP, R. M., "Reducibility Among Combinatorial Problems," in *Complexity of Computer Computation*, (R. E. Miller and J. W. Thatcher, eds.) Plenum Press, New York, 1972, 85–103.

11-35. KERNIGHAN, B. W., "Optimal Sequential Partitions of Graphs," *J. ACM*, Vol. 18, No. 1, 1971, 34–40.

11-36. KERNIGHAN, B. W., and S. LIN, "An Efficient Heuristic Procedure for Partitioning Graphs," *Bell System Tech. J.*, Vol. 49, Feb. 1970, 291–307.
11-37. KING, C. A., "A Graph-Theoretic Programming Language," in *Graph Theory and Computing*, (R. C. Read, ed.), Academic Press, New York, 1972, 63–74. See also King's doctoral thesis, University of West Indies, 1970.
11-38. KNOEDEL, W., *Graphentheoretische Methoden und ihre Anwendungen*, Springer-Verlag, Berlin, 1969.
11-39. KNUTH, D. E., *The Art of Computer Programming*, Vol. 1, *Fundamental Algorithms*, Addison-Wesley Publishing Company, Inc., Reading, Mass., 1968.
11-40. LEHMER, D. H., "Teaching Combinatorial Tricks to a Computer," *Proc. Symp. Appl. Math.*, Vol. 10, *Combinatorial Analysis*, American Mathematical Society, 1960, 179–193.
11-41. MATULA, D. W., G. MARBLE, and J. D. ISAACSON, "Graph Coloring Algorithms," in *Graph Theory and Computing* (R. C. Read, ed.), Academic Press, New York, 1972, 109–122.
11-42. MCILROY, M. D. "Algorithm 354: Generator of Spanning Trees," *Comm. ACM*, Vol. 12, No. 9, Sept. 1969, 511.
11-43. MINTY, G. J., "A Simple Algorithm for Listing All the Trees of a Graph," *IEEE Trans. Circuit Theory*, Vol. CT-12, No. 1, March 1965, 120.
11-44. MULLIGAN, G. D., "Algorithms for Finding Cliques of a Graph," Technical Report No. 41, Department of Computer Science, University of Toronto, Canada, May 1972.
11-45. MUNRO, J. I., "Efficient Determination of the Transitive Closure of a Directed Graph," *Information Processing Letters*, Vol. 1, 1971, 56–58.
11-46. NORMAN, R. Z., and M. O. RABIN, "An Algorithm for a Minimal Cover of a Graph," *Proc. Amer. Math. Soc.*, Vol. 10, 1959, 315–319.
11-47. PATON, K., "An Algorithm for Finding a Fundamental Set of Cycles of a Graph," *Comm. ACM*, Vol. 12, No. 9, Sept. 1969, 514–518.
11-48. PATON, K., "An Algorithm for the Blocks and Cutnodes of a Graph," *Comm. ACM*, Vol. 14, No. 7, July 1971, 468–476.
11-49. POHL, I., "A Method for Finding Hamilton Paths and Knight's Tour," *Comm. ACM*, Vol. 10, No. 7, July 1967, 446–449.
11-50. PRABHAKER, M., "Analysis of Algorithms for Finding all Circuits of a Graph," Master's Thesis, Department of Electrical Engineering, Indian Institute of Technology, Kanpur, India, August 1972.
11-51. PRATT, T. W., and D. P. FRIEDMAN, "A Language Extension for Graph Processing and Its Formal Semantics," *Comm. ACM*, Vol. 14, No. 7, 1971, 460–467.
11-52. READ, R. C., "Teaching Graph Theory to a Computer," in *Recent Progress in Combinatorics* (W. T. Tutte, ed.), Academic Press, Inc., New York, 1969, 161–173.
11-53. READ, R. C., "Graph Theory Algorithms," in *Graph Theory and Its Applications* (B. Harris, ed.), Academic Press, Inc., New York, 1970, 51–78.
11-54. READ, R. C., "The Coding of Various Kinds of Unlabeled Trees," in *Graph Theory and Computing* (R. C. Read, ed.) Academic Press, New York, 1972, 153–182.
11-55. REINGOLD, E. M., "Establishing Lower Bounds on Algorithms—A Survey," *Proc. AFIPS*, S.J.C.C., 1972, 471–481.
11-56. ROBERTS, S. M., and B. FLORES, "Systematic Generation of Hamiltonian Circuits," *Comm. ACM*, Vol. 9, No. 9, 1966, 690–694.
11-57. ROVNER, P. D., and D. A. HENDERSON, JR., "On the Implementation of AMBIT/G: A Graphical Programming Language," *Proc. Intern. Joint Conf. on Artificial Intelligence*, 1969, 9–20.
11-58. SEPPÄNEN, J. J., "Algorithm 399: Spanning Tree," *Comm. ACM*, Vol. 13, No. 10, 1970, 621–622.

11-59. SUSSENGUTH, E. H., JR., "A Graph Theoretic Algorithm for Matching Chemical Structures," *J. Chem. Doc.*, Vol. 5, No. 1, Feb. 1965, 36–43.

11-60. TABOURIER, Y., "All Shortest Distances in a Graph. An Improvement to Dantzig's Inductive Algorithm," *Discrete Mathematics*, Vol. 4, No. 1, Jan. 1973, 83–87.

11-61. TARJAN, R., "Depth-First Search and Linear Graph Algorithms," *SIAM J. Comput.*, Vol. 1, No. 2, June 1972, 146–160.

11-62. TARJAN, R., "An Efficient Planarity Algorithm," Computer Science Department, Report No. CS-244-71, Stanford University, November 1971. (A simplified version of this report was published as Cornell University Computer Science Technical Report No. TR73-165, in April 1973 by Hopcroft and Tarjan.)

11-63. TIERNAN, J. C., "An Efficient Search Algorithm to Find the Elementary Circuits of a Graph," *Comm. ACM*, Vol. 13, No. 12, Dec. 1970, 722–726.

11-64. UNGER, S. H., "GIT—A Heuristic Program for Testing Pairs of Directed Line Graphs for Isomorphism," *Comm. ACM*, Vol. 7, 1964, 26–34.

11-65. WARSHALL, S., "A Theorem on Boolean Matrices," *J. ACM*, Vol. 9, Jan. 1962, 11–12.

11-66. WEINBERG, L., "A Simple and Efficient Algorithm for Determining Isomorphism of Planar Triply Connected Graphs," *IEEE Trans. Circuit Theory*, Vol. CT-13, No. 2, 1966, 142–148.

11-67. WEINBLATT, H., "A New Search Algorithm for Finding the Simple Cycles of a Finite Directed Graph," *J. ACM*, Vol. 19, No. 1, Jan. 1972, 43–56.

11-68. WELCH, J. T., JR., "A Mechanical Analysis of the Cyclic Structure of Undirected Linear Graphs," *J. ACM*, Vol. 13, No. 2, April 1966, 205–210.

11-69. WELLS, M. B., *Elements of Combinatorial Computing*, Pergamon Press, Inc., Elmsford, N.Y., 1971.

11-70. WHITNEY, V. K. M., "Algorithm 422: Minimal Spanning Tree," *Comm. ACM*, Vol. 15, No. 4, April 1972, 273.

11-71. WOLFBERG, M. S., "An Interactive Graph Theory System," Ph.D. Thesis, The Moore School of Electrical Engineering, University of Pennsylvania, Philadelphia, 1969. Also Moore School Report No. 69-125.

11-72. YEN, J. Y., "Finding the *K*-Shortest Loopless Paths in a Network," *Management Sci.*, Vol. 17, No. 11, 1971, 712–716.

11-73. YEN, J. Y., "Finding the Lengths of all Shortest Paths in *N*-Node Nonnegative Distance Complete Network Using $\frac{1}{2}N^3$ Additions and N^3 Comparisons," *J. ACM*, Vol. 19, July 1972, 423–424.

PROBLEMS

11-1. For studying the behavior of an algorithm, random graphs are often used. A graph is random if its edges are drawn at random from the set of all distinct pairs of vertices. Write a subroutine for generating simple random graphs of a given n and density, where density is defined as the ratio $e/n(n-1)$ for a directed graph and $2e/n(n-1)$ for an undirected graph. (*Hint:* Use an appropriate pseudo-random-number generator, and obtain the adjacency matrix.)

11-2. Write subroutines for converting the following graph representations:
(a) Adjacency matrix to incidence matrix.
(b) Incidence matrix to adjacency matrix.
(c) Adjacency matrix to edge listing.
(d) Edge listing to successor listing.
(e) Successor listing to adjacency matrix.

11-3. Write a program in the assembly language of the machine you may have to pack a given adjacency matrix $X(n, n)$. Assume that the given matrix uses full words for each entry whether it is 0 or 1.

11-4. Write a FUNCTION subprogram ADJ, which when supplied with the subscripts (i, j) gives the ijth entry of the (packed) adjacency matrix as its value. Assume that the packed adjacency matrix is in a COMMON area.

11-5. The algorithm for connectedness given in the text modifies the adjacency matrix. How will you restore it? (*Hint:* Observe that only one row of the adjacency matrix is getting changed at a time.)

11-6. Analyze to determine when it would be profitable to determine the vertex of maximum degree in each component before fusion in Algorithm 1. Assume the graph is given in the form of an adjacency matrix.

11-7. Will Algorithm 1 require any modification to find the components of a digraph? Also write a program for identifying all fragments of a digraph.

11-8. In Algorithm 3, after an edge (z, p) has been considered it is deleted from the graph; that is, in adjacency matrix X entries $x_{z,p}$ and $x_{p,z}$ are made zero. What could be done if one wanted to avoid modifying X?

11-9. While considering the edge (z, p) if we found that p was already in the tree, we went ahead to discover the fundamental circuit. Let $L = \text{LEVEL}(z) - \text{LEVEL}(p)$. Prove that the fundamental circuit will contain all and only edges on the path of length L from z to the root of the tree, apart from the two edges (z, p) and $(p, \text{PRED}(p))$.

11-10. Use the following convention in drawing the tree developed by Algorithm 3. Draw the tree downward from the root, and add vertex p at depth $\text{LEVEL}(p)$ below the root. When vertex z is being examined, the edges (z, p) are added from left to right. If z is the vertex under examination, define the trunk as the path in the tree T from z to the root. Then show that the vertices in T fall into four classes:

1. Vertex z that is being examined.
2. Vertices below z—unexamined and added during the examination of z.
3. Vertices to the "left of" the trunk—unexamined and at a distance of one from the trunk.
4. Vertices on and "to the right of" the trunk—examined.

11-11. In Algorithm 4 let f_1 be a fundamental circuit found while examining vertex z_1, and f_2 be a fundamental circuit found later while examining z_2. If $f_1 \cap f_2 \neq \emptyset$, show that the tree edge $(z_1, \text{PRED}(z_1))$ is in both f_1 and f_2. Is this why it was not necessary to label the chords?

11-12. In Algorithm 5 suggest a quick way of testing whether a vertex v, which is being considered for extension, is already in the path array P. [Getting every $p(i)$ for $i = 1, \ldots, k$, where k is the length of the path built so far, and comparing with v is obviously bad.] Give a flow chart for the algorithm incorporating this.

11-13. In Algorithm 6 suppose you suspect that the graph is disconnected and the starting and the terminal vertex may not be in the same component. The flow chart given in Fig. 11-9 works for this case also, but it is inefficient. Suggest a method of speeding up the detection in this case. (*Hint:* You will have to add a test box at an appropriate place in the flow chart, on the outcome of which you would decide whether or not to continue.)

11-14. Modify Algorithm 6 so that it lists all shortest paths from s to t.

11-15. Given the optimal-policy matrix Z as mentioned in Algorithm 7, write a subroutine that returns the shortest path $(i, \ldots, j)$ for a specified i and j, as a sequence of vertex numbers.

11-16. Write a program similar to Algorithm 6 to obtain the longest distance from a vertex s to all vertices accessible from s in a given acyclic digraph.

11-17. Algorithm 7 can also be used for detecting whether or not a given graph is connected. (An ∞ in matrix D_k represents nonexistence of a path.) As presented, the algorithm is inefficient for this purpose. Rewrite Algorithm 7 solely for the purpose of identifying various components in a graph. (*Hint:* Use X rather than D and logical operations rather than arithmetic ones.) Compare its efficiency with that of Algorithm 1.

11-18. Give an algorithm to find a cut-set with respect to a given pair of vertices a, b. Assume that the graph is given in terms of its *F-H* representation. You are allowed to scan the *F-H* arrays only twice. [*Hint:* If there is no edge (a, b), add one. Choose the edge (a, b). Shrink an edge e in G not parallel to (a, b) by fusing its end vertices. Some edges might now become parallel; continue shrinking.]

11-19. Given an edge e^*, write an efficient algorithm to determine if e^* is a bridge. (*Hint:* Edge e^* is a bridge if and only if $\{e^*\}$ is a cut-set. Use the result of Problem 11-18.)

11-20. Write a program for generating all spanning trees, based on Minty's method described in Section 11-4.

APPENDIX OF PROGRAMS

Program listings of some of the algorithms described are given in the following pages. Program 11-1 is written in APL\360; the rest are in FORTRAN IV in the form of subroutines. The variables in the subroutines are dimensioned such that a graph with no more than 100 vertices can be given as input.

Program 11-1: Connectedness and Components

```
[1]     SL←R←ιN
[2]     C←1
[3]     LOOP1: I←D⌊/D←+/X
[4]     VA←X[I;]
[5]     VA[I]←1
[6]     LOOP2: VN←~VA
[7]     VA←VAv(v/X[;VA/SL])
[8]     →(0≠+/VN∧VA)/LOOP2
[9]     VN←~VA
[10]    'COMP ';C;'; VERT: ';VA/R
[11]    →(0=+/VN)/HALT
[12]    R←VN/R
[13]    NA←VN/SL
[14]    X←X[NA;NA]
[15]    SL←ιρR
[16]    C←C+1
[17]    →LOOP1
[18]    HALT: 'HALT'
```

X	Adjacency matrix of g
N	Number of vertices in G
C	Label of a component
I	Vertex with maximum degree in g
VA	Ith row of X
VN	Logical complement of VA
R	List of vertices in g
NA	Vertices not adjacent to I
SL	Relabeled list of vertices in g

Program 11-2: Spanning Tree/Forest

```
      SUBROUTINE  SPTREE ( F , H , N , E , EDGE , C )
      INTEGER  C , E , EDGE(E) , F(E) , H(E) , VERTEX(100)   V1 , V2
      DO 4 L = 1 , N
    4 VERTEX(L) = 0
      DO 6 L = 1 , E
    6 EDGE(L) = 0
      C = 0
      M = 0
      K = 0
   10 K = K + 1
      V1 = F(K)
      I = VERTEX(V1)
      IF ( I .EQ. 0 ) GO TO 39
      V2 = H(K)
      J = VERTEX(V2)
      IF ( J .EQ. 0 ) GO TO 36
      IF ( I - J) 21, 50, 18
   18 IJI = J
      J = I
      I = IJI
   21 DO 26 L = 1 , N
      IF ( VERTEX(L) - J )  26, 23, 25
   23 VERTEX(L) = I
      GO TO 26
   25 VERTEX(L) = VERTEX(L) - 1
   26 CONTINUE
      DO 32 L = 1 , E
      IF ( EDGE(L) - J )  32, 29, 31
   29 EDGE(L) = I
      GO TO 32
   31 EDGE(L) = EDGE(L) - 1
   32 CONTINUE
      C = C - 1
      EDGE(K) = I
      GO TO 49
```

```
      36  EDGE(K) = I
          VERTEX(V2) = I
          GO TO 49
      39  V2 = H(K)
          J = VERTEX(V2)
          IF ( J .EQ. 0 ) GO TO 45
          EDGE(K) = J
          VERTEX(V1) = J
          GO TO 49
      45  C = C + 1
          EDGE(K) = C
          VERTEX(V1) = C
          VERTEX(V2) = C
      49  M = M + 1
      50  IF (M .EQ. (N - 1) .OR. K .EQ. E) RETURN
          GO TO 10
          END
```

Program 11-3: Fundamental Circuits

```
          SUBROUTINE   FCRKTS ( X , N , NULTY )
          INTEGER  CIRKIT(100) , LEVEL(100) , P , PRED(100) , PREDOP
        1          , TW(100) , X(N,N) ,Z
          NULTY = 0
          DO 5 L = 1 , N
       5  LEVEL(L) = - 1
          NROOT = 1
       7  ITW = 1
          TW(1) = NROOT
          LEVEL(NROOT) = 0
      10  IF ( ITW .EQ. 0 ) GO TO 38
          Z = TW(ITW)
          LVLSUC = LEVEL(Z) + 1
          DO 35 P = 1 , N
          IF ( X(Z,P) ) 35 , 35 , 15
      15  IF ( LEVEL(P) + 1 .NE. 0 ) GO TO 21
          TW(ITW) = P
          ITW = ITW + 1
          PRED(P) = Z
          LEVEL(P) = LVLSUC
          GO TO 33
      21  NULTY = NULTY + 1
          PREDOP = PRED(P)
          M = 1
          CIRKIT(1) = Z
          J = Z
```

```
     26  J = PRED(J)
         M = M + 1
         CIRKIT(M) = J
         IF ( J .NE. PREDOP ) GO TO 26
         M = M + 1
         CIRKIT(M) = P
         PRINT 1000 , NULTY , ( CIRKIT(J) , J = 1 , M ) , CIRKIT(1)
     33  X(Z,P) = 0
         X(P,Z) = 0
     35  CONTINUE
         ITW = ITW - 1
         GO TO 10
     38  DO 39 NROOT = NROOT , N
     39  IF ( LEVEL(NROOT) .EQ. ( - 1 ) ) GO TO 7
     40  RETURN
   1000  FORMAT ( 4H THE I 4 , 24H FUNDAMENTAL CIRCUIT IS ( 20 I 4 ) )
         END
```

Program 11-4: Shortest Distance from *s* to *t*

```
              SUBROUTINE   DYSTRA ( D , N , S , T , LABELT )
        C
        C     9999999 IS OUR INFINITY
        C
              DIMENSION  LABEL(100)
              INTEGER  D(N,N) , P , S , T , VECT(100) , Z
              DO 6 L = 1 , N
              LABEL(L) = 9999999
            6 VECT(L) = 0
              LABEL(S) = 0
              VECT(S) = 1
              I = S
           10 M = 9999999
              DO 18 J = 1 , N
              IF ( VECT(J) .EQ. 1 ) GO TO 18
              Z = D(I,J) + LABEL(I)
              IF ( Z .LT. LABEL(J) ) LABEL(J) = Z
              IF ( LABEL(J) .GT. M ) GO TO 18
              M = LABEL(J)
              P = J
           18 CONTINUE
              VECT(P) = 1
              IF ( P .EQ. T ) GO TO 23
              I = P
              GO TO 10
           23 LABELT = LABEL(T)
              RETURN
              END
```

Program 11-5: Shortest Path Between Every Vertex Pair

```
          SUBROUTINE  FLOYD ( D , N )
C
C    9999999 IS OUR INFINITY
C
          INTEGER D(N,N) , S
          DATA INFNTY / 9999999 /
          DO 12 K = 1 , N
          DO 12 I = 1 , N
          IF ( D(I,K) .EQ. INFNTY ) GO TO 12
          DO 11 J = 1 , N
          IF ( D(K,J) .EQ. INFNTY ) GO TO 11
          S = D(I,K) + D(K,J)
          IF ( S .LT. D(I,J) ) D(I,J) = S
       11 CONTINUE
       12 CONTINUE
          RETURN
          END
```

12 GRAPHS IN SWITCHING AND CODING THEORY

The emphasis in the previous chapters has been on introducing more concepts of graph theory. Some applications were given, but mainly to make the concepts clearer. In the remaining chapters we shall discuss in detail some applications.

In Section 7-5 we saw how the configuration of a switching network inside a black box could be determined with the help of graph theory. Again, in Section 8-5 a minimal cover of a graph led to the minimization of a switching function. In this chapter graph theory will be applied to study switching networks further.

Switching theory came into being with the publication of Paul Ehrenfest's paper in 1910, in which he suggested that Boolean algebra could be applied to automatic telephone exchanges. The first mathematical formulation of the behavior of a contact network (a particular type of switching network) was given by C. Shannon in 1938. Since 1938, switching theory has developed rapidly. Originally, it was intended to provide the communications engineer with tools for analysis and synthesis of large-scale relay switching networks, such as a telephone exchange. In recent years, however, the enormous growth of switching theory has been mainly motivated by its use in the design of digital computers.

Unlike the signals in a classical electrical network (say, in a radio receiver), switching network signals have only two values—designated as 0 and 1. Switching networks are designed to process and store such binary signals.

A switching network can be classified as either a combinational network or a sequential network. A *combinational switching network* is one whose output at a given time depends only on the input at that time. A *sequential switching network*, on the other hand, is one whose output at a given time is a function of the input at that time and during its entire past history. In other words, a sequential network has memory, whereas a combinational network does not. All digital systems, from the largest multimillion-dollar computer to the smallest desk calculator, are constructed from these two basic types of circuitry—combinational and sequential.

A combinational switching network can further be classified as (1) a contact network, or (2) a gate network (see [12-5], page 77). It is in the study of contact networks that graphs appear as the most natural representation of the switching network, as we shall see in the next section. Although attempts have been made, little has been accomplished by the use of graph theory in gate networks (see Section 9-3 of [1-13]). We shall therefore confine ourselves to the contact-type networks in this chapter.

12-1. CONTACT NETWORKS

A *relay contact* (or a *contact*, for brevity) can be thought of as an ordinary household switch used for controlling the light. It is a two-terminal device having two states; in the *open* state there is no conductive path between the terminals; in the *closed* state there exists a path that will allow the electric current to flow in either direction. Thus a contact is a bilateral device. Usually, a contact is represented by one of the symbols shown in Fig. 12-1.

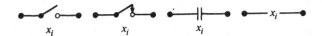

Fig. 12-1 Symbols used to represent a switch or contact.

A contact network is a network of interconnected contacts (see Fig. 12-2 for an example). Every contact network can be represented by a graph, in which the edges are the contacts and the vertices are the terminals. In fact, for our purpose, the following is the definition of a contact network: A contact network is an undirected, connected graph G (with no self-loops) in which each edge has a binary variable x_i associated with it, which can assume only two values, 1 or 0. The binary variable x_i assigned to a contact is 1 when the contact is closed and is 0 when the contact is open.

The input–output behavior of a contact network is usually expressed in the form of functions,

$$f_i(x_1, x_2, \ldots, x_k),$$

of the binary variables. Such a function f_i is called a *switching (or Boolean) function* and must itself assume a value of 0 or 1. *Boolean*† (or *switching*)

---

†Switching algebra, as defined here, is actually a special case of Boolean algebra. However, in switching theory these two terms are often used interchangeably, as it causes no confusion.

algebra, which is used in expressing and manipulating switching functions, is defined as follows:

A Boolean algebra (like rings and fields in Chapter 6) consists of a finite set $x_1, x_2, \ldots, x_k$ and two binary operations $+$ (called *Boolean addition*) and $\cdot$ (called *Boolean multiplication*) satisfying the following postulates:

1. Either $x_i = 1$ or $x_i = 0$.
2. For every x_i there exists another variable x_i', called the complement of x_i, such that if $x_i = 0$, $x_i' = 1$, and if $x_i = 1$, $x_i' = 0$.
3. (a) Sum $x_i + x_j = \begin{cases} 0, & \text{if } x_i = x_j = 0, \\ 1, & \text{otherwise;} \end{cases}$

   (b) Product $x_i x_j = \begin{cases} 1, & \text{if } x_i = x_j = 1, \\ 0, & \text{otherwise.} \end{cases}$

With these simple postulates a number of interesting results can be derived, which are very useful in the simplification of switching expressions. For example, it can be easily shown that $x_i + x_i x_j = x_i$.

In contact networks one encounters two types of problems—the problem of analysis and the problem of synthesis. In analysis we are given a contact network G and are asked to find conditions under which there will be an electrically conducting path between a pair of vertices (v_i, v_j) in G. In synthesis, on the other hand, we are asked to design (as cheaply as possible) a network that can meet the given requirements. We shall deal with the problem of analysis first and then with that of synthesis.

12-2. ANALYSIS OF CONTACT NETWORKS

Consider any two vertices in a contact network G. Since G is connected, there are one or more paths between these two vertices. Each of these paths can be identified by the Boolean product of the variables associated with the edges in the path. For example, in Fig. 12-2 the eight distinct paths between vertices a and b are

$$\begin{aligned}&(x_1 x_5), \quad (x_1 x_3' x_1), \quad (x_2 x_3 x_1), \quad (x_2 x_3 x_3' x_5), \\ &(x_2 x_1' x_1), \quad (x_2 x_1' x_3' x_5), \quad (x_3 x_4 x_1), \quad (x_3 x_4 x_3' x_5).\end{aligned} \quad (12\text{-}1)$$

Each of these products is called a *path product* between vertices a and b in the contact network G.

Clearly, the value of a path product is 1 if and only if each variable in the path product has a value of 1; otherwise, it is 0. The value 1 of a path product implies the existence of an electrically conducting path between a and b

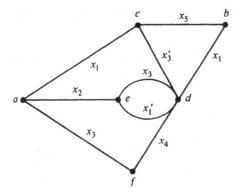

Fig. 12-2 Contact network with six vertices and nine contacts.

through the corresponding contacts in the network. For an electrical conduction between the two vertices, it is necessary and sufficient that at least one of the path products be 1. In other words, the Boolean sum of all path products between a specified pair of vertices (v_i, v_j) is 1 if and only if the terminals v_i and v_j are electrically connected in the contact network. Therefore, the Boolean sum of path products is referred to as the *transmission* of the contact network between the two specified vertices. For example, the transmission between vertices a and b in Fig. 12-2 is

$$F_{ab} = x_1 x_5 + x_1 x_3' x_1 + x_2 x_3 x_1 + x_2 x_3 x_3' x_5 + x_2 x_1' x_1 \\ + x_2 x_1' x_3' x_5 + x_3 x_4 x_1 + x_3 x_4 x_3' x_5. \quad (12\text{-}2)$$

Finding the transmission between specified vertices in a given contact network consists of enumerating all paths between the two vertices, and finding the Boolean sum of the path products. Furthermore, possible simplifications based on the postulates of the Boolean algebra are also performed. For example, in the path products listed in (12-1), the following identities are evident:

$$x_1 x_3' x_1 = x_1 x_3',$$
$$x_2 x_3 x_3' x_5 = 0,$$
$$x_2 x_1' x_1 = 0,$$

and

$$x_3 x_4 x_3' x_5 = 0.$$

Therefore, the switching function between vertices a and b in Fig. 12-2 is

$$F_{ab} = x_1 x_5 + x_1 x_3' + x_1 x_2 x_3 + x_1' x_2 x_3' x_5 + x_1 x_3 x_4. \quad (12\text{-}3)$$

Clearly, F_{ab} gives all different conditions under which a conductive path exists between a and b.

Normal Form: A switching function can be expressed in many different forms. For example, another way of expressing (12-3) is

$$F_{ab} = x_1(x_5 + x'_3 + x_3(x_2 + x_4)) + x'_1 x_2 x'_3 x_5. \qquad (12\text{-}4)$$

A Boolean function $F(x_1, x_2, \ldots, x_m)$ of m binary variables $x_1, x_2, \ldots, x_m$ when expressed as a sum of products (Boolean, of course) of the variables is said to be in the *normal* or *natural form*. Function F_{ab} in (12-3) is in normal form, but in (12-4) it is *not* in normal form.

Occasionally, one is interested in finding the transmissions between all pairs of vertices in a given contact network G. The result is best expressed as an n by n matrix called the *transmission matrix* $\mathsf{T} = [t_{ij}]$ of G, where n is the number of vertices, and t_{ij} is the transmission between vertices i and j in G. Clearly, T is a symmetric matrix with every diagonal entry $t_{ii} = 1$.

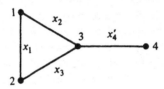

Fig. 12-3 Contact network.

The transmission matrix for the contact network shown in Fig. 12-3 is

$$\begin{array}{c} \\ 1 \\ 2 \\ 3 \\ 4 \end{array} \begin{bmatrix} 1 & 2 & 3 & 4 \\ 1 & x_1 + x_2 x_3 & x_2 + x_1 x_3 & x'_4(x_2 + x_1 x_3) \\ x_1 + x_2 x_3 & 1 & x_3 + x_1 x_2 & x'_4(x_3 + x_1 x_2) \\ x_2 + x_1 x_3 & x_3 + x_1 x_2 & 1 & x'_4 \\ x'_4(x_2 + x_1 x_3) & x'_4(x_3 + x_1 x_2) & x'_4 & 1 \end{bmatrix}.$$

The determination of a transmission matrix involves enumeration of all paths between every vertex pair in a network. A better method of determining a transmission matrix is from the primitive connection matrix, defined as follows:

The primitive connection matrix $\mathsf{Q} = [q_{ij}]$ of an n-vertex contact network G is an n by n matrix, whose elements q_{ij} are defined as

$q_{ii} = 1,$ for all i,

$q_{ij} = 0,$ if vertices i and j are not directly joined by a contact; otherwise,

$q_{ij} =$ Boolean sum of the variables associated with all edges directly joining vertices i and j.

The primitive connection matrices for the contact networks in Figs. 12-3 and

12-2, respectively, are

$$\begin{array}{c} \begin{array}{cccc} 1 & 2 & 3 & 4 \end{array} \\ \begin{array}{c} 1 \\ 2 \\ 3 \\ 4 \end{array} \left[\begin{array}{cccc} 1 & x_1 & x_2 & 0 \\ x_1 & 1 & x_3 & 0 \\ x_2 & x_3 & 1 & x_4' \\ 0 & 0 & x_4' & 1 \end{array} \right] \end{array},$$

$$\begin{array}{c} \begin{array}{cccccc} a & b & c & d & e & f \end{array} \\ \begin{array}{c} a \\ b \\ c \\ d \\ e \\ f \end{array} \left[\begin{array}{cccccc} 1 & 0 & x_1 & 0 & x_2 & x_3 \\ 0 & 1 & x_5 & x_1 & 0 & 0 \\ x_1 & x_5 & 1 & x_3' & 0 & 0 \\ 0 & x_1 & x_3' & 1 & x_1' + x_3 & x_4 \\ x_2 & 0 & 0 & x_1' + x_3 & 1 & 0 \\ x_3 & 0 & 0 & x_4 & 0 & 1 \end{array} \right] \end{array}.$$

The primitive connection matrix is also symmetric, and it contains the complete information about a contact network.

Let Q^k be the kth Boolean power of Q (i.e., Q multiplied by itself k times, using the rules of Boolean algebra, as defined in Section 12-1) for some positive integer k. Furthermore, let each entry in Q^k be simplified, such as

$$x + 1 = x + x' = 1, \qquad xx' = 0,$$

and

$$x + x = xx = x + xy = x.$$

Then examine the ijth entry in the simplified Q^k. What we have done amounts to tracing all edge sequences of length $1, 2, \ldots, k$ between the vertices i and j; and by employing the simplification process, we have eliminated all redundancies, including that of going over the same edge more than once ($xx = x$) and going over the same vertex more than once ($x + xy = x$). Thus the ijth entry in matrix Q^k represents all paths of length k or less between vertices i and j. Since in an n-vertex graph the longest path is of length $n - 1$, we have

THEOREM 12-1

The transmission matrix T of an n-vertex contact network, with primitive connection matrix Q, is given by

$$T = Q^{n-1}.$$

In case one is interested in evaluating the switching function only between a specified pair of vertices, Theorem 12-1, which computes the switching functions between all $n(n-1)/2$ pairs, is wasteful. Theorem 12-2 is more efficient:

THEOREM 12-2

Let Q_{ij} be the ijth minor of the primitive connection matrix Q (computed in Boolean algebra and simplified using Boolean identities). Then the switching function F_{ij} equals Q_{ij}.

Theorem 12-2 can be proved using arguments similar to those which led to Theorem 12-1. The details of the proof are left as an exercise.

Even the evaluation of the minor Q_{ij} is quite laborious and cumbersome. A simpler method, called the *node-removal method*, is often employed in evaluating F_{ij}. The interested reader is referred to [12-10], pages 315–323, for details on node-removal techniques.

12-3. SYNTHESIS OF CONTACT NETWORKS

Designing a network from given requirements is the general problem of network synthesis. We can assume that the requirements of the contact network to be designed are given in the form of switching functions. (If they are given in any other form, they can be converted into switching functions.) We can further assume that the switching functions are given in normal form (i.e., as a Boolean sum of products).

Two-Terminal Contact Networks: In a two-terminal synthesis we are given just one switching function $F(x_1, x_2, \ldots, x_m)$ of m variables, in normal form, and we are to design a network realizing this function as the transmission between two of its vertices.

This problem is trivial if we are not concerned with economizing the number of contacts, because any switching function in a natural form can be realized by a sufficiently large number of contacts. But such an extravagant realization is usually not acceptable. A realization to be useful should contain as few contacts as possible. It is this requirement that makes the synthesis problem difficult. In Fig. 12-4, for example, are shown three of many possible realizations of a very simple switching function (with only four variables). The simplest among the three networks is the one in Fig. 12-4(c). We may ask if this is *the* most economic realization possible. If so, how can we be sure? Are there any methods that will guarantee our arriving at a most efficient contact network for a given function?

The problem of finding a contact network that realizes an arbitrary switching function with a minimum number of contacts has not been solved yet (except by exhaustive enumeration) and is not likely to be solved in the

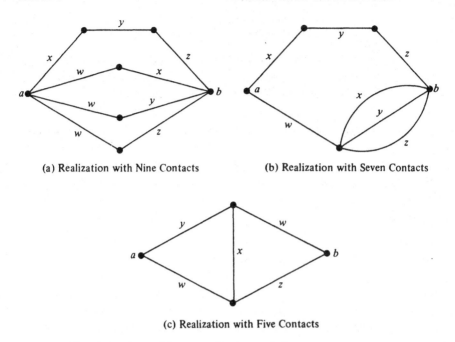

Fig. 12-4 Three different realizations of $F_{ab} = wx + wy + wz + xyz$.

near future. However, if we consider a restricted type of switching network, called single-contact networks, the problem becomes manageable.

Single-Contact Network: A contact network in which every binary variable x_i (either in uncomplemented or complemented form) is associated with only one edge is called a *single-contact (or SC) network*. Thus each contact in an SC network can be opened or closed independently. For example, the network in Fig. 12-3 is an SC network, but those in Figs. 12-2 and 12-4 are *not*.

Any transmission that can be realized by an SC network is called a *single-contact function* (or SC function). Since in an SC network a variable appears only once, it is not possible to simplify the sum of its path products any further [the type of simplification performed on expression (12-2) to produce expression (12-3)]. In other words, an SC function contains no redundant terms. Every product term represents a distinct path between the specified terminals, and every literal in a product term corresponds to a distinct edge in the path. For example, switching function $F = x_1 x_2 + x_2 x_3' + x_1' x_2 x_4$ is not an SC function.

Realization of an SC Function: Once we are assured that a given switching function F_{ab} is an SC function, we know unambiguously every path between vertices a and b in the network we intend to design. The following procedure shows how to design an SC network for an SC switching function. The network is unique up to 2-isomorphism. This is because the set of all paths between a pair of vertices specifies a graph uniquely within 2-isomorphism (see Section 4-8 and Problem 7-25).

Procedure for Realizing a Given SC Function of m Variables $x_1, x_2, \ldots, x_m$:

Step 1: From F_{ab} obtain the path matrix $P_{(a,b)}$ with respect to the vertex pair (a, b).

Step 2: Append a column of all 1's to $P_{(a,b)}$. This implies the addition of an edge (with associated variable, say, x_0) between a and b, thus converting every path into a circuit. Let the resulting circuit matrix be denoted as B.

Step 3: Use Jordan's method of elimination (mod 2) to eliminate all dependent rows in B. Rearrange the resulting fundamental circuit matrix B_f into the standard form (review Section 7-4, if necessary):

$$B_f = [I \mid B_t].$$

Step 4: From B_f obtain the fundamental cut-set matrix C_f, given by

$$C_f = [C_c \mid I_{n-1}] = [B_t^T \mid I_{n-1}].$$

Step 5: From C_f obtain A_f, the reduced incidence matrix, by appropriately performing modulo 2 sums of rows in C_f. This corresponds to obtaining a nonsingular transformation matrix R such that

$$A_f = R \cdot C_f,$$

where A_f has at most two 1's in each column. In step 5 we are essentially taking different ring sums of fundamental cut-sets so that they produce sets of edges incident at each vertex. This is the most laborious step in the entire procedure, and becomes prohibitive for large graphs (more on this later).

Step 6: Form the incidence matrix A by adding the missing row to A_f (so that each column has exactly two 1's). From matrix A draw the graph, and remove edge x_0.

Example: Let us apply this procedure to obtain a graph that realizes the SC function

$$\begin{aligned} F_{ab} = & x_1 x_2 x_3 x_5 x_7 + x_1 x_3 x_4 x_6 + x_1 x_5 x_6 x_8 + x_2 x_4 \\ & + x_2 x_3 x_5 x_8 + x_3 x_4 x_6 x_7 x_8 + x_5 x_6 x_7. \end{aligned} \quad (12\text{-}5)$$

Step 1: The path matrix is

$$P_{(a,b)} = \begin{bmatrix} x_1 & x_2 & x_3 & x_4 & x_5 & x_6 & x_7 & x_8 \\ 1 & 1 & 1 & 0 & 1 & 0 & 1 & 0 \\ 1 & 0 & 1 & 1 & 0 & 1 & 0 & 0 \\ 1 & 0 & 0 & 0 & 1 & 1 & 0 & 1 \\ 0 & 1 & 0 & 1 & 0 & 0 & 0 & 0 \\ 0 & 1 & 1 & 0 & 1 & 0 & 0 & 1 \\ 0 & 0 & 1 & 1 & 0 & 1 & 1 & 1 \\ 0 & 0 & 0 & 0 & 1 & 1 & 1 & 0 \end{bmatrix}.$$

Step 2: Appending a column of 1's at the end of matrix $P_{(a,b)}$ and identifying the column by variable x_0, we get circuit matrix B:

$$B = \begin{bmatrix} x_1 & x_2 & x_3 & x_4 & x_5 & x_6 & x_7 & x_8 & x_0 \\ 1 & 1 & 1 & 0 & 1 & 0 & 1 & 0 & 1 \\ 1 & 0 & 1 & 1 & 0 & 1 & 0 & 0 & 1 \\ 1 & 0 & 0 & 0 & 1 & 1 & 0 & 1 & 1 \\ 0 & 1 & 0 & 1 & 0 & 0 & 0 & 0 & 1 \\ 0 & 1 & 1 & 0 & 1 & 0 & 0 & 1 & 1 \\ 0 & 0 & 1 & 1 & 0 & 1 & 1 & 1 & 1 \\ 0 & 0 & 0 & 0 & 1 & 1 & 1 & 0 & 1 \end{bmatrix}.$$

Step 3: This step is somewhat involved. Jordan's method of elimination consists of adding (mod 2) rows to other rows so as to form an identity submatrix. For example, adding the first row to the second as well as to the third in B we get

$$B_1 = \begin{bmatrix} x_1 & x_2 & x_3 & x_4 & x_5 & x_6 & x_7 & x_8 & x_0 \\ 1 & 1 & 1 & 0 & 1 & 0 & 1 & 0 & 1 \\ 0 & 1 & 0 & 1 & 1 & 1 & 1 & 0 & 0 \\ 0 & 1 & 1 & 0 & 0 & 1 & 1 & 1 & 0 \\ 0 & 1 & 0 & 1 & 0 & 0 & 0 & 0 & 1 \\ 0 & 1 & 1 & 0 & 1 & 0 & 0 & 1 & 1 \\ 0 & 0 & 1 & 1 & 0 & 1 & 1 & 1 & 1 \\ 0 & 0 & 0 & 0 & 1 & 1 & 1 & 0 & 1 \end{bmatrix}.$$

In B_1 adding the second row to the fourth and sixth rows, we get

$$B_2 = \begin{array}{c} x_1\ x_2\ x_3\ x_4\ x_5\ x_6\ x_7\ x_8\ x_0 \\ \begin{bmatrix} 1 & 1 & 1 & 0 & 1 & 0 & 1 & 0 & 1 \\ 0 & 1 & 0 & 1 & 1 & 1 & 1 & 0 & 0 \\ 0 & 1 & 1 & 0 & 0 & 1 & 1 & 1 & 0 \\ 0 & 0 & 0 & 0 & 1 & 1 & 1 & 0 & 1 \\ 0 & 1 & 1 & 0 & 1 & 0 & 0 & 1 & 1 \\ 0 & 1 & 1 & 0 & 1 & 0 & 0 & 1 & 1 \\ 0 & 0 & 0 & 0 & 1 & 1 & 1 & 0 & 1 \end{bmatrix} \end{array}.$$

In this attempt to eliminate all but one 1 in a column (making sure the 1 in each column occurs in a different row), we ultimately get

$$B_3 = \begin{array}{c} x_1\ x_2\ x_3\ x_4\ x_5\ x_6\ x_7\ x_8\ x_0 \\ \begin{bmatrix} 1 & 1 & 1 & 0 & 1 & 0 & 1 & 0 & 1 \\ 0 & 1 & 0 & 1 & 0 & 0 & 0 & 0 & 1 \\ 0 & 1 & 1 & 0 & 1 & 0 & 0 & 1 & 1 \\ 0 & 0 & 0 & 0 & 1 & 1 & 1 & 0 & 1 \\ 0 & 0 & 0 & 0 & 0 & 0 & 0 & 0 & 0 \\ 0 & 0 & 0 & 0 & 0 & 0 & 0 & 0 & 0 \\ 0 & 0 & 0 & 0 & 0 & 0 & 0 & 0 & 0 \end{bmatrix} \end{array}.$$

From this matrix we get the fundamental circuit matrix in standard form:

$$B_f = \begin{array}{c} x_1\ x_4\ x_8\ x_6\ |\ x_2\ x_3\ x_5\ x_7\ x_0 \\ \begin{bmatrix} 1 & 0 & 0 & 0 & | & 1 & 1 & 1 & 1 & 1 \\ 0 & 1 & 0 & 0 & | & 1 & 0 & 0 & 0 & 1 \\ 0 & 0 & 1 & 0 & | & 1 & 1 & 1 & 0 & 1 \\ 0 & 0 & 0 & 1 & | & 0 & 0 & 1 & 1 & 1 \end{bmatrix} \end{array}$$

$$= [I_4 \mid B_t].$$

Now we have the following information about the desired network:

$$\text{rank of circuit matrix } B = \mu = e - n + 1 = 4,$$
$$\text{number of edges (including } x_0) \quad e = 9,$$

Therefore

$$\text{number of vertices } n = 6,$$
$$\text{rank of the cut-set matrix} = n - 1 = 5.$$

SEC. 12-3   SYNTHESIS OF CONTACT NETWORKS

Step 4: The fundamental cut-set matrix (with respect to the same tree as B_f in step 3) is immediately obtained as

$$C_f = [B_t^T \mid I_{n-1}] = \begin{array}{c} \\ \end{array} \begin{array}{cccc|ccccc} x_1 & x_4 & x_8 & x_6 & x_2 & x_3 & x_5 & x_7 & x_0 \\ \end{array} \\ \begin{bmatrix} 1 & 1 & 1 & 0 & 1 & 0 & 0 & 0 & 0 \\ 1 & 0 & 1 & 0 & 0 & 1 & 0 & 0 & 0 \\ 1 & 0 & 1 & 1 & 0 & 0 & 1 & 0 & 0 \\ 1 & 0 & 0 & 1 & 0 & 0 & 0 & 1 & 0 \\ 1 & 1 & 1 & 1 & 0 & 0 & 0 & 0 & 1 \end{bmatrix}.$$

Step 5: After many trials we find that if we perform the following three elementary row operations on C_f we get a matrix that contains at most two 1's per column. The operations are

Add (mod 2) row 5 to row 1,

Add (mod 2) row 3 to row 5,

Add (mod 2) row 4 to row 3.

The resulting reduced incidence matrix A_f is

$$A_f = \begin{array}{c} \end{array} \begin{array}{ccccccccc} x_1 & x_4 & x_8 & x_6 & x_2 & x_3 & x_5 & x_7 & x_0 \\ \end{array} \\ \begin{bmatrix} 0 & 0 & 0 & 1 & 1 & 0 & 0 & 0 & 1 \\ 1 & 0 & 1 & 0 & 0 & 1 & 0 & 0 & 0 \\ 0 & 0 & 1 & 0 & 0 & 0 & 1 & 1 & 0 \\ 1 & 0 & 0 & 1 & 0 & 0 & 0 & 1 & 0 \\ 0 & 1 & 0 & 0 & 0 & 0 & 1 & 0 & 1 \end{bmatrix}.$$

Step 6: We get the incidence matrix A by adding a row at the bottom such that every column now has exactly two 1's.

$$A = \begin{array}{c} \end{array} \begin{array}{ccccccccc} x_1 & x_4 & x_8 & x_6 & x_2 & x_3 & x_5 & x_7 & x_0 \\ \end{array} \\ \begin{bmatrix} 0 & 0 & 0 & 1 & 1 & 0 & 0 & 0 & 1 \\ 1 & 0 & 1 & 0 & 0 & 1 & 0 & 0 & 0 \\ 0 & 0 & 1 & 0 & 0 & 0 & 1 & 1 & 0 \\ 1 & 0 & 0 & 1 & 0 & 0 & 0 & 1 & 0 \\ 0 & 1 & 0 & 0 & 0 & 0 & 1 & 0 & 1 \\ 0 & 1 & 0 & 0 & 1 & 1 & 0 & 0 & 0 \end{bmatrix}.$$

Finally, the required contact network is constructed (see Fig. 12-5) from the incidence matrix A, and then edge x_0 is deleted.

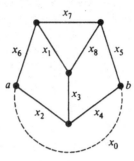

Fig. 12-5 Realization of F_{ab} in expression (12-5).

In this six-step synthesis of an SC switching function, we see that steps 1, 2, 4, and 6 are easy and require no conditions on F_{ab} for their completion. Step 3 involves some labor, but it is also guaranteed to terminate. Any matrix of rank k can be reduced by Jordan's process of elimination to one of the following forms (see [7-3]):

$$\begin{bmatrix} I_k \\ 0 \end{bmatrix}, \quad [I_k \mid 0], \quad \begin{bmatrix} I_k & 0 \\ \hline 0 & 0 \end{bmatrix}, \quad \text{or} \quad I_k.$$

Realizability: Steps 1, 2, 3, and 4 can always be performed whether or not the given switching function F is an SC function. The procedure will fail at step 5 if F is not an SC function, and we will not succeed in obtaining a reduced incidence matrix by elementary row operations. This leads us to an extremely important question in graph theory: When can a given (0, 1)-matrix be a cut-set matrix of some graph? An arbitrary (0, 1)-matrix M may or may not be a cut-set matrix. For example, the matrix

$$L = \begin{bmatrix} 1 & 0 & 1 & 1 & 1 & 0 & 0 \\ 1 & 1 & 0 & 1 & 0 & 1 & 0 \\ 1 & 1 & 1 & 0 & 0 & 0 & 1 \end{bmatrix}$$

cannot be a cut-set matrix of any graph. This can be verified by considering all seven possible (mod 2) sums of the three rows, and observing that this matrix cannot be transformed by elementary row operations into a matrix with at most two 1's per column. In other words, no incidence matrix can be found to correspond with L as a cut-set matrix.

The matrix L is unique. It can be shown that this is the smallest (0, 1)-matrix which cannot be a cut-set matrix of any graph. It is also clear that any matrix M, if it contains L as a submatrix, cannot be a cut-set matrix either.

Let us look at another facet of the situation. A cut-set matrix of a graph G is also the circuit matrix of its dual G^* if and only if G is planar. Suppose that matrix $M = [M_1 \mid I_k]$ contains a submatrix which we know to be the circuit matrix of some nonplanar graph H; then M cannot be a cut-set matrix; otherwise, we have a situation where a nonplanar subgraph has a dual, which is impossible. Thus we have a second necessary condition: if a matrix M = $[M_1 \mid I_k]$ is to be a cut-set matrix, it must not contain the circuit matrix of any nonplanar graph. From Theorem 5-9, we know that a graph is nonplanar if and only if it has as a subgraph either of the two Kuratowski graphs or any graph homeomorphic to either of them. Therefore, if M is to be a cut-set matrix, it must not contain a circuit matrix of either Kuratowski graph, or any graph homeomorphic to either of them.

It has been shown by Tutte in a remarkable paper that the two necessary conditions discussed so far are also sufficient. The proof of sufficiency is extremely long and is based on the theory of matroids. The realizability conditions for a cut-set matrix are precisely stated in Theorem 12-3 (for a proof see [12-11] or [12-14]).

THEOREM 12-3

Necessary and sufficient conditions for the (0, 1)-matrix M to be a cut-set matrix are that

1. M does not contain L or L^T as a submatrix.
2. M does not contain the circuit matrix of either Kuratowski graph, or any graph homeomorphic to either of them.

Realizability of M as a Circuit Matrix

Suppose that we want to find whether or not a matrix $M = [I_k \quad M_2]$ is the fundamental circuit matrix (rather than cut-set matrix) of some graph. The following result, the analog of Theorem 12-3 and proved by Tutte in the same paper, has the answer.

THEOREM 12-4

Necessary and sufficient conditions for the (0, 1)-matrix M, to be a circuit matrix are that

1. M does not contain L or L^T as a submatrix.
2. M does not contain the cut-set matrix of either Kuratowski graph, or any graph homeomorphic to either of them.

Note that an arbitrary (0, 1)-matrix M falls into one of four categories:

1. M is a fundamental cut-set matrix of some graph G and a fundamental circuit matrix of another graph G^* (graphs G and G^* are planar).
2. M is a fundamental cut-set matrix of some graph G, but is not a circuit matrix of any graph (G is nonplanar).

3. M is a circuit matrix of some graph G, but is not a cut-set matrix of any graph (G is nonplanar).
4. M is neither a cut-set matrix nor a circuit matrix of any graph.

12-4. SEQUENTIAL SWITCHING NETWORKS

So far, we have considered only combinational switching networks. Let us now study the *sequential switching networks* (better known as *sequential machines*†). As pointed out earlier in the chapter, the output of a sequential network depends not only on the present inputs but also on their past history. A sequential machine must, therefore, be able to retain information about the past inputs. This introduces the concept of "state" of a sequential network, where the "state" corresponds to the memory of the past inputs. Mathematically, a sequential machine is defined as follows:

A sequential machine is a mathematical system M, which consists of‡

1. A finite set $V = \{v_1, v_2, \ldots, v_n\}$ of *internal states* (or simply *states*).
2. A finite set $X = \{x_1, x_2, \ldots, x_m\}$ of inputs called the *input alphabet*.
3. A finite set $Z = \{z_1, z_2, \ldots, z_p\}$ of outputs called the *output alphabet*.
4. A function or mapping that assigns to every combination of the present state and the present input (v_i, x_j) a next state v_k. This function is called the *transition function* of M.
5. Another function, called the *output function*, assigns an output z_r to every combination (v_i, x_j) of the present state and the present input.

There are two equivalent methods of describing a sequential machine: (1) a tabular form, called the *state table*, and (2) a weighted, directed graph, called the *state graph* (or *state diagram*).

Each vertex in the state graph corresponds to a state of the sequential machine, and each directed edge represents a transition from the present state to the next. Every edge (v_i, v_j) has an ordered pair of weights x_k, z_q assigned to it. This weight pair represents the fact that if the present state of the machine is v_i and if the present input is x_k an output z_q results, and the next state will be v_j. The state table and the state graph of a sequential machine

†Sequential switching networks are also called *sequential networks*, *sequential machines*, *sequential nets*, or *sequential circuits*. The terms *finite-state machines* and *automata* are also used for sequential switching networks. The form sequential machine is perhaps the most commonly employed term and we shall use this term.

‡This definition of a sequential machine is somewhat restricted. It is the *Mealy model* of a *deterministic* and *completely specified* sequential machine.

with

$$\text{states} \quad V = \{A, B, C, D\},$$
$$\text{inputs} \quad X = \{1, 2\},$$
$$\text{outputs} \quad Z = \{a, b, c\}$$

are shown in Fig. 12-6. In the state graph the edge with weight pair from vertex A to B, for example, indicates that when the machine is in state A and the input 1 is applied the machine produces an output a and will go into state B.

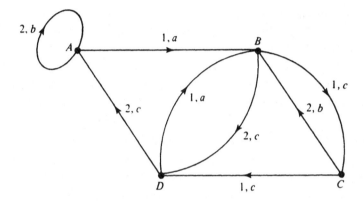

Present state	Next state/output		Inputs
	1	2	
A	B/a	A/b	
B	C/c	D/c	
C	D/c	B/b	
D	B/a	A/c	

Fig. 12-6 State graph and state table for a sequential machine.

Properties of State Graphs

The following observations can be made about the properties of state graphs:

1. In response to each specified input the machine in a given present state goes into a specific next state. Therefore, the out-degree of each vertex is m, one for each input; and the state graph has nm edges. Note that there is no similar restriction on the in-degrees.

2. Since an input may leave a sequential machine in its present state, self-loops may occur in a state graph.

3. A state graph may also have parallel edges, but they will have different weight pairs.

4. In most cases one of the states of a sequential machine is designated as a *starting state*, and the machine is required to be in this state before any input is applied. The state graph of a machine with a designated starting state is regarded as a rooted digraph, the root being the starting state.

5. A state (if any) that the machine cannot leave, no matter which input is applied, is called a *persistent state*. If a sequential machine has a persistent state, the corresponding vertex will have no directed edge going from it to another vertex.

6. A sequential machine is said to be *strongly connected* if its state graph is strongly connected. Thus a sequential machine M is strongly connected if and only if M can be brought to any state from any other state by an appropriate input sequence.

The state graph of a sequential machine contains all the information about the machine. Therefore, it is possible to study the properties of a given machine by studying its state graph. Some of the problems that arise in the theory of sequential machines are

1. *Analysis:* In analyzing the behavior of a machine, we may, for instance, be interested in determining the response (next states and outputs) of a given machine to a certain input sequence. Or we may be interested in drawing some conclusion about the internal behavior of a machine by applying a series of inputs and observing the outputs. If a machine has a designated starting state, the application of a given input sequence results in a unique output sequence.

2. *Synthesis:* To design a machine having a desired behavior, we start with the statement of the desired response and construct a state graph. Consider the following example:

Problem: Design a sequential machine to respond to an arbitrary input sequence of 0's and 1's. The machine should produce an output of 1 whenever there appears a set of four consecutive input bits of value greater than 9 in a serial 8-4-2-1 BCD code (the least significant bit comes to the machine first). Whenever the value of a four-bit sequence is 9 or less (i.e., 0000, 0001, 0010, ..., 1001), the output should be 0.

Solution: The machine should store the last three consecutive bits and should examine and respond to the next bit. Therefore, we should start with an eight-state ($2^3 = 8$) sequential machine. Let the eight states 000, 001, 010,

011, 100, 101, 110, and 111 be designated by A, B, C, D, E, F, G, and H, respectively. The input alphabet consists of $\{0, 1\}$, and the output alphabet also consists of $\{0, 1\}$.

When a new bit arrives at the left, the machine drops the rightmost bit and stores the new bit together with the two old ones. For example, if the machine is in state C (i.e., 010) and a 1 arrives, the next state is 101 (i.e., F) and the output is 1 (corresponding to 1010). The state table and the state graph of such a sequential machine can be easily constructed and are shown in Fig. 12-7.

Present state	Next State/Output	
	0	1
A	$A/0$	$E/0$
B	$A/0$	$E/0$
C	$B/0$	$F/1$
D	$B/0$	$F/1$
E	$C/0$	$G/1$
F	$C/0$	$G/1$
G	$D/0$	$H/1$
H	$D/0$	$H/1$

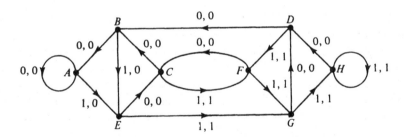

Fig. 12-7 State table and state graph.

3. *State equivalence and reduction:* The eight-state machine we just obtained is not necessarily the "simplest" one to perform the specified task. The next step, and a very important step, is to examine the state graph and see if it can be reduced to a "simpler" machine. The reduction can be accomplished if we can determine whether or not two states in a given machine (i.e.,

a pair of vertices in the state graph) are *equivalent*. If two states produce the same outputs and also go to a pair of equivalent next states for every input, they can be considered as one state and given the same label wherever they occur in the state table.

In the state graphs we can fuse the two equivalent vertices and remove any redundant edges (parallel edges with identical weights) that may result from the fusion. In Fig. 12-7 vertices A and B are equivalent, and therefore they can be fused. This fusing results in the seven-state machine shown in Fig. 12-8(a).

The process of reduction is shown in Fig. 12-8. When completed it yields the state graph of Fig. 12-8(e). State A in Fig. 12-8(e) is the replacement of A and B in the original state graph, state C is for C and D, and state E is for E, F, G, and H.

The three-state sequential machine in Fig. 12-8(e) performs the same task as the original eight-state machine in Fig. 12-7 did. This simple example illustrates the importance of the state-reduction process.

4. *State assignment:* The next step is the implementation of a sequential machine from the reduced state graph. Assuming that binary memory devices (i.e., two-state devices such as flip-flops or toggle switches) are used, an n-state machine will require q such devices, where

$$2^{q-1} < n \leq 2^q.$$

The q binary memory devices allow 2^q possible states. How to assign n of these 2^q states to the n vertices of the state graph such that we get the most economical machine is the problem of state assignment.

In graph theoretic terms the state assignment problem is the same as that of labeling the vertices of an n-vertex digraph with available 2^q ($\geq n$) labels, with certain optimizing criteria.

Finding an efficient algorithm to obtain the "best" assignment is an important unsolved problem in the theory of sequential machines. Listing all possible assignments and then picking out the best is impractical even for machines with 10 states. However, for a very small machine, such as the three-state machine in Fig. 12-8(e), it is possible to look at all distinct assignments and compare them. For $n = 3$ and $q = 2$, the number of distinct assignments is 3. The following table shows three distinct assignments (y_1 and y_2 are the two memory devices).

States	Assignment I		Assignment II		Assignment III	
	y_1	y_2	y_1	y_2	y_1	y_2
A	0	0	0	0	0	0
C	0	1	0	1	1	1
E	1	0	1	1	1	0

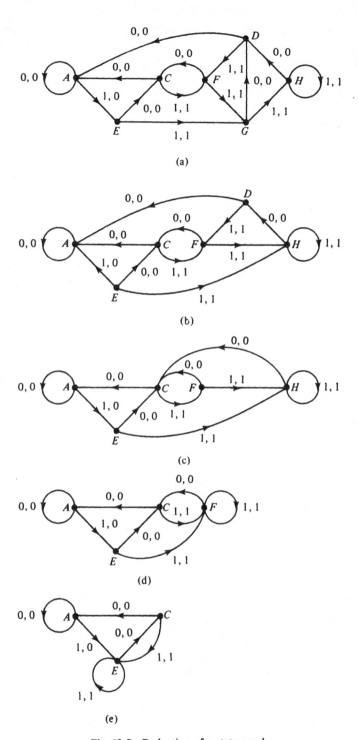

Fig. 12-8 Reduction of a state graph.

If you are familar with logical devices, design these machines completely, using flip-flops (or delay lines) and gates. You will notice that one assignment is decidedly superior to the other two.

There are a number of very important but difficult problems in sequential machine theory. Graph theory may have potential for solving many of these outstanding problems.

12-5. UNIT CUBE AND ITS GRAPH

Consider a set of m switching variables $x_1, x_2, \ldots, x_m$. Each x_i can take a value of either 0 or 1. Therefore, we can form 2^m distinct m-tuples. Each of these m-tuples can be represented by a vertex of the m-dimensional unit cube. Unit cubes for $m = 1, 2,$ and 3 are shown in Fig. 12-9. The extension to $m \geq 4$, although geometrically difficult, is simple enough to visualize.

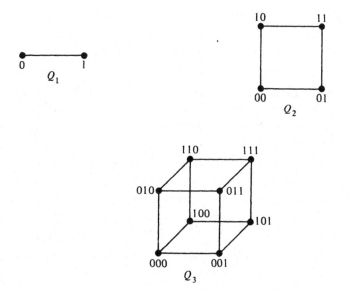

Fig. 12-9 One-, two-, and three-dimensional cubes.

The edges and vertices of an m-dimensional unit cube form a graph with 2^m vertices. Each vertex is labeled as a distinct binary sequence of m bits such that two vertices are adjacent if and only if they (i.e., their labels) differ in exactly one bit. Such a graph is called an *m-cube* and will be designated by Q_m. Once again, how we draw the m-cube is immaterial as long as we preserve the adjacency relationships of its vertices. For example, Q_3 is drawn in another way in Fig. 12-10(a). The 4-cube is sketched in Fig. 12-10(b). The m-cube is of interest in studying switching functions of m binary variables. The state-

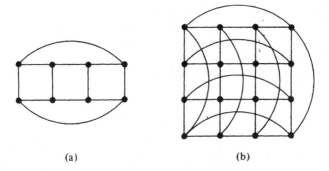

Fig. 12-10 Graphs of 3-cube and 4-cube.

assignment problem, discussed in the last section, can be looked upon as a problem of selecting and labeling vertices of an m-cube. See Chapter 13 of [12-2].

Some observations that can be made about the properties of an m-cube, Q_m, are

1. There are exactly m distinct labels that differ from a given label (of m bits) in one position. Therefore, each vertex in Q_m is of degree m. Thus Q_m is a regular graph of $n = 2^m$ vertices and $e = m \cdot 2^{m-1}$ edges.

2. The distance $\delta(v_i, v_j)$ (i.e., the number of edges in a shortest path) between two vertices v_i and v_j in an m-cube is equal to the number of positions in which the labels of v_i and v_j differ. For example, in Q_3 in Fig. 12-9 the distance between (011) and (101) is 2. This distance is known as the *Hamming distance* between the two binary words. It is easy to see that

   $\delta(v_i, v_j) = $ number of 1's in mod 2 vector sum of the labels of v_i and v_j.

3. The maximum distance possible between two vertices in an m-cube is m, because two m-bit sequences can differ at most in m positions.

Subcubes: A k-dimensional cube can be looked upon as a subcube of higher-dimensional cubes. Similarly, graph Q_k may be regarded as a subgraph of Q_m ($k \leq m$) such that Q_k consists of the 2^k vertices (of Q_m), whose labels have identical $m - k$ corresponding bits. For example, the vertices (011), (001), (111), and (101) in Fig. 12-9 have the same last bit, and constitute a subcube Q_2 in Q_3. Each vertex is a 0-cube, and any edge is a 1-cube.

Minterms: A Boolean product containing each of m variables $x_1, x_2, \ldots, x_m$ exactly once, either complemented or uncomplemented, is called a *minterm* (or *canonic product*) of m variables. For example, the minterms of three variables $= a, b, c$ are $(a'b'c')$, $(a'b'c)$, $(a'bc')$, $(a'bc)$, $(ab'c')$, $(ab'c)$, (abc'), and (abc).

There are 2^m distinct minterms of m variables, and they can be put into a one-to-one correspondence with the vertices of an m-cube. The minterm $(x'_1 x'_2 \ldots x'_m)$ corresponds to $(0\,0 \ldots 0)$ vertex, $(x'_1 x'_2 \ldots x'_{m-1} x_m)$ corresponds to $(0\,0 \ldots 0\,1)$ vertex, and so on; finally, the minterm $(x_1 x_2 \ldots x_m)$ corresponds to vertex $(1\,1 \ldots 1)$ of Q_m.

Switching Functions on the m-Cube: Any switching function $f(x_1, x_2, \ldots, x_m)$ of m variables can be expressed uniquely as a Boolean sum of a subset of 2^m minterms. This is termed as the *canonic form* of f. Clearly, the function f is 1 at those and only those vertices whose corresponding minterms are present in the canonic form of f. At all other vertices the function f is 0. The vertices† of Q_m at which f is 1 are called *true vertices* with respect to function f, and the vertices at which f is 0 are called the *false vertices* of Q_m with respect to function f. For example, consider the following function of three variables:

$$f(x_1, x_2, x_3) = x'_1 x'_2 x'_3 + x_1 x'_2 x'_3 + x'_1 x_2 x'_3 + x'_1 x_2 x_3.$$

The true vertices for this function on Q_3 are shown encircled in Fig. 12-11.

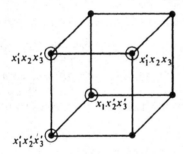

Fig. 12-11 True vertices on a Q_3 for a given function.

Thus every switching function of m variables uniquely partitions the vertices of the graph Q_m into two sets, one consisting of the true vertices and the other consisting of the false vertices. There are 2^{2^m} such partitions,‡ each corresponding to a distinct switching function of m variables. Thus the properties of switching functions can be determined by studying the properties of the subgraph of Q_m defined by the true vertices with respect to the given function.

---

†Those familiar with Karnaugh map will recognize that a vertex in Q_m corresponds to a square in Karnaugh map.

‡This includes two extreme cases when all 2^m vertices are true (i.e., $f = 1$), and all vertices are false (i.e., $f = 0$). Usually, partitions do not have empty subsets, but here we have called these two cases also partitions.

12-6. GRAPHS IN CODING THEORY

Gray Codes: Often, when information is converted from analog form to its digital equivalent, one requires a list of distinct binary m-tuples such that each differs from the one preceding it in just one coordinate. For example, to determine the angular position of a rotating shaft, the angles in adjacent quantum intervals are encoded into m-tuples (using m brushes on a commutator) of binary digits that differ in just one place. Taking $m = 3$, for instance, as the angle increases from 0 to 360°, the binary code for angles might go through the succession

$$
\begin{array}{lll}
000 & \text{for} & 0\text{--}45°, \\
001 & \text{for} & 45\text{--}90°, \\
011 & \text{for} & 90\text{--}135°, \\
010 & \text{for} & 135\text{--}180°, \\
110 & \text{for} & 180\text{--}225°, \\
111 & \text{for} & 225\text{--}270°, \\
101 & \text{for} & 270\text{--}315°, \\
100 & \text{for} & 315\text{--}360°,
\end{array}
$$

and back to

$$000 \quad \text{for} \quad 0\text{--}45°.$$

Such a code, which requires the changing of only one bit at a time, is called the *Gray* code, the *reflected binary* code, *circuit* code, or *cyclic* code. In contrast to the Gray code, other codes may require changing of several bits when going from one number to the next higher number. For example, going from 7 to 8 in 8-4-2-1 BCD (i.e., from 0111 to 1000) involves a change in all four bits simultaneously. Because of variations in the construction of the equipment, such multiple changes may not register simultaneously. Thus, during the change, false code combinations are supplied. Such false code words are eliminated in a Gray code, and this is why Gray codes are so important in analog-to-digital conversion of information.

An m-bit Gray code corresponds to a circuit in an m-cube. For instance, the 3-bit Gray code just illustrated for measuring the angular position of the rotating shaft is defined by the Hamiltonian circuit in Q_3, in Fig. 12-12 shown in heavy lines. The reason for the term cyclic or circuit code should be clear from this figure.

An m-bit code that uses all 2^m vertices is called a *complete code*. A circuit code need not be a complete code. For example, when 4-bit words are used to represent decimal digits, we use only 10 out of 16 vertices.

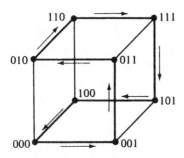

Fig. 12-12 Gray code on Q_3.

Snake-in-the-Box Codes: In selecting an incomplete code from 2^m available words, one would like to select a code that has certain error-checking properties. One such code has the desirable property that a single binary error (caused by malfunctioning of the equipment) in a word results in either (1) the next word, (2) the preceding word, or (3) a word that does not appear in the code at all. The last case indicates a *detected* error, and the first two cases introduce errors of relatively small magnitude. Such a code is called a *snake-in-the-box (SIB) code*, or *unit-distance error-checking code*.

An SIB code corresponds to a circuit in Q_m such that no two nonsuccessive vertices on the circuit are adjacent. A 6-word, SIB code in Q_3 is shown in Fig. 12-13.

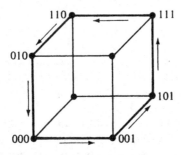

Fig. 12-13 Snake-in-the-box code on Q_3.

The SIB codes can be generalized to codes with additional error-checking properties, as follows: In graph Q_m, a circuit C_s is said to be of *spread s* if a person going around C_s cannot find a shortcut (i.e., a path with no edge from C_s) between two vertices of C_s consisting of fewer than s edges of Q_m. With this definition, every circuit in Q_m is of spread 1, and an SIB code corresponds to a circuit of spread 2.

For a given m and a specified s, one would like to find as large a circuit C_s as possible. At present no relationship is known that gives the size of the largest C_s in a Q_m for arbitrary m and s. For a survey of such problems on codes in Q_m, the reader is referred to a paper by Klee [12-8].

Huffman Graph-Theoretic Codes: We shall now briefly discuss the application of graphs to an entirely different type of coding. A *binary group code* is a set of binary code words with the property that the modulo 2 sum of any two code words in the set is also a code word in the set.† Binary group codes are of importance in information transmission, both for analytic and practical reasons. The group structure facilitates their mathematical study as well as their implementation. For more details on group codes see [12-12].

Since the ring sum of two cut-sets in a graph is another cut-set or an edge-disjoint union of cut-sets, it is evident that the set of all cut-sets and edge-disjoint union of cut-sets can be used to define a binary group code. In other words, the vectors (2^r of them, r being the rank of the graph) in the cut-set subspace W_S, over $GF(2)$, constitute a binary group code.

The rows of a fundamental cut-set matrix can be used to generate this binary group code. Such a code is called a *Huffman graph-theoretic* code.

For example, consider a graph and its fundamental cut-set matrix C_f in Fig. 12-14. The rows of C_f, their modulo 2 sums, and the zero vector yield the 5-bit, 8-word code shown in Fig. 12-14.

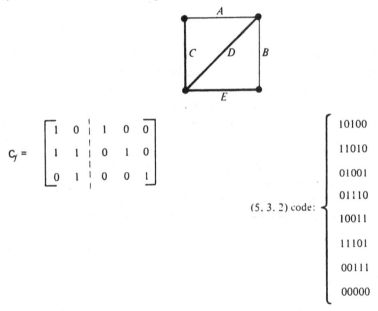

Fig. 12-14 Graph and its cut-set code.

Analogously, the fundamental circuit matrix of a graph also generates a binary group code. Thus we have two graph-theoretic codes associated with every graph.

†Note that we are and have been discussing only binary codes and only those binary codes in which each code word consists of the same number of bits. Such a code is called a *uniform binary code* or a *binary block code*. Gray codes and binary group codes are examples of binary block codes.

A graph-theoretic code is generally specified by three numbers—the number of edges e in the graph, the dimension of the associated subspace, W_s or W_r (i.e., rank r or nullity μ), and the smallest number of 1's in a nonzero code word. Thus the graph-theoretic code generated by the cut-sets of the graph in Fig. 12-14 is a (5, 3, 2) code.

Now that we know how to generate a code (in fact two codes) from any graph, we can investigate codes corresponding to important kinds of graphs—such as complete graphs, bipartite graphs, regular graphs, and planar graphs. Conversely, we can look for graphs that generate group codes with certain specified properties, such as efficiency and error-correcting capability. This is an area of current research. Some relationships between the properties of graphs and the properties of the associated codes have been investigated by Huffman [12-7], Frazer [12-3], Hakimi and Bredeson [12-4], and Saltzer [12-13].

SUMMARY

In this chapter graph theory was applied to switching circuits, automata theory, and coding theory. The applicability of graphs to digital systems and signals is not surprising, because both operate in GF(2).

For lack of space, only selected applications in switching theory were discussed. Many related topics, such as the study of series-parallel contact networks, planar and nonplanar contact networks, and regular expressions, were not even mentioned. Several other topics, such as the graphs of gate-type networks, generalized SIB codes, and properties of Huffman codes, were also left out. These are some of the areas of current research in switching theory. It is hoped that the serious reader will go to the references cited for a fuller account of this fascinating application of graph theory.

REFERENCES

Certain familiarity with switching theory was assumed in this chapter. For introductory switching theory, Caldwell [12-2], one of the earliest books in switching theory, is still one of the best, and Chapters 5, 8, 10, 12, and 13 are particularly relevant to the subject of this chapter. For a more abstract and formal treatment of switching theory, Harrison [12-5] is recommended. Chapter 5 of Miller [12-10] is excellent for graph-theoretic treatment of contact networks. Chapter 10 in Hill and Peterson [12-6] is good for understanding the problems in synthesis of sequential circuits. For coding theory, the classic book of Peterson [12-12] is recommended. Birkhoff and Bartee [12-1] may be read for an appreciation of why graph theory should be so readily applicable to switching theory and coding. Other sources referred to in the text are included in the following list of references.

12-1. BIRKHOFF, G., and T. C. BARTEE, *Modern Applied Algebra*, McGraw-Hill Book Company, New York, 1968.

12-2. CALDWELL, S. H., *Switching Circuits and Logical Design*, John Wiley & Sons, Inc., New York, 1958.
12-3. FRAZER, W. D., "A Graph-Theoretic Approach to Linear Codes," *Proc. Second Annual Allerton Conf. on Circuit and System Theory*, 1964, 888–898.
12-4. HAKIMI, S. L., and J. G. BREDESON, "Graph Theoretic Error-Correcting Codes," *IEEE Trans. Inform. Theory*, Vol. IT-14, No. 4, July 1968, 584–591.
12-5. HARRISON, M. A., *Introduction to Switching and Automata Theory*, McGraw-Hill Book Company, New York, 1965.
12-6. HILL, F. J., and G. R. PETERSON, *Introduction to Switching Theory and Logical Design*, John Wiley & Sons, Inc., New York, 1968.
12-7. HUFFMAN, D. A., "A Graph-Theoretic Formulation of Binary Group Codes," summaries of papers presented at 1964 ICMCI, pt. 3, 29–30.
12-8. KLEE, V., "The Use of Circuit Codes in Analog-to-Digital Conversion," in *Graph Theory and Its Applications* (B. Harris, ed.), Academic Press, Inc., New York, 1970, 121–131.
12-9. MAYEDA, W., "Synthesis of Switching Functions by Linear Graph Theory," *IBM J. Res. Develop.*, Vol. 4, July 1960, 320–328.
12-10. MILLER, R. E., *Switching Theory. Volume I: Combinational Circuits*, John Wiley & Sons, Inc., New York, 1965.
12-11. MINTY, G. J., "On the Axiomatic Foundations of the Theories of Directed Linear Graphs, Electrical Networks, and Network Programming," *J. Math. Mech.*, Vol. 15, 1966, 485–520.
12-12. PETERSON, W. W., *Error Correcting Codes*, The M.I.T. Press, Cambridge, Mass., 1961.
12-13. SALTZER, C., "Topological Codes," in *Error Correcting Codes* (H. B. Mann, ed.), John Wiley & Sons, Inc., New York, 1968.
12-14. TUTTE, W. T., *Introduction to the Theory of Matroids*, American Elsevier Publishing Company, Inc., New York, 1971.
12-15. WELSH, D. J. A., "Matroids and Their Applications," Seminar Notes, University of Michigan (to appear).

13 ELECTRICAL NETWORK ANALYSIS BY GRAPH THEORY

One of the reasons for the recent revival of interest in graph theory among students of electrical engineering is the application of graph theory to the analysis and design of electrical networks (more commonly known as electrical circuits). The idea of using graph theory for predicting the behavior of an electrical network is not new. It originated with G. Kirchhoff in 1847 and was improved upon by J. C. Maxwell in 1892. However, for hand computations (which were necessarily limited to small networks), the application of graph theory to network analysis offered little real advantage over the more elementary methods of node or loop analysis.

The picture has changed and is changing since the arrival of the high-speed digital computer. A milestone in graph-theoretic analysis of electrical networks was achieved by W. S. Percival, when he extended the Kirchhoff and Maxwell methods to networks with active elements. Computer programs are now available for analysis of large networks [13-2] based on the graph-theoretic approach. More efficient and less user-oriented computer programs for analyzing larger and more general types of networks are in the offing. In this chapter we shall present the underlying principle of graph-theoretic analysis of networks—which is how to use spanning trees (or chord sets) for evaluating determinants of a matrix.

Reminder on Terminology: Different disciplines using graph theory have developed somewhat different terminology. In electrical engineering, the term branch is used for edge, node for vertex, and loop for circuit. An electrical network is more commonly known as an electrical circuit. For the sake of consistency, however, the same graph theory terminology has been used throughout this book.

13-1. WHAT IS AN ELECTRICAL NETWORK?

An electrical network is a collection of interconnected electrical elements (or devices) such as resistors, capacitors, inductors, diodes, transistors, vacuum tubes, switches, storage batteries, transformers, delay lines, power sources, and the like. The behavior (such as the response to a unit impulse) of an electrical network is a function of two factors: (1) the characteristics of each of the electrical elements, and (2) how they are connected together, that is, their topology. It is the latter factor that brings graph theory into the picture.

An electrical element can be

1. Lumped or distributed.
2. One-port (i.e., two-terminal) or multiport.
3. Linear or nonlinear.
4. Time invariant or time varying.
5. Passive or active.
6. Bilateral or nonbilateral.

To avoid using partial differential equations, a distributed element, such as a transmission line, is either approximated by lumped elements or is considered separately. Thus an electrical network almost by definition implies a network consisting of lumped elements only. Also, a multiport device such as a transformer or a pentode can be replaced by a set of interconnected two-terminal elements, such as resistors, inductors, and dependent power sources (see Fig. 13-7). Thus we can confine ourselves to a network of lumped, two-terminal elements.

A two-terminal electrical element is represented by an edge e_k. Associated with each edge are two *edge variables*, $v_k(t)$ and $i_k(t)$. The variable $v_k(t)$ is called the *edge voltage* and may be regarded as a *cross variable*, because it exists across the two end vertices of the edge. The other variable $i_k(t)$ is called the *edge current* and may be thought of as a *through variable*, because it flows through the edge. Since the variables are directional, every edge is assigned an arbitrary orientation (see Fig. 13-1). The characteristics of each element are completely described in terms of these two variables. (The physics of an electrical element and its mathematical description form another subject in electrical engineering and are of little concern to us here.)

Thus an electrical network for us is a connected directed graph G in which each edge e_k is assigned two variables $v_k(t)$ and $i_k(t)$. The edge variables of each edge satisfy a relationship imposed by the nature of the corresponding element. Let the directed graph G have n vertices $1, 2, 3, \ldots, n$ and e edges

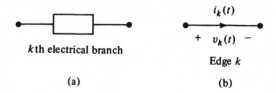

Fig. 13-1 Electrical element and its representation as an edge of a directed graph (the voltage + is always at tail of current arrow).

$b_1, b_2, \ldots, b_e$. Let the values of currents flowing through these edges at a given time be represented by a column vector (called the *edge-current vector*) i(t), where

$$\mathbf{i}(t) = \begin{bmatrix} i_1(t) \\ i_2(t) \\ \cdot \\ \cdot \\ \cdot \\ i_e(t) \end{bmatrix}.$$

Similarly, the edge voltages across the e edges are represented by another vector (called the *edge-voltage vector*) v(t), where

$$\mathbf{v}(t) = \begin{bmatrix} v_1(t) \\ v_2(t) \\ \cdot \\ \cdot \\ \cdot \\ v_e(t) \end{bmatrix}.$$

13-2. KIRCHHOFF'S CURRENT AND VOLTAGE LAWS

It was mentioned that each element in an electrical network is governed by a specific relationship imposed upon its two edge variables. When the elements are interconnected to form a network, is there any additional relationship imposed on these edge variables collectively? The answer, as every electrical engineer knows, is yes. The edge variables must also obey the two laws of Kirchhoff's:

Kirchhoff's Current Law (*KCL*): For any lumped electrical network, at any time the net sum (taking into account the orientations) of all the currents leaving any node (or vertex) is zero. That is, at the rth vertex of the corresponding directed graph G, we must have

$$\sum_{k=1}^{e} a_{rk} i_k(t) = 0, \qquad (13\text{-}1)$$

where a_{rk} is the rkth entry in the incidence matrix A of G, and $i_k(t)$ is the amount of current flowing through the kth edge of G. Since Eq. (13-1) holds simultaneously for $r = 1, 2, \ldots, n$, it can also be written in the matrix form

$$\text{A}\mathbf{i}(t) = 0. \tag{13-2}$$

Kirchhoff's Voltage Law (KVL): For any lumped electrical network, at any time the net sum (taking into account the orientations) of the voltages around a loop (i.e., circuit) is zero. In terms of the corresponding digraph, for the rth circuit we must have

$$\sum_{k=1}^{e} b_{rk} v_k(t) = 0, \tag{13-3}$$

where b_{rk} is the rkth entry in the circuit matrix B of G, and $v_k(t)$ is the amount of voltage across the kth edge. Since Eq. (13-3) holds simultaneously for every circuit in G, it can be represented in the matrix form as

$$\text{B}\mathbf{v}(t) = 0. \tag{13-4}$$

13-3. LOOP CURRENTS AND NODE VOLTAGES

Consider the vector space W_G (over the field of real numbers) associated with the directed graph G. Here G is a connected directed graph of e edges and n vertices, representing an electrical network. From Eq. (13-2), we see that the edge-current vector $\mathbf{i}(t)$ is orthogonal to each of the row vectors in the incidence matrix A. Since the row vectors in A span the entire cut-set subspace W_S (of dimension $n - 1$), $\mathbf{i}(t)$ is orthogonal to W_S. Therefore, $\mathbf{i}(t)$ lies in the circuit subspace W_Γ (of dimension $\mu = e - n + 1$) of G.

Since $\mathbf{i}(t)$ is contained in W_Γ, there must be a set of μ vectors in W_Γ whose linear combination will produce $\mathbf{i}(t)$. An obvious choice for this set of μ linearly independent vectors in W_Γ is the rows of the fundamental circuit matrix B_f with respect to some spanning tree. (Clearly, B_f is contained in B.) Let the coordinates (or coefficients) of $\mathbf{i}(t)$ in this basis formed by the rows $\mathbf{b}_1, \mathbf{b}_2, \ldots, \mathbf{b}_\mu$ of B_f be $i_{L1}(t), i_{L2}(t), \ldots, i_{L\mu}(t)$. In other words,

$$\mathbf{i}(t) = \begin{bmatrix} i_1(t) \\ i_2(t) \\ \vdots \\ i_e(t) \end{bmatrix} = [\mathbf{b}_1^T \mathbf{b}_2^T \ldots \mathbf{b}_\mu^T] \cdot \begin{bmatrix} i_{L1}(t) \\ i_{L2}(t) \\ \vdots \\ i_{L\mu}(t) \end{bmatrix} \tag{13-5}$$

or

$$\mathbf{i}(t) = B_f^T \mathbf{i}_L(t).$$

Thus each of the e edge currents can be expressed as a linear combination of μ quantities $i_{L1}(t), i_{L2}(t), \ldots, i_{L\mu}(t)$. These are called *loop currents* (or *mesh currents*); they represent current flowing in the μ independent circuits corresponding to the rows of B_f.

Substituting Eq. (13-5) into Eq. (13-2), we get

$$(AB_f^T)i_L(t) = 0. \tag{13-6}$$

Similarly, from Eq. (13-4) we see that the column vector $v(t)$ representing the edge voltages is orthogonal to the circuit subspace W_r and is, therefore, in cut-set subspace W_s. Thus $v(t)$ can be expressed as a linear combination of the $n-1$ rows of the reduced incidence matrix A_f. That is,

$$v(t) = \begin{bmatrix} v_1(t) \\ v_2(t) \\ \cdot \\ \cdot \\ \cdot \\ v_e(t) \end{bmatrix} = [a_1^T a_2^T \ldots a_{n-1}^T] \cdot \begin{bmatrix} v_{N1}(t) \\ v_{N2}(t) \\ \cdot \\ \cdot \\ \cdot \\ v_{N(n-1)}(t) \end{bmatrix}.$$

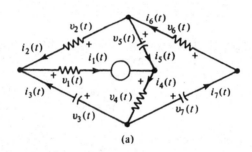

(a)

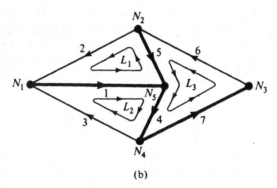

(b)

Fig. 13-2 Electrical network and its graph.

SEC. 13-3 LOOP CURRENTS AND NODE VOLTAGES

That is,

$$\mathbf{v}(t) = \mathbf{A}_f^T \mathbf{v}_N(t). \tag{13-7}$$

Thus each of e edge voltages can be expressed as a linear combination of $n-1$ quantities $v_{N_1}(t), v_{N_2}(t), \ldots, v_{N(n-1)}(t)$. These are called *node voltages*, and they represent the voltage at each of $n-1$ independent vertices with respect to the reference vertex.

Substituting Eq. (13-7) into Eq. (13-4), we get

$$(\mathbf{B}\mathbf{A}_f^T)\mathbf{v}_N(t) = \mathbf{0}. \tag{13-8}$$

Let us now illustrate with an example the loop currents and node voltages and how they are obtained from the edge currents and edge voltages, respectively. Figure 13-2(a) shows an electrical network with five vertices and seven edges. The corresponding directed graph is shown in Fig. 13-2(b). For this graph the reduced incidence matrix $\mathbf{A}_f$ with respect to vertex N_5 and the fundamental circuit matrix $\mathbf{B}_f$, with respect to the spanning tree $\{1, 4, 5, 7\}$ (shown in heavy lines), are

$$\mathbf{A}_f = \begin{array}{c} \\ N_1 \\ N_2 \\ N_3 \\ N_4 \end{array} \begin{array}{cccccccc} 1 & 2 & 3 & 4 & 5 & 6 & 7 \\ \left[\begin{array}{ccccccc} 1 & -1 & -1 & 0 & 0 & 0 & 0 \\ 0 & 1 & 0 & 0 & 1 & -1 & 0 \\ 0 & 0 & 0 & 0 & 0 & 1 & -1 \\ 0 & 0 & 1 & -1 & 0 & 0 & 1 \end{array}\right] \end{array},$$

$$\mathbf{B}_f = \begin{array}{c} L_1 \\ L_2 \\ L_3 \end{array} \left[\begin{array}{ccccccc} 1 & 1 & 0 & 0 & -1 & 0 & 0 \\ 1 & 0 & 1 & 1 & 0 & 0 & 0 \\ 0 & 0 & 0 & 1 & 1 & 1 & 1 \end{array}\right].$$

The edge-current vector expressed in terms of loop-current vector is

$$\mathbf{i}(t) = \mathbf{B}_f^T \mathbf{i}_L(t),$$

$$\begin{bmatrix} i_1(t) \\ i_2(t) \\ i_3(t) \\ i_4(t) \\ i_5(t) \\ i_6(t) \\ i_7(t) \end{bmatrix} = \begin{bmatrix} 1 & 1 & 0 \\ 1 & 0 & 0 \\ 0 & 1 & 0 \\ 0 & 1 & 1 \\ -1 & 0 & 1 \\ 0 & 0 & 1 \\ 0 & 0 & 1 \end{bmatrix} \cdot \begin{bmatrix} i_{L1}(t) \\ i_{L2}(t) \\ i_{L3}(t) \end{bmatrix}.$$

The edge voltages in terms of the node voltages (with respect to N_s) are

$$v(t) = A_f^T v_N(t),$$

$$\begin{bmatrix} v_1(t) \\ v_2(t) \\ v_3(t) \\ v_4(t) \\ v_5(t) \\ v_6(t) \\ v_7(t) \end{bmatrix} = \begin{bmatrix} 1 & 0 & 0 & 0 \\ -1 & 1 & 0 & 0 \\ -1 & 0 & 0 & 1 \\ 0 & 0 & 0 & -1 \\ 0 & 1 & 0 & 0 \\ 0 & -1 & 1 & 0 \\ 0 & 0 & -1 & 1 \end{bmatrix} \cdot \begin{bmatrix} v_{N1}(t) \\ v_{N2}(t) \\ v_{N3}(t) \\ v_{N4}(t) \end{bmatrix}.$$

13-4. *RLC* NETWORKS WITH INDEPENDENT SOURCES: NODAL ANALYSIS

In this section we shall restrict ourselves to electrical networks containing resistors, inductors, and capacitors (*RLC*) with independent voltage and current sources. In spite of its inherent simplicity, the *RLC* network covers a very large class of electrical networks in practice. In fact, it has been shown by Brune and Bott and Duffin that any time-invariant, two-terminal, linear, passive electrical element can be formed by a combination of *R*, *L*, and *C* (with real positive values of *R*, *L*, and *C*). A further stipulation may be made, without any loss of generality, that the voltage sources may only be connected in series with *RLC* elements and that current sources may only be connected in parallel with these elements. This stipulation allows us to convert all the energy sources either into a set of voltage sources or into a set of current sources.

Noda. Analysis: Consider an *RLC* network in which all energy sources have been converted into current sources. At each node combine all these current sources. Let the net current entering from the current sources into the *r*th node be $j_r(t)$. For the $n-1$ independent nodes, let the column vector

$$j(t) = \begin{bmatrix} j_1(t) \\ j_2(t) \\ \vdots \\ j_{n-1}(t) \end{bmatrix}.$$

The $n-1$ linearly independent equations from KCL can be expressed as

$$A_f i(t) = j(t), \tag{13-9}$$

where A_f is the reduced incidence matrix of the corresponding graph, and $i(t)$ is the e by 1 column vector of currents in each of the e *passive* edges.

Taking the Laplace transform of Eq. (13-9),

$$A_f I(s) = J(s). \tag{13-10}$$

But the voltage–current relation in the kth edge, consisting only of RLC elements, is given by

$$I_k(s) = Y_k(s) V_k(s), \tag{13-11}$$

where $I_k(s)$ is the Laplace transform of the current through the kth edge, $V_k(s)$ is the Laplace transform of the voltage across the kth edge, and $Y_k(s)$ is the admittance (or self-admittance) of the kth edge. Writing Eq. (13-11) for all the edges in matrix form,

$$\begin{bmatrix} I_1(s) \\ I_2(s) \\ \cdot \\ \cdot \\ \cdot \\ I_e(s) \end{bmatrix} = \begin{bmatrix} Y_1(s) & & & & \\ & Y_2(s) & & 0 & \\ & & \cdot & & \\ & 0 & & \cdot & \\ & & & & Y_e(s) \end{bmatrix} \begin{bmatrix} V_1(s) \\ V_2(s) \\ \cdot \\ \cdot \\ \cdot \\ V_e(s) \end{bmatrix}.$$

More compactly,

$$I(s) = Y(s) V(s), \tag{13-12}$$

where $I(s)$ is the Laplace-transformed column vector of the edge currents, $V(s)$ is the Laplace-transformed column vector of the edge voltages, and $Y(s)$ is the edge admittance matrix.

Substituting Eq. (13-12) into (13-10),

$$A_f Y(s) V(s) = J(s). \tag{13-13}$$

Eq. (13-7) provided a means of expressing the edge-voltage vector in terms of the node-voltage vector. Taking the Laplace transform of Eq. (13-7),

$$V(s) = A_f^T V_N(s), \tag{13-14}$$

and substituting Eq. (13-14) into (13-13),

$$[A_f Y(s) A_f^T] V_N(s) = J(s) \tag{13-15}$$

or
$$Y_N(s) V_N(s) = J(s).$$

The $(n-1)$ by $(n-1)$ matrix $A_f Y(s) A_f^T$ is called the *node admittance matrix* and is written as $Y_N(s)$. Note that in deriving Eq. (13-15) it was assumed that all

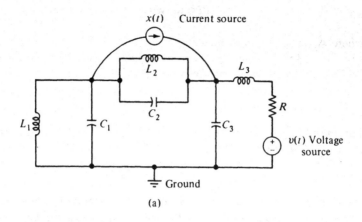

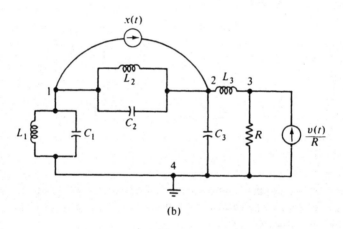

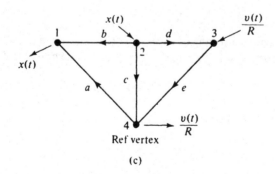

Fig. 13-3 Passive RLC network and its graph.

the initial conditions were zero. This too implies no loss of generality, because any energy stored in capacitors or inductors at time $t = 0$ can always be replaced by an appropriate energy source and hence incorporated into j(t).

Let us illustrate these concepts with an example. An *RLC* network with two independent sources—one voltage source and one current source—is given in Fig. 13-3(a). Figure 13-3(b) shows an equivalent network with only current sources. A directed graph of the network is shown in Fig. 13-3(c). The reduced incidence matrix A_f (with vertex 4 as reference) is

$$A_f = \begin{matrix} & a & b & c & d & e \\ 1 \\ 2 \\ 3 \end{matrix} \begin{bmatrix} -1 & -1 & 0 & 0 & 0 \\ 0 & 1 & 1 & 1 & 0 \\ 0 & 0 & 0 & -1 & 1 \end{bmatrix}.$$

The edge admittance matrix is

$$Y(s) = \begin{bmatrix} Y_a(s) & & & & 0 \\ & Y_b(s) & & & \\ & & Y_c(s) & & \\ & & & Y_d(s) & \\ 0 & & & & Y_e(s) \end{bmatrix} = \begin{bmatrix} \left(C_1 s + \frac{1}{L_1 s}\right) & & & & 0 \\ & \left(C_2 s + \frac{1}{L_2 s}\right) & & & \\ & & C_3 s & & \\ & & & \frac{1}{L_3 s} & \\ 0 & & & & \frac{1}{R} \end{bmatrix}.$$

The node admittance matrix $Y_N(s) = A_f Y(s) A_f^T$ is

$$Y_N(s) = \begin{bmatrix} Y_a(s) + Y_b(s) & -Y_b(s) & 0 \\ -Y_b(s) & Y_b(s) + Y_c(s) + Y_d(s) & -Y_d(s) \\ 0 & -Y_d(s) & Y_d(s) + Y_e(s) \end{bmatrix}.$$

J(s) column vector for this example is

$$J(s) = \begin{bmatrix} -X(s) \\ X(s) \\ \dfrac{V(s)}{R} \end{bmatrix}.$$

Network Analysis Problem: Let us pause for a moment and focus on the problem that we are solving. The general problem of network analysis can be formally stated as follows: Given a network whose structure determines matrix **A**, given its edge admittance matrix **Y**(s), and given the current source vector **J**(s), find the node voltages. [If edge voltages or edge currents are required, they can be readily obtained using Eqs. (13-7) and (13-12).]

This clearly requires solving Eq. (13-15), which involves inversion of the matrix **Y**$_N$(s). Inversion of a matrix (which must be nonsingular, of course) requires computation of its determinant and of all its cofactors. The conventional determinant technique is inefficient because of extra labor involved in computing many terms that eventually cancel out. Moreover, the entries in **Y**$_N$(s) consist of polynomials in s, and must be carried in literal form until after the matrix inversion. Therefore, the usual methods of matrix inversion are computationally difficult to implement.

Both these problems are circumvented by using graph theory to evaluate the determinant and cofactors. For this we invoke the Binet–Cauchy theorem (Appendix A) and use the fact that a major determinant (or simply major) of the reduced incidence matrix **A**$_f$ is nonzero if and only if it corresponds to a spanning tree.

Determinant of the Node Admittance Matrix: Let us denote by Δ_N the determinant of the node admittance matrix **Y**$_N$(s). That is,

$$\Delta_N = \det \mathbf{Y}_N(s) = \det[\mathbf{A}_f \mathbf{Y}(s) \mathbf{A}_f^T].$$

Using the Binet–Cauchy theorem,

$$\Delta_N = \text{sum of products of all pairs of corresponding majors of } [\mathbf{A}_f \mathbf{Y}(s)] \text{ and } \mathbf{A}_f^T. \quad (13\text{-}16)$$

Had every branch in the network been a 1-ohm resistor, **Y**(s) would be an identity matrix and det **Y**$_N$(s) would equal det ($\mathbf{A}_f \mathbf{A}_f^T$), which is equal to the total number of the spanning trees in the network (Chapter 9). But for an *RLC* network, in general, **Y**(s) is not an identity matrix. It is, however, diagonal, and therefore $\mathbf{A}_f \mathbf{Y}(s)$ has the same structure as $\mathbf{A}_f$ except that the kth column in $\mathbf{A}_f$ is multiplied by $Y_k(s)$. Every nonzero major determinant in $\mathbf{A}_f \mathbf{Y}(s)$, as well as $\mathbf{A}_f$, still corresponds to a spanning tree of the network.

If we call the product of all $n - 1$ edges of a specific spanning tree a *tree admittance product*, Eq. (13-16) becomes.

$$\Delta_N = \sum_{\text{all spanning trees}} \text{tree admittance product.} \quad (13\text{-}17)$$

Equation (13-17) was proposed by Maxwell and hence is known as *Maxwell's formula*. To calculate the node admittance determinant by Maxwell's formula, one must find all the spanning trees of the network, multiply the $n - 1$ edge admittances of each spanning tree, and then add the resulting products. Let us illustrate Maxwell's formula for the network of Fig. 13-3. The spanning trees of this graph are *abd*, *abe*, *acd*, *ace*, *ade*, *bcd*, *bce*, and *bde*. Multiplying the edge admittances in each spanning tree and adding them, we get the determinant of the node admittance matrix Δ_N:

$$\Delta_N = Y_a(s)Y_b(s)Y_d(s) + Y_a(s)Y_b(s)Y_e(s) + Y_a(s)Y_c(s)Y_d(s) + Y_a(s)Y_c(s)Y_e(s)$$
$$+ Y_a(s)Y_d(s)Y_e(s) + Y_b(s)Y_c(s)Y_d(s) + Y_b(s)Y_c(s)Y_e(s) + Y_b(s)Y_d(s)Y_e(s)$$

$$= \left(C_1 s + \frac{1}{L_1 s}\right)\left(C_2 s + \frac{1}{L_2 s}\right)\frac{1}{L_3 s} + \left(C_1 s + \frac{1}{L_1 s}\right)\left(C_2 s + \frac{1}{L_2 s}\right)\frac{1}{R}$$
$$+ \left(C_1 s + \frac{1}{L_1 s}\right)\frac{C_3}{L_3} + \left(C_1 s + \frac{1}{L_1 s}\right)\frac{C_3 s}{R} + \left(C_1 s + \frac{1}{L_1 s}\right)\frac{1}{RL_3 s}$$
$$+ \left(C_2 s + \frac{1}{L_2 s}\right)\frac{C_3}{L_3} + \left(C_2 s + \frac{1}{L_2 s}\right)\frac{C_3 s}{R} + \left(C_2 s + \frac{1}{L_2 s}\right)\frac{1}{RL_3 s}$$

$$= \frac{s^2}{R}(C_1 C_2 + C_1 C_3 + C_2 C_3) + \frac{s}{L_3}(C_1 C_2 + C_1 C_3 + C_2 C_3)$$
$$+ \frac{1}{R}\left(\frac{C_2 + C_3}{L_1} + \frac{C_1 + C_3}{L_2} + \frac{C_1 + C_2}{L_3}\right) + \frac{1}{sL_3}\left(\frac{C_2 + C_3}{L_1} + \frac{C_1 + C_3}{L_2}\right)$$
$$+ \frac{1}{s^2 R}\left(\frac{1}{L_1 L_2} + \frac{1}{L_2 L_3} + \frac{1}{L_3 L_1}\right) + \frac{1}{s^3 L_1 L_2 L_3}.$$

Note that to compute Δ_N we do not need to write $Y_N(s)$. Also note that no terms are canceled in this method of computing Δ_N. The reader is urged to compute det $Y_N(s)$ directly from matrix $Y_N(s)$ and verify that it equals the expression for Δ_N just obtained. Observe the large number of terms that cancel in the process of directly evaluating det $Y_N(s)$. Also note that Δ_N is independent of the reference vertex chosen because the trees of a graph do not depend on the reference vertex in writing A_f.

Cofactors of $Y_N(s)$ and 2-Trees

Evaluation of cofactors of the node admittance matrix $Y_N(s)$ is slightly more involved than det $Y_N(s)$. Let the cofactor of the *ij*th entry in $Y_N(s)$ be designated by Δ_{ij}. Then by definition

$$\Delta_{ij} = \left((-1)^{i+j}\right) \cdot \begin{pmatrix} \text{determinant of } (n-2) \text{ by } (n-2) \text{ submatrix of } Y_N(s) \\ \text{left after deleting its } i\text{th row and } j\text{th column.} \end{pmatrix}$$

Since $Y_N(s) = A_f Y(s) A_f^T$, deleting the *i*th row from A_f and the *j*th column from A_f^T will delete the *i*th row and *j*th column from $Y_N(s)$, respectively. Moreover, deleting the *j*th column from A_f^T is equivalent to deleting the *j*th row in A_f. Therefore,

$$\Delta_{ij} = (-1)^{i+j} \det [A_{f-i} Y(s)(A_{f-j})^T], \qquad (13\text{-}18)$$

where A_{f-i} denotes the submatrix of A_f remaining after its *i*th row has been deleted.

If A_f is the reduced incidence matrix of a graph G, what does matrix A_{f-i} represent? Matrix A_{f-i} is the reduced incidence matrix of the graph G_i obtained from G by fusing its ith vertex with the reference vertex and removing any self-loop resulting from the fusion (Problem 13-14).

Let us first evaluate symmetric cofactors Δ_{ii}, which according to Eq. (13-18) is

$$\Delta_{ii} = \det[A_{f-i}Y(s)(A_{f-i})^T],$$

and the right-hand side of this equation is simply the sum of the tree admittance products for the graph G_i. Therefore,

$$\Delta_{ii} = \text{sum of tree admittance products for all spanning trees of } G_i. \quad (13\text{-}19)$$

Now look at a spanning tree of G_i as a subgraph of the original graph G. This subgraph has $n - 2$ edges, n vertices, and no circuits. Therefore, it must consist of two components (one of which may possibly be an isolated vertex). Such a subgraph is called a *2-tree* of G. For example, in Fig. 13-4 the subgraph ad is a spanning tree of G_3 and is a 2-tree in G. (Note that G_3 is obtained by fusing vertex 3 to the reference vertex 4 and removing the resulting self-loop of edge e.)

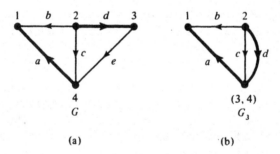

Fig. 13-4 Spanning tree of G_3 is a 2-tree (3, 4) of G.

Moreover, in this 2-tree of G, the vertex i and the reference vertex r must occur in different components; otherwise, fusing them would yield a circuit. Such 2-trees in which two specified vertices occur in different components are designated by 2-tree (i, r). For example, in Fig. 13-4 subgraph ad is a 2-tree $(3, 4)$. Thus Eq. (13-19) can be rewritten as

$$\Delta_{ii} = \text{sum of admittance products of all 2-trees } (i, r). \quad (13\text{-}20)$$

Let us now use Eq. (13-20) to evaluate the cofactor Δ_{33} of the network in

Fig. 13-3. It has five 2-trees of (3, 4) type, and these are, as seen from Fig. 13-4(b),

$$ab, \quad ac, \quad ad, \quad bc, \quad \text{and} \quad bd.$$

Therefore,

$$\Delta_{33} = Y_a(s)Y_b(s) + Y_a(s)Y_c(s) + Y_a(s)Y_d(s) + Y_b(s)Y_c(s) + Y_b(s)Y_d(s)$$
$$= s^2(C_1 C_2 + C_1 C_3 + C_2 C_3) + \left(\frac{C_1 + C_3}{L_2} + \frac{C_1 + C_2}{L_3} + \frac{C_2 + C_3}{L_1}\right)$$
$$+ \frac{1}{s^2}\left(\frac{1}{L_1 L_2} + \frac{1}{L_1 L_3} + \frac{1}{L_2 L_3}\right).$$

To evaluate Δ_{ij}, the cofactor of an off-diagonal entry, observe that in Eq. (13-18) the nonzero majors of $\mathbf{A}_{f-i}Y(s)$ correspond to 2-trees (i, r), where r is the reference node. The nonzero majors of $\mathbf{A}_{f-j}$ correspond to 2-trees (j, r). The terms that contribute to Δ_{ij} in Eq. (13-18) must be due to both 2-trees of (i, r) and (j, r). Since a 2-tree has only two components and vertex r must be in one of the components, both i and j vertices must be in the other. Such a 2-tree is designated by a 2-tree (ij, r). Thus

$$\Delta_{ij} = \left(\pm(-1)^{i+j}\right)\begin{pmatrix}\text{sum of admittance products of}\\ \text{all 2-trees } (ij, r).\end{pmatrix} \quad (13\text{-}21)$$

In Eq. (13-20) we did not have to worry about the sign of the nonzero majors, because corresponding majors of both $\mathbf{A}_{f-i}$ and $(\mathbf{A}_{f-i})^T$ had the same sign. The situation in Eq. (13-21), however, is different. Since $\mathbf{A}_{f-i}$ and $\mathbf{A}_{f-j}$ are different matrices, we have no assurance that the signs of the products of the corresponding majors will be positive. In fact, it can be shown (Problem 13-15) that

product of corresponding nonzero majors of $\mathbf{A}_{f-i}$ and $\mathbf{A}_{f-j} = (-1)^{i+j}$.
$$(13\text{-}22)$$

Therefore,

$$\Delta_{ij} = \text{sum of admittance products of all 2-trees } (ij, r). \quad (13\text{-}23)$$

Returning to the example of Fig. 13-3, once more

$$\Delta_{31} = \text{sum of admittance products of all 2-trees } (31, 4)$$
$$= Y_b(s)Y_d(s) = \frac{C_2}{L_3} + \frac{1}{L_2 L_3 s^2},$$
$$\Delta_{32} = Y_a(s)Y_d(s) + Y_b(s)Y_d(s)$$
$$= \frac{C_1 + C_2}{L_3} + \frac{1}{L_1 L_3 s^2} + \frac{1}{L_2 L_3 s^2}.$$

Node Voltages: Now we can compute any node voltage required. For example, the voltage at node 3 in Fig. 13-3 is given by

$$V_3(s) = \frac{-\Delta_{31}X(s) + \Delta_{32}X(s) + \Delta_{33}\frac{V(s)}{R}}{\Delta_N}.$$

Network Functions: Now that we have formulas for the determinant and every cofactor Δ_{ij} of the node admittance matrix, any network function that was originally expressed in terms of node admittance matrix can now be expressed in terms of various tree-admittance products. For example, the open-circuit transfer function of a three-terminal network in Fig. 13-5(a) (all driving currents zeroed except J_1), taking 4 as the reference, is

$$H(s) = \frac{V_3(s)}{V_1(s)} = \frac{\Delta_{31}(s)}{\Delta_{11}(s)} = \frac{\sum 2\text{-tree }(13, 4) \text{ admittance product}}{\sum 2\text{-tree }(1, 4) \text{ admittance product}}.$$

Formulas like these are called *topological formulas for networks*.

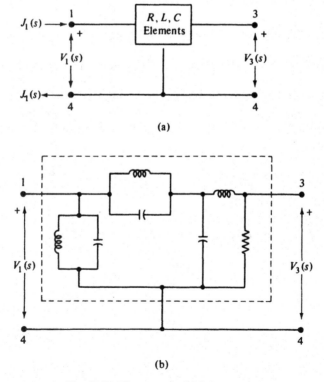

Fig. 13-5 Three-terminal *RLC* network.

Applying this topological formula to the network in Fig. 13-5(b), which is the same as the network in Fig. 13-3 with its driving sources removed, we get

$$H(s) = \frac{Y_b(s)Y_d(s)}{Y_b(s)Y_d(s) + Y_b(s)Y_e(s) + Y_c(s)Y_d(s) + Y_c(s)Y_e(s) + Y_d(s)Y_e(s)}$$

$$= \frac{s^2 + \dfrac{1}{L_2 C_2}}{s^3\left(\dfrac{L_3}{R_1} + \dfrac{L_3 C_3}{C_2 R_1}\right) + s^2\left(1 + \dfrac{C_3}{C_2}\right) + s\left(\dfrac{1}{RC_2} + \dfrac{L_3}{L_2 C_2 R}\right) + \dfrac{1}{L_2 C_2}}.$$

13-5. RLC NETWORKS WITH INDEPENDENT SOURCES: LOOP ANALYSIS

In Section 13-4, had we considered KVL instead of KCL (converting any current source into an equivalent voltage source), we would have obtained a set of $\mu = e - n + 1$ simultaneous loop equations,

$$\mathbf{B}_f \mathbf{Z}(s) \mathbf{B}_f^T \mathbf{I}_L(s) = \mathbf{E}(s), \tag{13-24}$$

where $\mathbf{B}_f$ is the fundamental circuit matrix of the network with respect to some spanning tree, and $\mathbf{B}_f^T$ is its transpose. The e by e matrix $\mathbf{Z}(s)$ is the edge impedance matrix, describing the electrical property of each of e edges in the network; that is,

$$\begin{bmatrix} V_1(s) \\ V_2(s) \\ \vdots \\ V_e(s) \end{bmatrix} = \begin{bmatrix} Z_1(s) & & & 0 \\ & Z_2(s) & & \\ & & \ddots & \\ 0 & & & Z_e(s) \end{bmatrix} \begin{bmatrix} I_1(s) \\ I_2(s) \\ \vdots \\ I_e(s) \end{bmatrix}.$$

Note that for an RLC network the edge impedance matrix $\mathbf{Z}(s)$ is the inverse of its edge admittance matrix $\mathbf{Y}(s)$. $\mathbf{I}_L(s)$ is the Laplace transform of the loop current vector $\mathbf{i}_L(t)$, and $\mathbf{E}(s)$ is the Laplace transform of the voltage sources (or equivalent voltage sources) applied externally in the μ fundamental circuits.

The step-by-step derivation of Eq. (13-24) is similar to the derivation of Eq. (13-15) and is left as an exercise (Problem 13-9).

The μ by μ matrix $\mathbf{B}_f \mathbf{Z}(s) \mathbf{B}_f^T$ in Eq. (13-24) is called the *loop impedance matrix* and is usually denoted by $\mathbf{Z}_L(s)$. Thus Eq. (13-24) is rewritten as

$$\mathbf{Z}_L(s) \mathbf{I}_L(s) = \mathbf{E}(s). \tag{13-25}$$

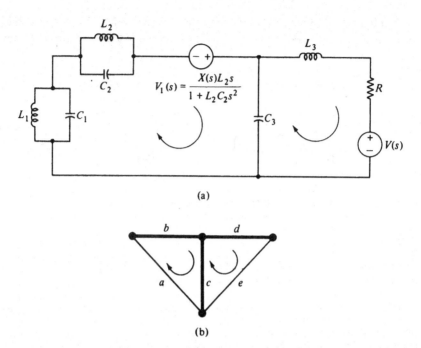

Fig. 13-6 Network of Fig. 13-3(a) for loop analysis.

For example, consider the electrical network of Fig. 13-3(a), once again. By replacing the current source $x(t)$ with an equivalent voltage source, we get the network as shown in Fig. 13-6(a) and its graph as in Fig. 13-6(b).

$$Z_L(s) = \begin{bmatrix} \left(\dfrac{L_1 s}{1 + C_1 L_1 s^2} + \dfrac{L_2 s}{1 + C_2 L_2 s^2} + \dfrac{1}{C_3 s}\right) & \left(-\dfrac{1}{C_3 s}\right) \\ \left(-\dfrac{1}{C_3 s}\right) & \left(\dfrac{1}{C_3 s} + L_3 s + R\right) \end{bmatrix}$$

$$I_L = \begin{bmatrix} I_{L1}(s) \\ I_{L2}(s) \end{bmatrix}, \quad \text{and} \quad E(s) = \begin{bmatrix} \dfrac{X(s)L_2 s}{1 + L_2 C_2 s^2} \\ -V(s) \end{bmatrix}.$$

The solution of Eq. (13-25) requires obtaining the determinant and cofactors of $Z_L(s)$. The expression for Δ_L, the determinant of $Z_L(s)$, according to the Binet–Cauchy theorem is given by

$$\begin{aligned}\Delta_L &= \det Z_L(s) = \det(B_f Z(s) B_f^T) \\ &= \text{sum of products of all pairs of corresponding} \\ &\quad \text{majors of } [B_f Z(s)] \text{ and } B_f^T.\end{aligned} \qquad (13\text{-}26)$$

Since a major of B_f is nonzero if and only if it corresponds to a chord set, Eq. (13-26) becomes

$$\Delta_L = \text{sum of chord impedance products for all spanning trees of the network.} \quad (13\text{-}27)$$

Equation (13-27) was originally given by Kirchhoff for a purely resistive network. For the network of Fig. 13-6(b), all possible chord sets are ce, cd, be, bd, bc, ae, ac, and ad. Therefore,

$$\begin{aligned}\Delta_L &= Z_c(s)Z_e(s) + Z_c(s)Z_d(s) + Z_b(s)Z_e(s) + Z_b(s)Z_d(s) \\ &\quad + Z_b(s)Z_c(s) + Z_a(s)Z_e(s) + Z_a(s)Z_d(s) + Z_a(s)Z_c(s) \\ &= \frac{R}{C_3 s} + \frac{L_3}{C_3} + \frac{RL_2 s}{1 + C_2 L_2 s^2} + \frac{L_2 L_3 s^2}{1 + C_2 L_2 s^2} + \frac{RL_1 s}{1 + C_1 L_1 s^2} \\ &\quad + \frac{L_1}{C_3(1 + C_1 L_1 s^2)} + \frac{L_2}{C_3(1 + C_2 L_2 s^2)} + \frac{L_1 L_3 s^2}{1 + C_1 L_1 s^2}.\end{aligned}$$

The expressions for the cofactors of $Z_L(s)$ both symmetrical and asymmetrical can be obtained in a fashion similar to those for $Y_N(s)$ (Problems 13-11 and 13-12).

Note the duality between the nodal and loop analyses (Problem 13-16).

13-6. GENERAL LUMPED, LINEAR, FIXED NETWORKS

Topological formulas for Δ_N, Δ_L, Δ_{ij}, and so on, derived in the last two sections were dependent on two important restrictions on the network:

1. Existence of edge admittance matrix $Y(s)$ [or edge impedance matrix $Z(s)$], which implied that the network elements were lumped, linear, and time invariant.

2. The edge admittance matrix $Y(s)$ [and therefore also $Z(s)$] was diagonal. This implied that there was no mutual coupling between edges of the network. Thus three- or four-terminal devices (which produce couplings between two vertex pairs), such as transformers, transistors, tubes, and gyrators, could not have been included.

In this section we shall still retain restriction 1, but do away with 2. This will allow us to handle a general linear network containing lumped, linear, time-invariant, r-terminal ($r \geq 2$) elements—passive devices like transformers (which are bilateral also) and gyrators (which are nonbilateral), as well as

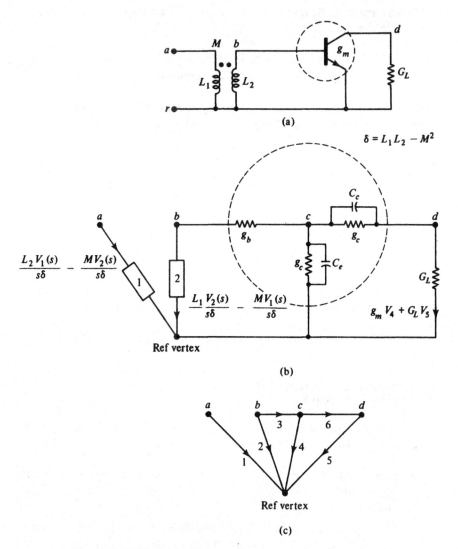

Fig. 13-7 (a) Network with a transformer and a transistor; (b) Its equivalent network; (c) Graph representation of (b).

active devices, such as tubes and transistors. An example of such a network is shown in Fig. 13-7.

In the network in Fig. 13-7 (which has six edges and five vertices), we observe that the current through edge 5 is dependent not only on V_5, but also on V_4, the voltage across edge 6. Similarly, the currents through 1 and 2 are dependent on the voltage across each other. Thus edges 1 and 2 are mutually coupled and so are 5 and 4. (Edges such as 3 and 6 that have no coupling with any other edge are called *ordinary edges*.) The edge admittance matrix $Y(s)$ is shown in the following equation, $I(s) = Y(s)V(s)$ for the network:

$$\begin{bmatrix} I_1(s) \\ I_2(s) \\ I_3(s) \\ I_4(s) \\ I_5(s) \\ I_6(s) \end{bmatrix} = \begin{bmatrix} Y_1 & Y_{12} & 0 & 0 & 0 & 0 \\ Y_{12} & Y_2 & 0 & 0 & 0 & 0 \\ 0 & 0 & g_b & 0 & 0 & 0 \\ 0 & 0 & 0 & Y_4 & 0 & 0 \\ 0 & 0 & 0 & g_m & G_L & 0 \\ 0 & 0 & 0 & 0 & 0 & Y_6 \end{bmatrix} \begin{bmatrix} V_1(s) \\ V_2(s) \\ V_3(s) \\ V_4(s) \\ V_5(s) \\ V_6(s) \end{bmatrix},$$

where

$$Y_1 = \frac{L_2}{s\delta}, \quad Y_{12} = -\frac{M}{s\delta}, \quad Y_2 = \frac{L_1}{s\delta},$$

$$Y_4 = g_e + C_e s, \quad \text{and} \quad Y_6 = g_c + C_c s.$$

Clearly, $Y(s)$ is not diagonal.

Node Admittance Matrix

Just as in Section 13-4, Kirchhoff's current law in its Laplace transformed form will yield

$$A_f Y(s) A_f^T V_N(s) = J_N(s),$$

and therefore the node admittance matrix is

$$Y_N(s) = A_f Y(s) A_f^T. \tag{13-28}$$

The difference between Eqs. (13-15) and (13-28) is only that in (13-15) matrix $Y(s)$ was diagonal, whereas it is not diagonal in Eq. (13-28).

For the network and its graph shown in Fig. 13-7,

$$A_f = \begin{array}{c} \\ a \\ b \\ c \\ d \end{array} \begin{array}{c} 1 \ 2 \ \ 3 \ \ 4 \ 5 \ \ 6 \\ \begin{bmatrix} 1 & 0 & 0 & 0 & 0 & 0 \\ 0 & 1 & 1 & 0 & 0 & 0 \\ 0 & 0 & -1 & 1 & 0 & 1 \\ 0 & 0 & 0 & 0 & 1 & -1 \end{bmatrix} \end{array},$$

and the node admittance matrix $Y_N(s) = A_f Y(s) A_f^T$ is

$$Y_N(s) = \begin{bmatrix} Y_1 & Y_{12} & 0 & 0 \\ Y_{12} & Y_2 + g_b & -g_b & 0 \\ 0 & -g_b & g_b + Y_4 + Y_6 & -Y_6 \\ 0 & 0 & g_m - Y_6 & G_L + Y_6 \end{bmatrix}$$

Determinant Δ_N

Again, our aim is to evaluate the determinant and cofactors of the node admittance matrix $Y_N(s)$. We write

$$\Delta_N = \det Y_N(s) = \det A_f Y(s) A_f^T.$$

Using the Binet-Cauchy theorem,

$$\begin{aligned}\Delta_N &= \sum \text{ products of corresponding majors of } A_f Y(s) \text{ and } A_f^T \\ &= \sum_\alpha \det [A_f Y(s)]_\alpha \cdot \det [A_f]_\alpha,\end{aligned} \quad (13\text{-}29)$$

where the subscript α denotes a set of $n-1$ columns of $A_f Y(s)$ and A_f (same as a set of $n-1$ rows of A_f^T). Thus α also denotes a set of $n-1$ edges of the corresponding graph. In Eq. (13-29), $Y(s)$ is not diagonal; therefore, the product $A_f Y(s)$ is not as simply related to A_f as it was in Eq. (13-15). So we apply again the Binet-Cauchy theorem to evaluate $\det [A_f Y(s)]_\alpha$. And since

$$[A_f Y(s)]_\alpha = A_f [Y(s)]_\alpha,$$

we get

$$\det [A_f Y(s)]_\alpha = \sum_\beta \det [A_f]_\beta \cdot \det [Y(s)]_\alpha^\beta, \quad (13\text{-}30)$$

where $[A_f]_\beta$ is a set of $n-1$ columns of A_f and $[Y(s)]_\alpha^\beta$ is the corresponding set of $n-1$ rows of $[Y(s)]_\alpha$. Thus $[Y(s)]_\alpha^\beta$ is an $(n-1)$ by $(n-1)$ submatrix of $Y(s)$. Substituting Eq. (13-30) into (13-29), we get

$$\begin{aligned}\Delta_N &= \sum_\alpha \left(\sum_\beta \det [A_f]_\beta \cdot \det [Y(s)]_\alpha^\beta \right) \det [A_f]_\alpha \\ &= \sum_{\alpha,\beta} \det [A_f]_\alpha \cdot \det [A_f]_\beta \cdot \det [Y(s)]_\alpha^\beta.\end{aligned} \quad (13\text{-}31)$$

In Eq. (13-31) the summation is over all possible pairs of sets of $n-1$ edges of the graph, but $\det[A_f]_\alpha$ and $\det[A_f]_\beta$ are zero unless α and β correspond to spanning trees of the network, in which case they are $+1$ or -1. Therefore,

$$\Delta_N = \sum_{\alpha,\beta} \epsilon_{\alpha\beta} \det [Y(\alpha)]_{\alpha,\beta} \quad (13\text{-}32)$$

summed over all possible spanning tree pairs (α, β). The term $\epsilon_{\alpha\beta}$ is the product of the signs of spanning trees α and β.

In general, α and β can represent different spanning trees. If $Y(s)$ is diagonal, $\det [Y(s)]_\alpha^\beta = 0$, unless $\alpha = \beta$. But if $\alpha = \beta$, $\epsilon_{\alpha\beta} = 1$ and Eq. (13-32) reduces to (13-17).

But if $Y(s)$ is not diagonal, a spanning tree α, besides making a tree pair

SEC. 13-6　GENERAL LUMPED, LINEAR, FIXED NETWORKS

with itself, may be able to "pair-up" with some other spanning trees. These terms will be contributions to Δ_N due to the couplings between edges.

Pairs of Spanning Trees: The following method of picking out all pairs of spanning trees (α, β) for which $\det [Y(s)]_\alpha^\beta \neq 0$ depends on the fact that for any (lumped, linear, time-invariant) electrical network the edge admittance matrix $Y(s)$ can be expressed as

$$Y(s) = \begin{bmatrix} Y_1(s) & & & & \\ & Y_2(s) & & 0 & \\ & & \cdot & & \\ & & & \cdot & \\ & 0 & & & \cdot \\ & & & & Y_h(s) \end{bmatrix},$$

where the nonzero submatrices $Y_1(s)$, $Y_2(s)$, ..., $Y_h(s)$ are relatively small square matrices. See, for example, the edge admittance matrix of the network in Fig. 13-7.

Assuming that we have the list of all spanning trees of the graph, the following principle determines which spanning tree β pairs with a given spanning tree α, such that $\det[Y(s)]_\alpha^\beta \neq 0$.

The set of rows β must be selected such that $[Y(s)]_\alpha^\beta$ contains no row or column entirely of zeros. Therefore, if α contains edge p, and column p in $Y(s)$ contains nonzero entries in rows $x, y, \ldots,$ then β must contain one (or more) of the edges $x, y, \ldots$. Thus, if column p falls in the submatrix $Y_k(s)$ of $Y(s)$, at least one of the rows must also be in the submatrix $Y_k(s)$.

A corollary of the observation just made is that if spanning tree α contains an ordinary edge u, β must also contain that ordinary edge u.

Let us illustrate the selection principle by means of the example of the network in Fig. 13-7. The graph has eight spanning trees:

(1, 2, 3, 5),　(1, 2, 3, 6),　(1, 2, 4, 5),　(1, 2, 4, 6),　(1, 2, 5, 6),
(1, 3, 4, 5),　(1, 3, 4, 6),　(1, 3, 5, 6).

Since 3 and 6 are ordinary edges, the following five are the only candidates for possible pairings out of the total of $(8 \times 7)/2 = 28$ pairs of spanning trees:

1. {(1, 2, 3, 5) and (1, 3, 4, 5)}: both have edge 3.
2. {(1, 2, 4, 6) and (1, 2, 5, 6)}: both have edge 6.
3. {(1, 2, 3, 6) and (1, 3, 4, 6)}: both have 3 and 6.
4. {(1, 2, 3, 6) and (1, 3, 5, 6)}: both with 3 and 6.
5. {(1, 3, 4, 6) and (1, 3, 5, 6)}: both with 3 and 6.

The existence of the same set of ordinary edges is only a necessary and not a sufficient condition for pairing.

Let us now apply the tree-pair selection principle to nonordinary edges:

1. If $Y_k(s)$ is a 2 by 2 square submatrix, it corresponds to a transformer or a gyrator, and its contribution to Δ_N is

$$Y_k(s) = \begin{bmatrix} a & b \\ c & d \end{bmatrix} \begin{matrix} \leftarrow \text{row } p \\ \leftarrow \text{row } (p+1) \end{matrix}$$

$$\begin{matrix} \nearrow & \nwarrow \\ \text{column } p & \text{column } (p+1) \end{matrix}$$

		or		or	
If α Contains	Column p		Column $(p+1)$		Both p and $(p+1)$
β must contain row	p | $(p+1)$		p | $(p+1)$		Both p and $(p+1)$
Contribution to Δ_N due to $Y_k(s)$ is	a | b		c | d		$ad - bc$
	or		or		

2. If $Y_k(s)$ is a 2 by 2 triangular matrix, it corresponds to a transistor or a vacuum tube. In that case

$$Y_k(s) = \begin{bmatrix} a & 0 \\ c & d \end{bmatrix} \begin{matrix} \leftarrow \text{row } p \\ \leftarrow \text{row } (p+1) \end{matrix}$$

$$\begin{matrix} \nearrow & \nwarrow \\ \text{column } p & \text{column } (p+1) \end{matrix}$$

		or		or	
If α Contains	Column p		Column $(p+1)$		Both p and $(p+1)$
β must contain row	p | $(p+1)$		$(p+1)$		Both p and $(p+1)$
Contribution to Δ_N due to $Y_k(s)$ is	a | c		d		ad
	or				

In light of these two tables, let us look at the five tree pairs that are possible candidates in the network of Fig. 13-7.

Three of the five pairs, 1, 3, and 4, do not form valid tree pairs, because in each of the three one spanning tree contains both edges 1 and 2, while the other one contains only edge 1.

The remaining two pairs [(1, 2, 5, 6), (1, 2, 4, 6)] and [(1, 3, 5, 6), (1, 3, 4, 6)] satisfy the tree-pair-solution criterion, and their contributions to Δ_N are

$$(Y_1Y_2 - Y_{12}^2)g_mY_6 \quad \text{and} \quad Y_1g_bg_mY_6,$$

respectively.

The criterion of selection of pairs of spanning trees can be easily extended to $Y_k(s)$ of sizes larger than 2 by 2 (see [13-7]).

Signs of Tree Pairs: In the case of a spanning tree α consisting of ordinary edges only, the spanning tree pairs only with itself, and we need not know if $\det [A_f]_\alpha = +1$ or -1 because

$$\epsilon_{\alpha\alpha} = \det [A_f]_\alpha \cdot \det [A_f]_\alpha = +1.$$

But for tree pairs (α, β) consisting of nonordinary edges (and therefore $\alpha \neq \beta$), we must know the relative (not absolute) signs of the spanning trees in each pair.

According to the method of sign determination discussed in Chapter 9, for the tree pair in Fig. 13-7,

$$[(1, 2, 5, 6), (1, 2, 4, 6)], \quad \epsilon_{\alpha\beta} = +1,$$

and for

$$[(1, 3, 5, 6), (1, 3, 4\ 6)], \quad \epsilon_{\alpha\beta} = +1.$$

Thus Δ_N as a sum of the eight spanning admittance products (α, α pairings) and two additional terms due to (α, β) pairings is expressed as

$$\begin{aligned}\Delta_N = &(Y_1Y_2 - Y_{12}^2)g_bG_L + (Y_1Y_2 - Y_{12}^2)g_bY_6 + (Y_1Y_2 - Y_{12}^2)Y_4G_L \\ &+ (Y_1Y_2 - Y_{12}^2)Y_4Y_6 + (Y_1Y_2 - Y_{12}^2)G_LY_6 + Y_1g_bY_4G_L \\ &+ Y_1g_bY_4Y_6 + Y_1g_bG_LY_6 + (Y_1Y_2 - Y_{12}^2)g_mY_6 + Y_1g_bg_mY_6.\end{aligned}$$

Note once again that there is no cancellation of terms.

The derivation of cofactors Δ_{ij} for active networks can be carried on similarly by a combination of the technique discussed in Section 13-4 and the use of spanning-tree pairs.

SUMMARY

The technique developed in this chapter can be extended to solve any linear-system problem. Roughly speaking, any linear-system problem can be expressed in the following form:

$$\Lambda X^* = Y^*,$$

where Λ is a linear operator, Y^* a known vector, and X^* an unknown vector for which the solution is sought.

A standard method of solving this equation is to find an operator Λ^{-1} (assuming it exists and is unique), the inverse of Λ, and then to premultiply both sides to obtain the required vector

$$X^* = \Lambda^{-1}Y^*.$$

In an electrical network consisting of lumped, linear, time-invariant devices, the problem consists of solving a set of simultaneous, linear, differential equations with constant coefficients. Application of the Laplace transform converts these differential equations into linear algebraic equations. Thus the operator Λ is a matrix whose entries are functions of s, the Laplace variable, and Y^* is the vector of independent driving voltages (or currents).

Thus the electrical network problem (like most linear-system problems) consists of matrix inversion, which is the same as finding the determinants and cofactors. And all that has been done in this chapter is to show how graph theory can be used (rather than algebra) to evaluate determinants and cofactors of the nonsingular matrix Λ, if Λ could be expressed as a triple matrix product

$$\Lambda = \mathsf{PMP}^T,$$

where P is a unimodular (0, 1)-matrix—a reduced incidence (or fundamental cut-set or fundamental circuit matrix) of a graph—describing the "structure" of Λ; and M is a matrix describing the values of the nonzero entries in Λ.

The same approach can be used for solution of any lumped, linear, time-invariant system, provided a "system graph" can be found. This has a direct bearing on the realizability problem discussed in Section 12-5, as to when a given unimodular matrix P can be the cut-set or circuit matrix of a graph.

Whether there is any computational advantage in using graph theory for network analysis is totally dependent on whether one can generate all spanning trees, 2-trees, and the like, of a large graph rapidly and without duplication. A graph of moderate size (20 vertices and 50 edges) could have several million spanning trees. Even the storing of all the trees in a computer memory can be a problem. The algorithm should therefore be such that spanning trees are rapidly generated, one at a time, and its admittance product is added to or subtracted from (depending on the sign) the cumulative sum. The algorithm should guarantee that no spanning tree will be generated twice, so that one does not have to check every newly obtained tree against all the trees previously generated. Moreover, the algorithm must also guarantee that no spanning tree in the graph is left out.

As discussed in Chapter 11, a number of algorithms for generating all spanning trees of a graph have been proposed in the literature. The best ones do generate one spanning tree at a time without duplication and generate all spanning trees. But the algorithms are still not as efficient as one would like them to be.

REFERENCES

The bare essentials of topological analysis of networks have been presented in this chapter for the purpose of introducing the reader to the application of graph theory to electrical network analysis.

Nothing, for instance, was said about the application of graphs to the synthesis of electrical networks. Nor did we discuss the extension of these techniques to nonlinear networks. We confined ourselves to the frequency-domain analysis, and did not consider the time-domain analysis via state-space techniques. Many other topics, such as the duality in electrical networks or stability of electrical networks, were also not covered.

For these and more, the reader may go to one of several books and scores of research and tutorial papers available on these specific subjects. Seshu and Reed [1-13] is the classic and one of the best books. Kim and Chien [13-5] is another excellent book. Chan's book [13-3] includes the state-space approach, which is not dealt with in Seshu and Reed or in Kim and Chien.

Some of the excellent survey papers are by Bryant [13-1], Dawson [13-4], and Kuo [13-6]. These papers also include a large bibliography on the subject.

For the sake of simplicity in the case of networks with active devices, Talbot's [13-7] approach of a single graph was used rather than Mayeda's method of dealing with two different graphs—voltage and current graphs.

A list of classical papers, such as those of Kirchhoff, Maxwell, Percival, Mayeda, Bashkow, Bryant, and others, can be found in almost any of th references cited.

13-1. BRYANT, P. R., "Graph Theory Applied to Electrical Networks," Chapter 3 in *Graph Theory and Theoretical Physics* (F. Harary, ed.), Academic Press, Inc., New York, 1967.
13-2. CALAHAN, D. A., *Computer-Aided Network Design*, Revised Edition, McGraw-Hill, Inc., New York, N. Y., 1972.
13-3. CHAN, S. P., *Introductory Topological Analysis of Electrical Networks*, Holt, Rinehart and Winston, Inc., New York, 1969.
13-4. DAWSON, D. F., "The Topological Approach to Computer-Aided Analysis," Chapter 2 in *Computer Oriented Circuit Design* (F. F. Kuo and W. G. Magnuson, eds.), Prentice-Hall, Inc., Englewood Cliffs, N.J., 1969.
13-5. KIM, W. H., and R. T. CHIEN, *Topological Analysis and Synthesis of Communication Networks*, Columbia University Press, New York, 1962.
13-6. KUO, F. F., "Network Analysis by Digital Computer," *Proc. IEEE*, Vol. 54, June 1966, 820–829.
13-7. TALBOT, A., "Topological Analysis of General Linear Networks," *IEEE Trans. Circuit Theory*, Vol. CT-12, June 1965, 170–180.

PROBLEMS

13-1. Show that Kirchhoff's voltage and current laws imply "conservation of power." [*Hint:* Using Eqs. (13-5) and (13-7), show that $\sum_{k=1}^{e} v_k(t) i_k(t) = 0$.]

13-2. An electrical network with e edges has $2e$ unknowns (the current through and voltage across each edge). Identify the $2e$ independent equations, and discuss the existence and uniqueness of the solutions.

13-3. Kirchhoff's current law may be expressed in more general form as follows: The net sum (taking into account the orientations) of all currents flowing across a cut-set is zero. Using a cut-set matrix (instead of incidence matrix) and this form of KCL, develop equations parallel to Eqs. (13-2), (13-7), (13-8), (13-15), and (13-17).

13-4. In Fig. 13-2(b) list all spanning trees and all 2-trees $(N_3 N_1, N_5)$.

13-5. In Fig. 13-4(a) sketch all 2-trees (2, 4) and all 2-trees (23, 4).

13-6. Of Kirchhoff's and Maxwell's formulas, which one will you prefer for evaluating Δ_N and Δ_L? Why?

13-7. In Fig. 13-2, assume the resistance value R_i or capacitance value C_i (as the case may be) in the ith edge of the network. Let $x(t)$ be the value of the independent voltage source shown, and let N_5 be the reference node. Convert $x(t)$ into an equivalent current source in parallel with R_1. Use Maxwell's formula to evaluate Δ_N and Δ_{31}. Using these two quantities, evaluate the voltage at node N_3.

13-8. In Problem 13-7, keep the voltage source in series with R_1. Write the loop-impedance matrix $Z(s)$ of the network. Write $Z_L(s)$. Evaluate Δ_L using Eq. (13-27). Evaluate the appropriate cofactor of $Z_L(s)$ required for obtaining $V_2(s)$. Finally, obtain $V_2(s)$, and compare the result with that of Problem 13-7.

13-9. In a step-by-step fashion derive Eq. (13-24).

13-10. For an RLC network, prove that
$$\frac{\Delta_N}{\Delta_L} = Y_1(s) Y_2(s) \ldots Y_e(s).$$

[*Hint:* $Y_i(s) Z_i(s) = 1$, and therefore for each spanning tree T in the network,

$$\frac{\text{admittance product of the spanning tree } T}{\text{impedance product of chord sets with respect to } T} = $$

$[Y_1(s) \cdot Y_2(s) \ldots Y_e(s).]$

13-11. Similarly to Eq. (13-20), show that the (i, i)th cofactor of the loop impedance matrix $Z_L(s)$ of an RLC network G is equal to the sum of the chord impedance products for all spanning trees of the network G', obtained from G by deleting the ith chord.

13-12. Attempt an expression for the (i, j)th cofactor of the loop impedance matrix Z_L of an RLC network.

13-13. In deriving expressions for Δ_N and Δ_L, we tacitly assumed the nonsingularities of $Y_N(s)$ and $Z_L(s)$. Discuss the requirements imposed on an electrical network because of the nonsingularity requirements. [*Hint:* The network should have (1) each voltage source only in series with some passive element, (2) each current source only in parallel with some passive element, and (3) no perfectly coupled transformer; i.e., $L_1 L_2 > M_{12}^2$.]

13-14. Let A_f be the $(n-1)$ by e reduced incidence matrix of a connected (directed or undirected) graph G of n vertices and e edges, with respect to some reference vertex r. And let G_i be the graph obtained from G by fusing its ith vertex with the reference vertex r, and removing any self-loops produced in the process. Prove that A_{f-i}, the $(n-2)$ by e matrix obtained from A_f by deleting its ith row, is the reduced incidence matrix (with the fused vertex as the reference vertex) of G_i. [*Hint:* In G_i the $n-2$ vertices have exactly the same incidences as they had in G.

between r and i are gone, but the edges that were incident on either r or i but not on both have one end incident on the fused vertex.]

13-15. In Problem 13-14, show that the product of any two corresponding nonzero majors of $(n-2)$ by e unimodular matrices A_{f-i} and A_{f-j} is equal to $(-1)^{i+j}$, providing the rows and columns of both these matrices are arranged in the same order.

13-16. Draw a dual electrical network to the one in Fig. 13-2, and then study the dual relationship between various quantities between the two networks, such as the loop equations in one being the node equation in the other.

13-17. For a one-port RLC network, shown in Fig. 13-8(a), show that the driving point admittance at terminals $(1, r)$ is

$$\frac{\sum \text{tree admittance product}}{\sum 2\text{-tree } (1, r) \text{ admittance product}}$$

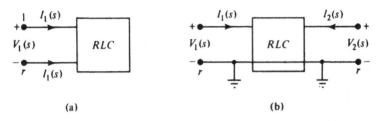

(a)   (b)

Fig. 13-8   One- and two-port RLC networks.

13-18. A two-port network has four short-circuit admittance functions. Derive the topological formula for each in the two-port RLC network with common reference vertex r shown in Fig. 13-8(b).

14 GRAPH THEORY IN OPERATIONS RESEARCH

Graph theory is a very natural and powerful tool in combinatorial operations research. In earlier chapters we have already applied graph theory to operations-research problems. The traveling-salesman problem (Chapter 2), finding the shortest spanning tree in a weighted graph (Chapter 3), obtaining an optimal matching of jobs and men (Chapter 8), and locating the shortest path between two vertices in a graph (Chapter 11) are some examples of the uses of graph theory in operations research. This chapter will be devoted entirely to solving problems in operations research using graph-theoretic tools. We shall consider three related areas of operations research in which graph theory is used most frequently and profitably. They are transport networks, activity networks, and the theory of games.

14-1. TRANSPORT NETWORKS

In Section 4-6 we saw how a graph can be used as a model for a network of pipelines through which some commodity is transported from one place to another. The general problem in such a transport network (also called a flow network) is to maximize the flow or minimize the cost of a prescribed flow. This is an operations-research problem and can be solved by linear programming, but the graph-theoretic approach has been found to be computationally more efficient. In this section we shall see how network-flow problems can be formulated and solved using graphs. Let us first define some terms.

Transport Network: A simple, connected, weighted, digraph G is called a *transport (or flow) network* if the weight associated with every directed edge in G is a nonnegative number. In a transport network this number represents the *capacity* of the edge and is designated as c_{ij} for the edge directed from vertex i to vertex j. A transport network is shown in Fig. 14-1, where the numbers written beside the edge are the edge capacities.

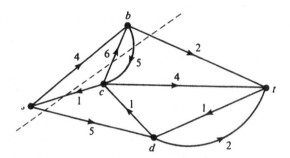

Fig. 14-1 Transport network.

The capacity c_{ij} of an edge (i, j) can be thought of as the maximal amount of some commodity (such as water, gas, electrical energy, number of cars, bits of information, etc.) that can be transported from station i to j, along the edge (i, j), per unit of time in a steady state. Then a natural question is: What is the maximal amount of the commodity flow from a given vertex s to another specified vertex t via the entire network? Let us first formulate the question mathematically.

Maximal flow: In a given transport network G, a *flow* (or *a static flow*) is an assignment of a nonnegative number f_{ij} to every directed edge (i, j) such that the following conditions are satisfied:

1. For every directed edge (i, j) in G

$$f_{ij} \leq c_{ij}. \tag{14-1}$$

2. There is a specified vertex s in G, called the *source*, for which

$$\sum_i f_{si} - \sum_i f_{is} = w, \tag{14-2}$$

where the summations are taken over all vertices in G. Quantity w is called the *value of the flow*.

3. There is another specified vertex t in G, called the *sink*, for which

$$\sum_i f_{ti} - \sum_i f_{it} = -w. \tag{14-3}$$

4. All other vertices are called intermediate vertices. For each intermediate vertex j,

$$\sum_i f_{ji} - \sum_i f_{ij} = 0. \tag{14-4}$$

Condition (14-1) states that the flow through any edge does not exceed its capacity. The other three conditions state that the net flow out of the source is w, the net flow into the sink is w, and the flow is conserved at each intermediate vertex. This is why w is called the value of the flow from s to t. Condition (14-3) can, in fact, be derived from (14-2) and (14-4), and is therefore not independent. It is understood that if there is no edge from vertex p to q, $f_{pq} = 0$. An edge (i, j) for which $f_{ij} = c_{ij}$ is said to be *saturated*.

A set of flows f_{ij}'s for all (i, j)'s in G is called a *flow pattern*. A flow pattern that maximizes the quantity w is called a *maximal flow pattern*. The first problem one encounters in a transport network is: Given G, s, and t, find a maximal flow pattern.

Linear Programming Formulation: Those familiar with linear programming (LP) will recognize this as an LP problem. As an example, take the transport network in Fig. 14-1. The variables are the flows through each of the 10 edges. Although $w = f_{sb} + f_{sd} - f_{cs}$, we can regard w as another variable. Let the flow pattern be denoted by a column vector $\mathbf{f}$:

$$\mathbf{f} = \begin{bmatrix} f_{sb} \\ f_{sd} \\ f_{cs} \\ f_{bc} \\ f_{bt} \\ f_{cb} \\ f_{ct} \\ f_{dt} \\ f_{dc} \\ f_{td} \end{bmatrix},$$

and let

$$\mathbf{f}' = \begin{bmatrix} w \\ \mathbf{f} \end{bmatrix}$$

denote the variable vector of the LP problem. Let $\mathbf{h}$ denote the row vector

$$(1, 0, 0, 0, 0, 0, 0, 0, 0, 0, 0).$$

Then the problem is to maximize $\mathbf{h} \cdot \mathbf{f}'$ subject to the constraints

$$\mathbf{A}' \cdot \mathbf{f}' = 0, \quad \mathbf{f} \le \mathbf{c}, \quad \text{and} \quad \mathbf{f}' \ge 0,$$

where

$$\mathbf{A}' = \begin{bmatrix} -1 & 1 & 1 & -1 & 0 & 0 & 0 & 0 & 0 & 0 & 0 \\ 0 & -1 & 0 & 0 & 1 & 1 & -1 & 0 & 0 & 0 & 0 \\ 0 & 0 & 0 & 1 & -1 & 0 & 1 & 1 & 0 & -1 & 0 \\ 0 & 0 & -1 & 0 & 0 & 0 & 0 & 0 & 1 & 1 & -1 \\ 1 & 0 & 0 & 0 & 0 & -1 & 0 & -1 & -1 & 0 & 1 \end{bmatrix}$$

and

$$\mathbf{c} = \begin{bmatrix} 4 \\ 5 \\ 1 \\ 5 \\ 2 \\ 6 \\ 4 \\ 2 \\ 1 \\ 1 \end{bmatrix}.$$

Observe that $\mathbf{A}'$ is the incidence matrix of a digraph obtained by adding an edge from t to s in the transport network of Fig. 14-1. Also note that the edges in $\mathbf{f}$, $\mathbf{c}$, and $\mathbf{A}'$ must appear in the same order.

Clearly, a maximal flow can be obtained by solving this LP problem, but, as mentioned earlier, the graph-theoretic approach is more efficient. Using the graph-theoretic concept, we shall now state and prove the max-flow min-cut theorem, the most important result in the theory of transport networks.

Cut and Its Capacity: Ignoring the directions of edges in a transport network, let us consider a cut-set with respect to vertices s and t, that is, a cut-set which separates the source s from sink t. Such a set of edges in a transport network is called a *cut*. The notation $(P, \bar{P})$ is used to denote a cut that partitions the vertices into two subsets P and $\bar{P}$, where P contains s and $\bar{P}$ contains t. The *capacity of a cut* denoted by $c(P, \bar{P})$ is defined to be the sum of the capacities of those edges directed from the vertices in set P to the vertices in $\bar{P}$; that is,

$$\sum_{\substack{i \in P \\ j \in \bar{P}}} c_{ij} = c(P, \bar{P}).$$

For example, in Fig. 14-1 the cut (dashed line) separating $P = \{s, b\}$ from $\bar{P} = \{c, d, t\}$ has a capacity of $5 + 5 + 2 = 12$.

Theorem 14-1

In a given transport network G, the value of flow w from source s to sink t is less than or equal to the capacity of any cut separating s from t.

Proof: Let $(P, \bar{P})$ be an arbitrary cut such that the source s is in vertex set P and the sink t is in vertex set $\bar{P}$. Let us write Eq. (14-4) for all intermediate vertices in P and add them to Eq. (14-2). This yields

$$\sum_{\substack{p \in P \\ i \in G}} f_{pi} - \sum_{\substack{p \in P \\ i \in G}} f_{ip} = w,$$

which can be rewritten as

$$\sum_{\substack{p \in P \\ i \in \bar{P}}} f_{pi} + \sum_{\substack{p \in P \\ i \in P}} f_{pi} - \sum_{\substack{p \in P \\ i \in \bar{P}}} f_{ip} - \sum_{\substack{p \in P \\ i \in P}} f_{ip} = w.$$

But

$$\sum_{\substack{p \in P \\ i \in P}} f_{pi} - \sum_{\substack{p \in P \\ i \in P}} f_{ip} = 0.$$

Therefore,

$$\sum_{\substack{p \in P \\ i \in \bar{P}}} f_{pi} - \sum_{\substack{p \in P \\ i \in \bar{P}}} f_{ip} = w. \tag{14-5}$$

Since $\sum_{\substack{p \in P \\ i \in \bar{P}}} f_{ip}$ is always a nonnegative quantity, we have

$$w \leq \sum_{\substack{p \in P \\ i \in \bar{P}}} f_{pi} \leq \sum_{\substack{p \in P \\ i \in \bar{P}}} c(p, i) = c(P, \bar{P}). \blacksquare$$

In the following theorem we shall prove that it is possible to achieve a value of the flow which equals the capacity of the smallest cut separating s from t.

Theorem 14-2 (Max-Flow Min-Cut Theorem)

In a given transport network G, the maximum value of a flow from s to t is equal to the minimum value of the capacities of all the cuts in G that separate s from t.

Proof: In view of Theorem 14-1 we need only to prove that there exists a flow pattern in G such that the value of the flow w_0 from s to t is equal to $c(P_0, \bar{P}_0)$, the capacity of some cut $(P_0, \bar{P}_0)$ separating s from t.

Let there be some flow pattern in G such that the value of the flow from s to t is at its maximum possible value w_0. Define a vertex set P in G recursively as follows:

(a) $s \in P$.
(b) If vertex $i \in P$ and either $f_{ij} < c_{ij}$ or $f_{ji} > 0$, then $j \in P$. Any vertex not in P belongs to $\bar{P}$.

Now vertex t cannot be in P. If it were, there would be a path p (see Fig. 14-2) from s to t, say, $s, v_1, v_2, \ldots, v_j, v_{j+1}, \ldots, v_k, t$, for which in every edge either flow $f_{v_j v_{j+1}} < c_{v_j v_{j+1}}$, or $f_{v_{j+1} v_j} > 0$. In path p an edge (v_j, v_{j+1}) directed from v_j to v_{j+1} is called a *forward edge* and an edge (v_{j+1}, v_j) directed from v_{j+1} to v_j is called a *backward edge* (Fig. 14-2).

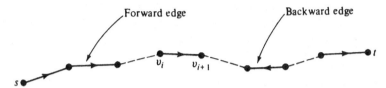

Fig. 14-2 Path p in the proof of Theorem 14-2.

In path p let δ_1 be the minimum of all differences $[c_{v_j v_{j+1}} - f_{v_j v_{j+1}}]$ in forward edges and δ_2 be the minimum of all flows in backward edges. Both δ_1 and δ_2 are positive quantities. Let $\delta = \min(\delta_1, \delta_2)$. Then the flow in the network G can be increased by increasing the flow in each forward edge and decreasing the flow in each backward edge by an amount δ. [Conditions (14-1), (14-2), (14-3), and (14-4) are still satisfied.] This contradicts the assumption that w_0 was the maximum flow.

Thus t must be in the vertex set $\bar{P}$. In other words, the cut $(P, \bar{P})$ separates s from t. Furthermore, according to condition (b), for each vertex p in P and i in $\bar{P}$, we have

$$f_{pi} = c_{pi} \quad \text{and} \quad f_{ip} = 0.$$

Therefore, from Eq. (14-5) we get the value of the flow:

$$w_0 = \sum_{\substack{p \in P \\ i \in \bar{P}}} f_{pi} - \sum_{\substack{p \in P \\ i \in \bar{P}}} f_{ip}$$

$$= \sum_{\substack{p \in P \\ i \in \bar{P}}} c_{pi} = c(P, \bar{P}),$$

which proves the theorem. ∎

As an example, let us consider the transport network of Fig. 14-1, once again. It has eight (2^3) cuts that separate s from t. These cuts (identified by vertex set P) and their capacities are

Vertex Set P	$c(P, \bar{P})$
$\{s\}$	9
$\{s, b\}$	12
$\{s, c\}$	19
$\{s, d\}$	7
$\{s, b, c\}$	11
$\{s, b, d\}$	10
$\{s, c, d\}$	16
$\{s, b, c, d\}$	8

The cut with minimum capacity among these is the one in which $P = \{s, d\}$ and $\bar{P}$ is $\{b, c, t\}$. The maximum flow possible in s to t in the network is therefore 7 units.

The proof does not include an algorithm for finding the actual value of the maximal flow w_{max}. Nor does it give a flow pattern that realizes this maximal flow. If we were interested only in finding w_{max}, we would take some algorithm for generating a minimal cut (see, for instance, Plisch [14-19] for an efficient computer code to generate all minimal cuts in a given transport network), and then compute its capacity.

For those wanting to construct a maximal flow pattern, an algorithm based on the foregoing proof of the max-flow min-cut theorem is also available. This is an efficient algorithm, and it uses a vertex-labeling process for constructing a maximal flow pattern. However, the proof that this algorithm terminates in a finite number of steps depends on the edge capacities being integers. For more on this labeling algorithm and its modifications, see [14-7], [14-9], or [14-12].

14-2. EXTENSIONS OF MAX-FLOW MIN-CUT THEOREM

The max-flow min-cut theorem as stated is applicable to a transport network (simple, weighted, connected digraph) with one source and one sink. There are, however, many other types of network-flow problems that can be solved by extending the max-flow min-cut theorem appropriately. Some of these extensions are straightforward and others are quite involved. Let us consider them in increasing order of difficulty.

1. *Multiple Sources and Sinks:* If there are several sources $s_1, s_2, \ldots, s_k$ and several sinks $t_1, t_2, \ldots, t_r$, and if the flow from any source can be sent to any sink, then this problem can be converted immediately into a one-source and one-sink problem as follows: Introduce a *supersource* s with edges (of unlimited capacity) directed to $s_1, s_2, \ldots, s_k$ and a *supersink* t with edges (also of unlimited capacity) directed from $t_1, t_2, \ldots, t_r$, as shown in Fig. 14-3. The problem of maximizing the total value of the flow from all sources is then the same as that of maximizing the value of the flow from s to t.

However, if the restriction is made that the flow from a specified source

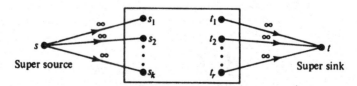

Fig. 14-3 Multi-source multi-sink transport network.

s_i must be sent to a specified sink t_i, the problem becomes much more difficult. Such a flow, known as the multicommodity flow, will be discussed shortly as a separate topic.

2. *Vertices with Specified Capacity:* Suppose that we have a transport network in which some (or all) vertices also have specified capacities. The total flow into a vertex v must not exceed its capacity $c(v)$, a real positive number. This network can be converted into an ordinary transport network by replacing each such vertex v with two vertices v' and v'' and an edge from v' to v'' with capacity $c(v)$. All edges originally incident into v are made incident into v', and all edges originally incident out of v are made incident out of v'', as illustrated in Fig. 14-4.

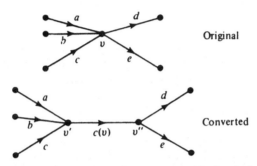

Fig. 14-4 Replacement of a vertex v with v' and v''.

3. *Networks Containing Undirected Edges:* Often one encounters the problem of maximizing a flow through a network in which some or all of the edges are undirected. In such a network an undirected edge between vertices p and q of capacity c_{pq} implies that the flow can occur in either direction, and

$$f_{pq} \leq c_{pq},$$
$$f_{qp} \leq c_{pq}.$$

Moreover, since simultaneous flows in opposite directions cancel each other, the flow is assumed to be in only one direction. That is,

$$f_{pq} \cdot f_{qp} = 0.$$

Thus the maximum-flow problem in a network containing undirected edges can be solved by replacing each undirected edge with a pair of oppositely directed edges, each having a capacity of the original edge.†

---

†In the case of a multicommodity flow, where two or more commodities flow through the network simultaneously, this replacement is not valid. The product $f_{pq} \cdot f_{qp} \neq 0$, because different commodities can flow in opposite directions without canceling each other. This is one of the major difficulties in multicommodity flow problems.

4. Lower Bound on Edge Flows: So far we have assumed that the lower bound on a flow through an edge in a transport network is zero. Occasionally, one encounters a practical situation that requires a minimum flow b_{ij} through an edge (for instance, an oil pipeline in Alaska may need a specified minimum flow to keep it from freezing). That is, conditions

$$b_{ij} \leq f_{ij} \leq c_{ij} \tag{14-6}$$

replace (14-1), where b_{ij} is a nonnegative real number no larger than c_{ij}.

For some network there may not even exist a feasible flow pattern, that is, one which satisfies the constraints (14-2), (14-3), (14-4), and (14-6). For example, all $f_{ij} = 0$ is not a feasible flow pattern, unlike in the case with no lower bounds on the edge flows. Therefore, we have to first determine if indeed there is a flow pattern in G that satisfies all the upper and lower bounds, and if so how do we get a flow.

It can be shown using the arguments of Theorem 14-1 that the value w for any feasible flow pattern must satisfy the following simultaneous requirements for every cut $(P, \bar{P})$ in G separating s and t:

$$w \leq c(P, \bar{P}) - b(\bar{P}, P),$$
$$w \geq b(P, \bar{P}) - c(\bar{P}, P),$$

where

$$b(P, \bar{P}) = \sum_{\substack{i \in P \\ j \in \bar{P}}} b_{ij} \quad \text{and} \quad b(\bar{P}, P) = \sum_{\substack{i \in \bar{P} \\ j \in P}} b_{ij}.$$

Furthermore, analogous to Theorem 14-2, it can be shown that if there exists a flow pattern satisfying the lower and upper bounds, a maximum flow can be achieved, and the value of the flow equals the minimum value of the quantity

$$c(P, \bar{P}) - b(\bar{P}, P)$$

taken over all cuts $(P, \bar{P})$ separating vertices s and t. Similarly, the minimum value of a flow equals the maximum value of

$$b(Q, \bar{Q}) - c(\bar{Q}, Q)$$

taken over all cuts $(Q, \bar{Q})$ separating vertices s and t.

The problem of determining conditions under which a flow pattern exists satisfying constraints (14-2), (14-3), (14-4), and (14-6) is slightly more involved. The reader is referred to Liu [8-3], pages 270–275, or Chapter 2 of Ford and Fulkerson [4-3] for further discussions.

5. Lossy Networks: So far we have assumed that the flow does not vary along an edge. In many practical transport networks, however, the flow does suffer loss during transmission, due to leakage, evaporation, and so forth, Such networks are called *lossy transport networks* (or *lossy networks*).

A lossy network has an additional parameter, called efficiency, λ_{ij}, associated with each directed edge (i, j). For each edge (i, j) there are two flows: flow f_{ij} entering the edge and flow f_{ij}^* leaving the edge. These quantities are related as follows:

$$f_{ij}^* = \lambda_{ij} \cdot f_{ij}, \quad \text{for each edge } (i,j) \text{ in } G.$$

The efficiency λ_{ij} is a positive number. It is less than unity if there is a loss during transmission and is more than unity of there is a gain (for instance, improvement in the signal due to repeaters in a communication line).

At each intermediate vertex the total outgoing flow must still be equated to the total incoming flow. The larger of the two quantities f_{ij} and f_{ij}^* must still not exceed c_{ij}, the capacity of the edge (i, j). As in the case of ordinary transport networks (in which $\lambda_{ij} = 1$, for every edge), the goal is to maximize the flow arriving at the sink t. Moreover, for the same value of the flow arriving at the sink, we may have different values of flow leaving the source. Therefore, another goal is to find a flow pattern that gives the maximum flow arriving at the sink for a minimum amount leaving the source. This is called an optimal flow in a lossy network.

The max-flow min-cut theorem has been extended to lossy networks. Conditions for optimality have been obtained, and algorithms for optimal flows have been devised. For details, see the paper by Onaga [14-18] or pages 277–288 in [14-7].

14-3. MINIMAL-COST FLOWS

Suppose that associated with each edge (i, j) in a transport network G there is an additional number d_{ij}, which may be thought of as the cost of unit flow through (i, j). It is desired to construct a flow pattern sending a specified value w from source s to sink t satisfying constraints (14-1), (14-2), (14-3), and (14-4), which minimizes the total flow cost,

$$\sum_{(i,j) \in G} d_{ij} \cdot f_{ij}, \tag{14-7}$$

over all flows that send w units from s to t.

This is one of the most practical problems in network flows. It is also a classic problem in linear programming and is known as the *transportation problem*. Many problems in operations research can be formulated as a transportation problem.

To find a flow pattern that minimizes the cost, we start with a minimal-cost directed path from s to t and saturate this path (i.e., assign a flow to the path such that at least one edge in the path reaches its capacity). Then by using the following theorem recursively we obtain the minimal-cost flow pattern of desired value. Let us call a path from s to t unsaturated for a given flow in G if $f_{ij} \leq c_{ij}$ for every forward edge (i, j) and $f_{ij} \geq 0$ for every backward edge (see Fig. 14-2).

THEOREM 14-3

Let f be the minimal-cost flow pattern of value w from s to t. Then the flow pattern f', obtained from f by adding $\delta \leq 0$ to the flow in forward edges of a minimal-cost unsaturated path, and subtracting δ from the flow in the backward edges of the path, is a minimal-cost flow of value $w + \delta$.

This theorem is of central importance in constructing minimal-cost flow patterns. For a formal proof of this intuitively obvious result, the reader is referred to Ford and Fulkerson [4-3], pages 121–122. Theorem 14-3 states that at every stage of construction each additional unit of flow is to be sent through the least-cost available path. All unsaturated paths from s to t are available paths, and in computing the cost of an available path p one takes into account not only the cost of adding the flow to the forward edges in p but also the savings due to reduction of existing flows in the backward edges of p. Let us illustrate the application of the theorem with an example.

In Fig. 14-5 we have a transport network. Of the pair of numbers written next to an edge, the first number is the capacity c_{ij} and the second one is the cost d_{ij} of a unit flow. To find a minimal-cost maximal flow from s to t, we go through the following steps.

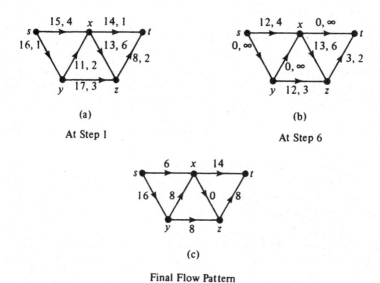

Final Flow Pattern

Fig. 14-5 Minimal-cost flow.

1. The minimal-cost path is *syxt*, and total path cost is 4. We can send a maximum possible flow of 11 units through this path, thus saturating edge (y, x) in this path.
2. We modify the network by subtracting 11 from the current capacities of all edges in *syxt*. Set $d_{yx} = \infty$.
3. In the modified network, the minimal-cost path from s to t is *sxt*. The cost is 5. We sent the maximum possible flow of 3 units through *sxt*, which saturates edge (x, t) in the path.
4. We further update the network by sending the capacities in the path *sxt* and setting $d_{xt} = \infty$.
5. In the resulting network the minimal-cost path is *syzt* of cost 6 and capacity 5. Sending 5 units of flow through *syzt* saturates (s, y).
6. Appropriate updating yields the network in Fig. 14-5(b).
7. In Fig. 14-5(b) the minimal-cost path is *sxyzt* with a cost of $4 - 2 + 3 + 2 = 7$. Sending 3 units along this path saturates a cut-set and thus the algorithm terminates. The desired flow pattern obtained is given in Fig. 14-5(c). The value of the flow from s to t is $11 + 3 + 5 + 3 = 22$ units, and the cost is $4 \times 11 + 5 \times 3 + 6 \times 5 + 7 \times 3 = 110$.

14-4. MULTICOMMODITY FLOW

In some practical situations it becomes necessary to deal with several distinct commodities flowing simultaneously through a given transport network. Each commodity has its own source and its own sink. All flows share the edge capacity, and therefore, as in the single-commodity case, the sum of all flows through an edge must not exceed the capacity of the edge. For each commodity the flow is preserved at every intermediate vertex.

For illustration, let us consider the transport network in Fig. 14-6 through which commodities 1 and 2 are flowing. Commodity 1 is to be transported from s_1 to t_1 and commodity 2 from s_2 to t_2.

For a two-commodity case, let f_{ij}^1 and f_{ij}^2 be the flows of commodities 1

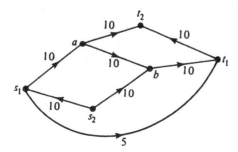

Fig. 14-6 Two-commodity transport network.

and 2, respectively, through an edge (i, j) in G. Then, analogous to the single-commodity case, the constraints in Fig. 14-6 are

$$f_{ij}^1 + f_{ij}^2 \leq c_{ij}$$

and

$$\sum_{j \in G} f_{ij}^1 = \begin{cases} w_1 & \text{for } i = s_1, \\ -w_1 & \text{for } i = t_1, \\ 0 & \text{for all other vertices,} \end{cases}$$

$$\sum_{j \in G} f_{ij}^2 = \begin{cases} w_2 & \text{for } i = s_2, \\ -w_2 & \text{for } i = t_2, \\ 0 & \text{for all other vertices.} \end{cases}$$

These constraints can be easily written down for a k-commodity flow. In such flows two problems are usually raised: (1) Construct patterns for all k-commodities such that the total sum of the flow values $w_1 + w_2 + \cdots + w_k$ is maximized; (2) Given the flow values $w_1, w_2, \ldots, w_k$ for each commodity and a network G, find out if these values of flows can be achieved simultaneously.

Simply sending the maximum amount of each commodity will not in general maximize the total value. This can be seen even in the simple case of Fig. 14-6. If we maximize w_1 alone, we get $w_1 = 15$ and $w_2 = 0$. On the other hand, the maximum value of $w_1 + w_2$ is obtained with $w_1 = 5$ and $w_2 = 20$. Thus to maximize the total value, we must know how to allocate commodities to each edge.

There is no result similar to the max-flow min-cut theorem for the multicommodity flow in general. Only in some special cases (such as when G is undirected and there are only two commodities) has it been possible to get a theorem analogous to the max-flow min-cut theorem.

For further reading in this specialized and rather involved topic, the interested reader is referred to Chapter 11 of Hu's book [14-12], Chapter 3 of Frank and Frisch [14-7], and the Ph.D. dissertation of Sakarovitch [14-20].

14-5. ADDITIONAL APPLICATIONS

We have been discussing how various types of shipping problems can be solved by means of network-flow techniques. In addition to these, there are a surprisingly large number of combinatorial problems in operations research that can be formulated (and then solved) as network-flow problems. Take for instance the matching or assignment problem discussed in Section 8-4. We have p men $M_1, M_2, \ldots, M_p$ and q jobs $J_1, J_2, \ldots, J_q$, and it is known which men are qualified for which jobs. When is it possible to fill all jobs with qualified men or when is it possible to assign each man a job he is qualified for?

The problem can be formulated as a network-flow problem, as shown in Fig. 14-7. Construct a p-source q-sink flow network, such that an edge (M_i, J_k) exists if and only if man M_i is qualified for job J_k. Join all sources to a supersource s and all sinks to a supersink t. Assign capacities of one unit to each (s, M_i) and to each (J_i, t). The capacities of the remaining edges are made infinite. Then the optimal assignment problem becomes that of constructing a flow pattern with maximum value from s to t.

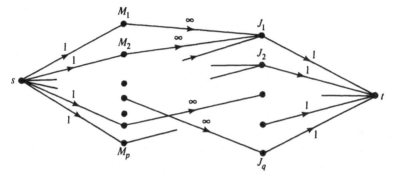

Fig. 14-7 Flow network for an assignment problem.

Observe that in such a flow pattern

$f_{sM_i} = 1,$ if ith man is assigned to a job,
$\quad\quad = 0,$ otherwise,
$f_{J_k t} = 1,$ if kth job has been assigned,
$\quad\quad = 0,$ otherwise,
$f_{M_i J_k} = 1,$ if ith man is assigned to the kth job,
$\quad\quad = 0,$ otherwise.

More complicated personnel assignment problems have been formulated in terms of network flow. Numerous other types of problems have also been solved as flow problems. For these the reader is referred to the bibliography in the survey paper by Fulkerson [14-9].

14-6. MORE ON FLOW PROBLEMS

Flow problems may be looked upon as a generalization of connectivity problems, studied in Chapter 4. The study of connectivity involves a search for paths between pairs of vertices in a graph. A path from a vertex x to a vertex y implies that some amount of flow can be sent from x to y. To find how much, we have to consider the capacities of the edges in the path.

The maximum number of edge-disjoint paths between a pair of vertices x, y is equal to the minimum number of edges that when removed from the

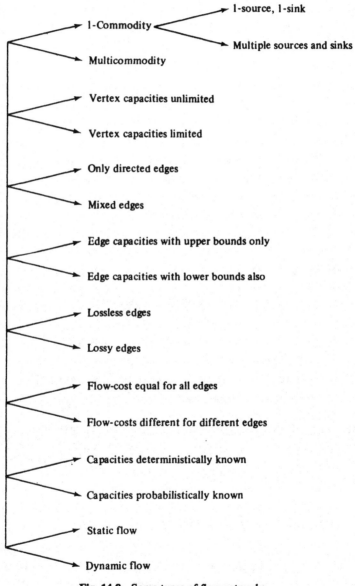

Fig. 14-8 Some types of flow networks.

graph leave no path between x and y. This number is precisely the number of edges in the smallest cut-set with respect to x and y. This concept, when applied to a graph with edge capacities, becomes the max-flow min-cut theorem. It equates the sum of maximum capacities of paths between x and y to the capacity of the minimum cut-set with respect to x and y.

It is natural to seek results as elegant as the max-flow min-cut theorem for more general networks. Some generalizations are easily made. Others, such as for the multicommodity flow, have not been possible so far. A summary of some common types of flow networks is given in Fig. 14-8.

It is interesting to compare transport networks with electrical networks, studied in Chapter 13. A transport network can be thought of as a special type of resistor network that obeys Kirchhoff's current law (KCL), but not the voltage law (KVL). Moreover, the resistors have no resistance for currents (i.e., flows) up to a certain value c_{ij} and then have an infinite resistance for current larger than that. In such a network no voltage (potential, pressure, or tension) exists across any branch.

Conversely, an electrical network problem can also be formulated as a flow problem. Consider a resistor network G with current sources in which we wish to find currents (i.e., flows) f_{ij} flowing through every edge (i, j). The upper and lower bounds on the currents are

$$c_{ij} = \infty,$$
$$b_{ij} = -\infty.$$

The flow pattern must satisfy KCL; that is,

$$\sum_i f_{ji} - \sum_i f_{ij} = 0 \qquad (14\text{-}4)$$

for every vertex j in G. The flow pattern must also satisfy KVL. It was observed by J. C. Maxwell in 1893 that among all flow patterns satisfying (14-4) the one that minimizes the power dissipation

$$\sum_{\substack{\text{for all} \\ (i,j) \in G}} r_{ij} \cdot f_{ij}^2 \qquad (14\text{-}8)$$

is the one that satisfies Kirchhoff's voltage law also. Quantity r_{ij} is the electrical resistance of the edge (i, j). [That minimization of (14-8) is equivalent to satisfying KVL, assuming KCL, in a resistive network with current sources is left as an exercise.]

Thus an electrical network problem can be viewed as a flow problem, which minimizes a *quadratic* flow-cost function (14-8) subject to linear constraints (14-4). Obviously, then, an electrical network problem (subject to KCL and KVL) is not an LP problem.

14-7. ACTIVITY NETWORKS IN PROJECT PLANNING

One of the most popular and successful applications of networks in operations research is in the planning and scheduling of large complicated projects. The two best-known names in this connection are CPM (Critical Path Method) and PERT (Program Evaluation and Review Technique). A project is divided into many well-defined and nonoverlapping individual jobs, called *activities*. Due to technical restrictions, some jobs must be finished before others can be started (such as washing before drying, putting foundation before erecting walls, etc.). In addition to this precedence relationship among the activities, each activity also requires a certain time, called the *duration* of the activity. Given the list of activities in a project, the list of immediate prerequisites (i.e., predecessors) for each activity, and the durations, a weighted digraph can be drawn to depict the project, as follows: Each edge represents an activity, and its weight represents the duration of the activity. The vertices represent beginnings and endings of activities and are called *events* or *milestones* in the project. An activity (i, j) cannot be started before all activities leading to the event i have been completed. Each event in the project is a well-defined occurrence in time (such as walls erected, shipment arrived, etc.). Such a weighted, connected digraph representing activities in a project is called an *activity network*.

Let us take an extremely simple example. Suppose that we have a project consisting of six activities A, B, C, D, E, and F, with the restriction that A must precede C and D; B and D must precede E; and C must precede F. The durations for the activities A, B, C, D, E, and F are 5, 7, 6, 4, 15, and 2 days, respectively. The activity network of this project is shown in Fig. 14-9.

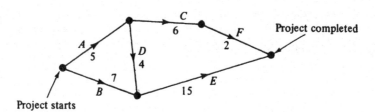

Fig. 14-9 Activity network.

Observe that an activity network must be acyclic; otherwise, we would have an impossible situation in which no activity in the directed circuit could be initiated—a vicious cycle. Also observe that the vertex denoting the start of the project must have zero in-degree, since no activity precedes this vertex. Likewise, the vertex denoting the termination of the project must have zero out-degree, as no activity follows this vertex.

Dummy Activity: In the example of the activity network considered in Fig. 14-9, suppose we had an additional restriction that activity *F* could not be started before *B* and *D* were completed. We can incorporate this precedence relationship by drawing an edge from vertex *x* to *y* (Fig. 14-10). Such an

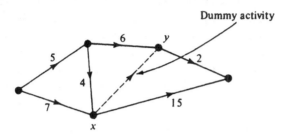

Fig. 14-10 Dummy activity in a network.

edge, which represents only a precedence relationship and not any job in the project, is called a *dummy activity*. Dummy activities become necessary when the existing activities are not enough to portray all precedence relationships accurately. All dummy activities are of zero duration and are usually shown in broken lines.

Two parallel edges (i.e., activities having the same immediate predecessor and the same immediate successor) may be replaced by a single edge, combining both activities into one [Fig. 14-11(a)]. If, however, the activities are to be kept track of separately, then a dummy activity and a dummy event

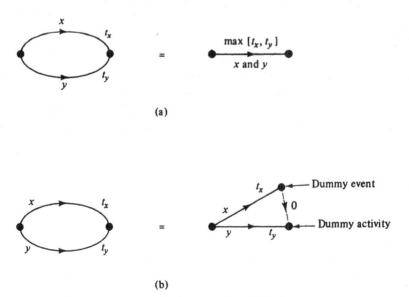

Fig. 14-11 Replacement of parallel edges.

must be created [Fig. 14-11(b)]. And, as there can be no self-loop in an activity network, we have only simple digraphs for an activity network.

An activity network can be assumed to have exactly one vertex with zero in-degree and exactly one vertex with zero out-degree. If there is more than one vertex having zero in-degree, one arbitrarily selects one of these for the start event and draws dummy activities from this to the other vertices. The vertices with zero out-degrees are handled similarly.

In brief, an activity network is a representation of two aspects of a project: (1) precedence relationships among the activities, and (2) their durations. It is a connected, weighted, simple, acyclic digraph with exactly one vertex of zero in-degree and exactly one vertex of zero out-degree.

14-8. ANALYSIS OF AN ACTIVITY NETWORK

A newly constructed network should first be checked for any directed circuit. A directed circuit implies inconsistency in the network, which must be corrected. Although Theorem 9-17 gives an algorithm for finding whether or not a digraph has a directed circuit, a more efficient method is provided by topological sorting of vertices, defined as follows:

Topological Sorting: The vertices of a digraph G are said to be in *topological order* if they are labeled $1, 2, 3, \ldots, n$ such that every edge in G leads from a smaller numbered vertex to a larger one. That is, for every edge (i, j) in G we have $i < j$. The process of relabeling the vertices such that they are in a topological order is called *topological sorting*. Clearly, if a digraph contains a directed circuit, it is not possible to put its vertices in a topological order. The following construction procedure shows that the vertices of every acyclic digraph G can be put in a topological order.

Start with a vertex with zero in-degree and label it 1. Delete vertex 1 from G, and in the remaining digraph $(G - 1)$ find a vertex with zero in-degree and label it 2. From $(G - 1)$ delete vertex 2 and repeat the process, till either (1) every vertex is labeled, or (2) we find a subdigraph g in which there is no vertex with zero in-degree. In view of Theorem 9-15, case (2) is possible only if g contains a directed circuit. Thus we can state

THEOREM 14-4

The vertices in a digraph can be arranged in a topological order if and only if the digraph is acyclic.

Topological sorting performs two functions in an activity network: (1) it detects directed circuits, if any, in the network, and (2) it puts the events in a topological order $1, 2, 3, \ldots, n$, where 1 is the start event and n is the completion event of the project.

Analysis of an Activity Network

Topological sorting is an important process in many problems besides activity network analysis. For example, if we want to arrange the words in a glossary so that no term is used before it has been defined, we resort to a topological sorting. We shall therefore present an algorithm for this important process in a step-by-step fashion.

Algorithm for Topological Sorting

1. Set $i \leftarrow 1$.
2. Find an unlabeled vertex with zero in-degree, and label this vertex i. If no such vertex exists, go to step 4.
3. Set $i \leftarrow i + 1$; and go to step 2.
4. If every vertex in G has been labeled, stop. Otherwise, go to step 5.
5. If the out-degree of any vertex labeled so far is nonzero, remove all edges incident out of every labeled vertex and go to 2. If there are some unlabeled vertices and the out-degree of each of the labeled vertices is zero, we have a directed circuit in the network, stop.

Note that there may be more than one topological ordering of the vertices in a given acyclic digraph, because at step 2 in the algorithm it is possible to have more than one vertex with zero in-degree.

Critical Path: Having made sure that the activity network G contains no directed circuits and (in the process) having placed the vertices of G in a topological order $1, 2, \ldots, n$, our next task is to determine the project duration. The minimum time required to complete the entire project is equal to the length (i.e., sum of the activity durations) of the longest directed path in G. (The longest directed path is, of course, from 1 to n.) The longest directed path is called a *critical path* (CP). The vertices and edges in a CP are called the *critical events* and the *critical activities*, because any delay in them will delay the entire project. In Fig. 14-9 the critical path is ADE and the project duration is $5 + 4 + 15 = 24$ days. There may be more than one critical path in a given activity network.

Instead of determining the longest path only from 1 to n, let us determine the longest paths from vertex 1 to every vertex k in G, where $k = 2, 3, \ldots, n$. The length of the longest path from 1 to k is called the *earliest event time* for event k, because this is the earliest possible time at which event k can be realized.

Since digraph G is acyclic, the method of obtaining shortest paths from a specified vertex to all others, given in Chapter 11, can be easily modified for finding the longest paths. In fact, since the vertices are already topologically ordered, the task is even simpler. Let

$$t_{ij} = \text{duration of activity } (i,j) \text{ in } G,$$

and let

$$T(k) = \text{length (i.e., time) of longest path from 1 to } k, \quad \text{for } k = 1, 2, 3, \ldots, n.$$

Clearly, $T(1) = 0$. Vertex 2 can be reached only from vertex 1 (because of the topological order), and therefore

$$T(2) = T(1) + t_{12} = t_{12}.$$

Vertex 3 cannot be reached from any vertex except from 1 and 2. Therefore,

$$T(3) = \max[T(1) + t_{13}, T(2) + t_{23}].$$

Similarly, vertex 4 can possibly be reached only from 1, 2, and 3. Therefore,

$$T(4) = \max[T(1) + t_{14}, T(2) + t_{24}, T(3) + t_{34}].$$

And so on. The general expression can thus be written as

$$T(k) = \max_{i<k}[T(i) + t_{ik}], \tag{14-9}$$

where the maximum is over all vertices i from which there is a directed edge (i, k) to vertex k.

The solutions of these equations can be performed one by one, and $T(1)$, $T(2), \ldots, T(n)$ obtained successively. Let us take a simple example:

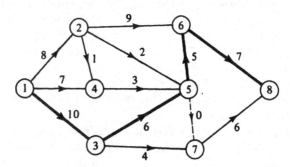

Fig. 14-12 Activity network.

An activity network consisting of 8 events, 12 activities, and 1 dummy activity is shown in Fig. 14-12. The event labels are shown inside the small circles representing the vertices. They are in topological order $1, 2, 3, \ldots, 8$. (Since the vertices are identified and the digraph is simple, the activity labels have been dispensed with.) The durations of activities (in some unit of time) are shown next to the edges. Let us compute the earliest event times $T(i)$ for $i = 1, 2, 3, \ldots, 8$.

$$T(1) = 0,$$
$$T(2) = t_{12} = 8,$$
$$T(3) = 10.$$

With successive application of Eq. (14-9), we get

$$T(4) = \max[t_{14}, T(2) + t_{24}]$$
$$= \max[7, 9] = 9,$$
$$T(5) = \max[T(2) + t_{25}, T(3) + t_{35}, T(4) + t_{45}]$$
$$= \max[10, 16, 12] = 16.$$

Similarly,

$$T(6) = 21, \quad T(7) = 16, \quad T(8) = 28.$$

Thus the project duration $T(8) = 28$. The critical path is 1, 3, 5, 6, 8, and is shown in heavy lines in Fig. 14-12.

This computation of earliest times of topological ordered events by tracing the longest paths from vertex 1 to vertices $2, 3, \ldots, n$, successively, is referred to as *forward calculation*.

Latest Event Time: To ensure that the project is finished at time $T(n)$, we have to make sure that none of the critical activities is delayed. There is, however, a certain amount of latitude in scheduling noncritical activities. A noncritical activity may be allowed to slip (and thereby save money or nerves) to a certain extent without delaying the project completion. The latest time by which an event k must be realized without increasing the project duration is called the *latest event time $T'(k)$*. For example, in Fig. 14-12 event 7 may be realized latest by time unit 22 (and no later) without affecting the completion time of the project. It is not difficult to see that the latest event time is given by the relation

$$T'(k) = T(n) - \text{time taken along longest path from vertex } k \text{ to } n.$$

If we reverse the direction of every edge in the network, the vertices will still be topologically sorted, but the order would be reversed, $n, n-1, \ldots, 2, 1$. Starting from vertex n, one could move toward vertex 1 and compute the times taken along the longest paths, using Eq. (14-9) successively in the reversed network. Starting with relation

$$T'(n) = T(n),$$

we get the following recursive relation for the latest event time for vertex k:

$$T'(k) = \min[T'(i) - t_{ki}], \qquad (14\text{-}10)$$

where the minimization is over vertices i to which there is directed edge (k, i) from vertex k. For example, in Fig. 14-12,

$$T'(7) = 28 - 6 = 22,$$
$$T'(6) = 28 - 7 = 21,$$
$$T'(5) = \min[T'(7) - 0, T'(6) - 5] = 16,$$

and so on. Table 14-1 shows all earliest and latest event times in the activity network of Fig. 14-12. Note that $T(i) = T'(i)$ if and only if vertex i is in a critical path.

Event i	$T(i)$	$T'(i)$
1	0	0
2	8	12
3	10	10
4	9	13
5	16	16
6	21	21
7	16	22
8	28	28

Table 14-1 Earliest and Latest Event Times

Slacks: As a measure of maximum latitude available in a noncritical activity, let us look at the following quantity called *total slack* (or *float*) of activity (i, j).

$$s_{ij} = T'(j) - T(i) - t_{ij}. \qquad (14\text{-}11)$$

Quantity s_{ij} represents the maximum permissible delay in activity (i, j), which is possible when i is realized as early as possible, and j is delayed as much as possible.

Since each activity has two end vertices, each one of which has two time values, it is possible to define four different slacks for each activity. We have considered only the most important one, the total slack. The second most important slack is the *free slack*, v_{ij}, defined as

$$v_{ij} = T(j) - T(i) - t_{ij}. \qquad (14\text{-}12)$$

This is the amount by which an activity (i, j) can be delayed without delaying the early start of any other activity. Total slacks and free slacks for all activities in the network of Fig. 14-12 are shown in Table 14-2.

Observe that the total slack $s_{ij} = 0$ if and only if (i, j) is a critical activity, whereas v_{ij} may be zero even if (i, j) is not a critical activity. Also note that $s_{ij} \geq v_{ij} \geq 0$.

Activity	Total Slack	Free Slack
(1, 2)	4	0
(1, 3)	0	0
(1, 4)	6	2
(2, 4)	4	0
(2, 5)	6	6
(2, 6)	4	4
(3, 5)	0	0
(3, 7)	8	2
(4, 5)	4	4
(5, 6)	0	0
(5, 7)	6	0
(6, 8)	0	0
(7, 8)	6	6

Table 14-2 Total and Free Slacks of Activities in Network of Fig. 14-12

In the foregoing analysis of a network, called the critical path method (CPM), we have accomplished the following:

1. Checked for directed circuits.
2. Arranged events in topological order.
3. Identified critical path (or paths) and computed the project duration.
4. Computed earliest event time $T(k)$ for each event.
5. Computed latest event time $T'(k)$ for each event.
6. Computed slacks for each activity.

Having identified the critical activities, we can concentrate only on these and by expediting them reduce the total project duration. Second, the slacks can be utilized to reduce the peak demands for certain machines or skilled workers.

Project Cost Curve: Although we have assumed a constant duration for each activity, in practice allocation of more money can usually get a job done faster. Given a fixed budget for the project, how should the money be allocated among the activities so that the project is completed at the earliest possible date? If for each activity the time–cost relation is linear, this problem can be shown to be a minimal-cost flow problem (see pages 151–162 of [4-3] or [14-8]). The solution of the problem will be a curve showing project cost versus project duration, and, depending on the budget (or the target completion date), one would pick a point on this curve. Such a curve is known as the *project cost curve*.

Often in activity networks, in addition to time and cost, there may be other parameters, such as personnel required, shop facilities necessary, and so forth, associated with each edge.

In CPM networks activity durations were assumed to be precisely known. If the activity durations are random variables with given probability distributions, the network goes by the acronym PERT (Program Evaluation and Review Technique). Whereas CPM focuses on optimizing the total project cost, PERT is more concerned with estimates of completion dates, scheduling requirements, and so forth. Activity networks with variations of these two are also encountered.

14-9. FURTHER COMMENTS ON ACTIVITY NETWORKS

In this chapter vertices were used to represent events and the edges to represent activities. There is another representation often used in the literature in which the vertices denote activities and the edges represent only the precedence relationships among the activities. Obviously, for a given set of activities these two will yield different graphs. For example, the project activities of Fig. 14-9 are shown in both representations in Fig. 14-13.

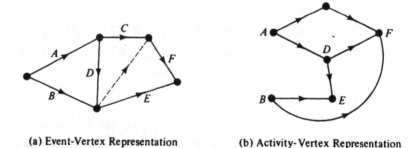

(a) Event-Vertex Representation (b) Activity-Vertex Representation

Fig. 14-13 Two representations for the same activities.

It is not difficult to transform the event-vertex representation into the activity-vertex representation and vice versa (see [14-5]). In fact, the activity-vertex representation is the edge digraph (see Problem 9-16) of the event-vertex representation, if we disregard the dummy activities in the latter. There is no need of dummy activities in the activity-vertex representation. Both types of representations are widely used in the literature. Each has its own slight advantage over the other, but there is no basic difference between the two as far as their analyses are concerned.

The activity network and its application in project planning have been in existence only since 1957. In these few years its success has been spectacular.

Computer programs for analyzing CPM and PERT networks are part of the standard program library of almost every computing center. Large networks consisting of thousands of activities are often analyzed.

Generally, the network is constructed from a list of activities and a precedence table. Generation of a precedence table is a manual job because it involves an intimate knowledge of the processes in the project. The construction of the activity network, including the dummy activities, from the precedence table can be relegated to the computer, although it is still done mostly manually. Construction of a composite network from subnetworks can also be programmed [14-22].

A typical computer program for critical-path analysis consists of three phases: (1) cycle-checking and topological-sorting phase, (2) forward-time-calculation phase, and (3) backward-time-calculation phase. Shortcuts have been suggested that can complete the critical-path analysis in a single phase, and save computation time in the case of large networks [14-17].

We have presented the bare essentials of the activity network analysis. Much more can be done with graph theory in project planning.

14-10. GRAPHS IN GAME THEORY

The theory of games has become an important field of mathematical research since the publication of the first book on the subject by John von Neumann and Oskar Morgenstern in 1944. Game theory is applied to problems in engineering, economics, and war science to find the optimal way of performing certain tasks in a competitive environment.

The general idea of game theory is the same as the one we associate with parlor games such as chess, bridge, and checkers. The distinction between a puzzle and a game is that in a game one plays against one or more human opponents, whereas a puzzle involves a solitary effort to solve a problem.

A game may be played between two persons, such as chess, or among more than two persons, such as poker. The former is called a *two-person* game and the latter an *n-person* game. Another classification of games is based on whether or not an element of randomness is introduced, such as by dice or cards. A third element in categorizing a game is whether or not a player has complete information on the position of a game at every move. A game such as chess, in which each player knows exactly where the game stands is called a *perfect-information* game. Bridge, in which one does not know what cards the other players have, is an *imperfect-information* game. A game is called *finite* if each player has a finite number of choices available at each move and the game must end after a finite number of moves. An *infinite* game is one in which a player chooses a move from an infinite set of moves.

We shall confine ourselves to the study of two-person, perfect-information, finite games without chance moves. A digraph is a natural representation of such a game. The vertices represent the *positions* (also called *states*) in the game and the edges represent the moves. There is a directed edge from vertex v_i to v_j if and only if the game can be transformed from position (state) v_i to v_j by a move permissible under the rules of the game. As an example, let us look at a very simple game. It is a simplified version of a game called *nim*.

Simplified Nim: Two piles of sticks are given and players A and B take turns, each taking any number of sticks from any one pile. The player who takes the last stick wins, and since the finite quantity of sticks will eventually be exhausted, it is obvious that the game allows no draw. As a further simplification, let us start with two piles containing two sticks each. The complete game is described by the digraph in Fig. 14-14. Each state of the game is described by an ordered pair of labels (x, y), indicating the number of sticks in the first and the second pile, respectively.

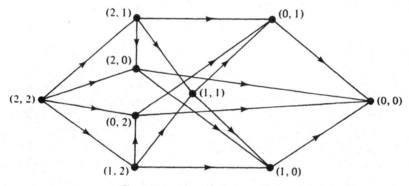

Fig. 14-14 Simplified game of nim.

Let us observe some properties of such a *game digraph* (a digraph representing a two-person, perfect-information, finite game without chance moves):

1. The digraph has a unique vertex with a zero in-degree. This vertex represents the starting position in the game and is therefore called the *starting vertex*. Vertex (2, 2) in Fig. 14-14 is the starting vertex.
2. There are one or more vertices with zero out-degree. These correspond to the closing positions in the game, and are called the *closing vertices*. Vertex (0, 0) is the closing vertex in Fig. 14-14.
3. A game digraph is a connected, acyclic digraph. A directed circuit would imply that the game could go on indefinitely. (In practice, in a game such as chess, where the game may return to a state, endless

matches are prevented by means of a rule that after a certain number of repetitions of the same move, the game is declared stalemated.)

4. Each directed path from the starting vertex to a closing vertex represents one complete play of the game. This path consists of edges representing the moves of the two players alternately.

The most important question in a game is the following: When and how can a player choose his moves so that he is certain of winning? We shall first answer this question for the specific game in Fig. 14-14, and then generalize it.

Let us call a position "won" if the player who brought the game to this position can force a victory. Conversely, a position is dubbed "lost" if the player who brought the game to this position can be forced to lose. In keeping with this characterization of vertices, the closing vertex in Fig. 14-14 is to be marked as won, because the player who brought the game to this position is the winner. Having marked this vertex as won, let us use the following procedure to mark the remaining vertices as won or lost.

Mark an unmarked vertex won if *all* its successors are marked lost, and mark an unmarked vertex lost if at least *one* of its successors is marked won. (This is because it is assumed that each player is intelligent and makes the best possible move at each stage.) This results in vertices (0, 0), (1, 1), and (2, 2) being marked as won and the remaining as lost. And thus the player who makes the second move has the winning strategy, since he can force his opponent to move to the vertices marked as lost.

To generalize the foregoing method of finding a winning strategy, let us introduce the concept of kernel in a digraph.

Kernel of a Digraph: A set of vertices K in a digraph G is called a *kernel* (or *nucleus*) of G if

1. No two vertices in K are joined by an edge.
2. Every vertex v not in K has an edge directed from v to some vertex in K.

Conditions 1 and 2 correspond respectively to definitions of an independent set and a dominating set in an undirected graph (recall Chapter 8). What are some types of digraphs that have kernels? Theorem 14-5 characterizes one such type.

THEOREM 14-5

Every acyclic digraph has a unique kernel.

Proof: The theorem will be proved by a constructive procedure, at the end of which all vertices forming the kernel will be painted red. Let G be the given acyclic digraph. According to Theorem 9-15, G must have at least one vertex with zero out-degree. Let V_1 be the set of all vertices in G with zero out-degree. Since these vertices must all be in the kernel of G, paint them red.

Next, let W_1 be the set of all those vertices in G from which there is at least one directed edge to some vertex in V_1. Clearly, no vertex in W_1 can be included in the kernel. Delete from G all vertices in W_1 (together with the edges incident on them, of course), and thus obtain subgraph $(G - W_1)$.

Subgraph $(G - W_1)$ is also acyclic. Let V_2 be the set of all vertices with zero out-degree in $(G - W_1)$. Since in the original digraph G no vertex in set V_2 had a zero out-degree, every vertex in V_2 had to have at least one edge going to some vertex in W_1. Moreover, in digraph G no vertex in V_2 could have been adjacent to any vertex in V_1. Nor could any vertex in V_2 have been adjacent to any other vertex in V_2, because the out-degree of each vertex in set V_2 of $(G - W_1)$ is zero.

Thus we conclude that every vertex in V_2 must also be included in the kernel of G and therefore be painted red.

This procedure is continued till every vertex in G is either deleted or painted red. The unique set of vertices painted red constitutes the kernel of the digraph. ■

As an illustration, let us find the kernel in the acyclic digraph in Fig. 14-14. It is easily seen that the set of three vertices marked (0, 0), (1, 1), and (2,2) is the kernel.

Let A be the player who makes the first move in the game and B be the player who makes the second move. Assuming that the rule of the game is such that the player who is able to make the last possible move in the game is always the winner, we have the following important result.

Theorem 14-6

In the game digraph if the starting vertex is not in the kernel K, then player A is assured of a win, and A can win by always selecting vertices in K.

Proof: Since the starting vertex is not in set K, player A can move the game to a vertex x in K. If this vertex x is a closing vertex, A is the winner. If not, the second player B will have to move to some vertex y, which is not in the kernel K. In his next move A can take the game to some vertex in K. The game continues, with B forced to take it out of K and A bringing it back into K. Eventually, the play will be brought to a closing vertex by A, because all closing vertices are in K. Thus A wins the game. ■

Corollary

It follows from the proof of this theorem that if the starting vertex is in the kernel, the second player B has the winning strategy; and B can win by always selecting vertices in the kernel.

In the foregoing analysis the rules of the game were assumed to be such that the player who made the last possible move was always the winner. In many games (such as chess or tic-tac-toe) some of the closing vertices represent a draw and others a win. In such a game choosing vertices from the kernel will only assure a player a win or a draw.

There are also games in which the rule is such that the player who is forced to make the last move is the loser rather than the winner. From such a game digraph, if we remove all edges corresponding to the last moves, the game can be converted to the type in which the winner is the player making the last move. In other words, the game is decided at the time the second-to-the-last moves are made. For example, let us modify the nim game of Fig. 14-14 such that the player forced to take the last stick is the loser. Then erasing the last moves from the digraph, we get Fig. 14-15. The three vertices marked won constitute the kernel in this digraph. Even in this game player B has the winning strategy.

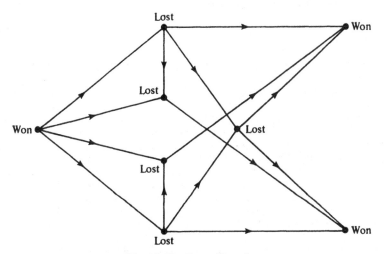

Fig. 14-15 Game digraph.

Thus, at least in theory, it is possible to construct the game digraph for any two-person, perfect-information, finite game with no chance moves, and, since the digraph is acyclic, it is also possible to obtain its kernel. Therefore, if each player plays according to his best strategy, the outcome is predetermined. It will be either a draw or a certain win for the player who makes the first move if the starting vertex is not in the kernel, or for the player who makes the second move if the starting vertex is in the kernel. In this sense, every game of this type is either "unfair" or "futile."

In reality the situation is not so bleak as it appears. In most nontrivial games, such as checkers or chess, the number of positions (i.e., the vertices in the game digraph) is so enormous that the game digraph cannot even be stored in the memory unit of any existing or contemplated computer.

This is precisely why in real problems in operations research the theory of games provides an approach rather than a complete analysis. Moreover, graph theory is applicable only to a very special but important class of games.

SUMMARY

We have considered three important classes of problems in combinatorial operations research: transportation problems, activity networks, and game theory. These problems can be expressed and solved elegantly as graph-theory problems involving connected and weighted (mostly acyclic) digraphs. From a practical point of view, all these problems are trivial (and so is any combinatorial problem) if the network is small. Many real-life situations, however, consist of huge networks, and therefore it is important to look at these network problems in terms of solving them on computers.

REFERENCES

There is an impressive literature on the subject of flows in networks. Several books are devoted entirely to network-flow problems. Ford and Fulkerson's book [4-3], the first one to be written on the subject, is a classic. Following the publication of this book, a good deal of work has been reported in the literature, particularly on the problems of two-commodity and multicommodity flows, probabilistic flows, flows through lossy network, flows with queues, and synthesis of flow networks. Hu [14-12] and Frank and Frisch [14-7] cover much of these areas. Other books on the subject are Berge and Ghouila-Houri [14-4] and Iri [14-14]. Of several excellent survey papers, Fulkerson [14-9], Hu [14-13], and Beckenbach [14-2] are particularly recommended. These papers provide a list of important references. Chapters on flow problems in Berge [1-1], Busacker and Saaty [1-2], and Liu [8-3] are also recommended. As an example of a good computer program for a typical flow problem see [14-3], which determines the least-cost flow for a network with both upper and lower bounds on edge capacities.

Many books and hundreds of articles have been written on critical-path methods. Some of the books recommended are by Battersby [14-1], Elmaghraby [14-6], and Moder and Phillips [14-16]. Papers by Klein [14-15], Montalbano [14-17], and Schurmann [14-22] discuss computer algorithms and programs for activity networks. Busacker and Saaty [1-2] also contains a section (pages 128–135) on activity networks.

For a brief and succinct historical survey of *nim-like games* (also called *disjunctive* or *take-away games*), see Gardner [14-10]. A lucid general article on games is [14-23]. Two other papers on games, [14-11] and [14-21], are also recommended. Berge [1-1], Busacker and Saaty [1-2], Kaufmann [9-4], and Iri [14-14] each contain a section dealing with graph representation of games.

14-1. BATTERSBY, A., *Network Analysis for Planning and Scheduling*, 2nd ed., St. Martins Press, Inc., New York, 1967.
14-2. BECKENBACH, E. F., "Network Flow Problems," Chapter 12 in *Applied Combinatorial Mathematics* (E. F. Beckenbach, ed.), John Wiley & Sons, Inc., New York, 1964.
14-3. BRAY, T. A., and C. WITZGALL, "Algorithm 336: Netflow," *Comm. ACM*, Vol. 11, Sept. 1968, 631–632; corrected in Vol. 12, 1969.

14-4. BERGE, C., and A. GHOUILA-HOURI, *Programming, Games and Transportation Networks* (translated by M. Merrington and C. Ramanujacharyula), John Wiley & Sons, Inc., New York, 1965.
14-5. DIMSDALE, B., "Computer Construction of Minimal Project Networks," *IBM Systems J.*, Vol. 2, March 1963, 24–36.
14-6. ELMAGHRABY, S. E., *Some Network Models in Management Science*, Springer-Verlag New York, Inc., New York, 1970. Also appeared as articles in *Management Sci.*, Vol. 17, Sept. and Oct. 1970.
14-7. FRANK, H., and I. T. FRISCH, *Communications, Transmission, and Transportation Networks*, Addison-Wesley Publishing Company, Inc., Reading, Mass., 1971.
14-8. FULKERSON, D. R., "A Network Flow Computation for Project Cost Curves," *Management Sci.*, Vol. 7, 1961, 167–178.
14-9. FULKERSON, D. R., "Flow Networks and Combinatorial Operations Research," *Am. Math. Monthly*, Vol. 73, 1966, 115–138.
14-10. GARDNER, M., "Mathematical Games," *Sci. Am.*, Vol. 226, No. 1, Jan. 1972, 104–107.
14-11. GRUNDIG, P. M., and C. A. SMITH, "Disjunctive Games with the Last Player Losing," *Proc. Cambridge Phil. Soc.*, Vol. 52, part 2, 1956, 527–533.
14-12. HU, T. C., *Integer Programming and Network Flows*, Addison-Wesley Publishing Company, Inc., Reading, Mass., 1969.
14-13. HU, T. C., "The Development of Network Flow and Related Areas in Programming," University of Wisconsin M.R.C. Technical Report No. 1096, Aug. 1970; also in the *Proceedings of the 7th International Mathematical Programming Symposium at the Hague*, Sept. 1970.
14-14. IRI, M., *Network Flow, Transportation and Scheduling*, Academic Press, Inc., New York, 1969.
14-15. KLEIN, M. M., "Scheduling Project Networks," *Comm. ACM*, Vol. 10, 1967, 225–231.
14-16. MODER, J. J., and C. R. PHILLIPS, *Project Management with CPM and PERT*, Van Nostrand Reinhold Company, New York, 1970.
14-17. MONTALBANO, M., "High-Speed Calculation of the Critical Paths of Large Networks," *IBM Systems J.*, Vol. 6, 1967, 163–191.
14-18. ONAGA, K., "Optimal Flows in General Communication Networks," *J. Franklin Inst.*, Vol. 283, No. 4, April 1967, 308–327.
14-19. PLISCH, D. C., "New Results Concerning Separation Theory of Graphs," Ph.D. Thesis, University of Wisconsin, Madison Wisc., 1970.
14-20. SAKAROVITCH, M., "The Multicommodity Flow Problem," Ph.D. Dissertation, Operations Research Centre, University of California, Berkeley, 1966.
14-21. SMITH, C. A., "Graphs and Composite Games," *J. Combinatorial Theory*, Vol. 1, 1966, 51–81.
14-22. SCHURMANN, A., "GAN, a System for Generating and Analyzing Activity Networks," *Comm. ACM*, Vol. 11, No. 10, Oct. 1968, 675–679.
14-23. WANG, H., "Games, Logic and Computers," *Sci. Am.*, Vol. 213, No. 5, Nov. 1965, 98–106.

15 SURVEY OF OTHER APPLICATIONS

In the last three chapters we have explored in considerable detail the application of graph theory to three disciplines, switching and coding theory, electrical networks, and operations research. In this final chapter we shall briefly describe how graph theory is used in a number of other areas. The first three sections are somewhat related. They all deal with representation of a system structure by means of a weighted, connected digraph and subsequent analysis of the system through an appropriate study of the digraph. In Section 15-1 a linear system is modeled as a weighted digraph, which has proved to be a convenient tool for analysis. Section 15-2 deals with representation of a stochastic process (a discrete Markov process) by a digraph and makes use of Section 15-1 for its analysis. Section 15-3 uses weighted digraphs for the analysis of computer programs. A discrete Markov process is an appropriate model for many programs, and thus Section 15-2 is made use of in Section 15-3.

Section 15-4, in which graph theory is used as a tool for identification of chemical compounds, is an isolated section. It is, however, an important application of graph theory. Finally, Section 15-5 lists some miscellaneous applications, with relevant references.

15-1. SIGNAL-FLOW GRAPHS

Most problems in analysis of a linear system are eventually reduced to solving a set of simultaneous, linear algebraic equations. This problem, usually solved by matrix methods, can also be solved via graph theory. The graph-theoretic approach is often faster, and, more importantly, it displays cause–effect relationships between the variables—something totally obscured in the matrix approach. This graph-theoretic analysis of a linear system consists of two parts: (1) constructing a labeled, weighted digraph called the *signal-flow graph*, and (2) solving for the required dependent variable from the signal-flow graph.

In a signal-flow graph each vertex represents a variable and is labeled so. A directed edge from x_i to x_j implies that variable x_j depends on variable x_i (but not the reverse). The coefficients in the equations are assigned as the weights of the edges such that the variable x_k is equal to the sum of all products $w_{ik}x_i$, where w_{ik} is the weight of the edge coming into x_k from x_i. As an example, let us construct a signal-flow graph for the system given by the set of three equations,

$$c_{11}x_1 + c_{12}x_2 + c_{13}x_3 = y_1,$$
$$c_{21}x_1 + c_{22}x_2 + c_{23}x_3 = y_2, \qquad (15\text{-}1)$$
$$c_{31}x_1 + c_{32}x_2 + c_{33}x_3 = y_3,$$

which can be rewritten as

$$(c_{11} + 1)x_1 + c_{12}x_2 + c_{13}x_3 - y_1 = x_1,$$
$$c_{21}x_1 + (c_{22} + 1)x_2 + c_{23}x_3 - y_2 = x_2, \qquad (15\text{-}2)$$
$$c_{31}x_1 + c_{32}x_2 + (c_{33} + 1)x_3 - y_3 = x_3.$$

The signal-flow graph representing Eqs. (15-2) is given in Fig. 15-1.

Clearly, the in-degree of a vertex v in a signal-flow graph is zero if and

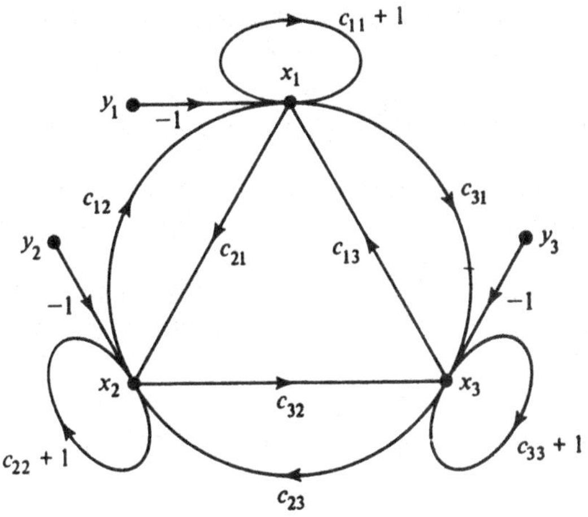

Fig. 15-1 Signal-flow graph for Eqs. (15-2).

only if v represents an independent variable. Also note that a signal-flow graph is connected; otherwise, we have two uncoupled (unrelated) systems thrown together.

A signal-flow graph can be compared to a signal transmission network, in which the vertices corresponding to the independent variables are signal sources, and the other vertices are repeaters, which act as receiving, summing, and transmitting devices. The signals travel along the edges and are multiplied (amplified or attenuated) by the weights of the edges traversed. The label x_i of a vertex equals the sum of all incoming signals, and is the strength of the signal in each outgoing edge from x_i. It is from this analogy that the name "signal-flow graph" comes. For the same reason the edge weights are called *edge gains*, and independent variable vertices are referred to as *source vertices*.

Note that a signal-flow graph contains the same information as the equations from which it is derived; but there does not exist a one-to-one correspondence between the system of equations and the digraph. From the same set of n equations we can obtain $n!$ different signal-flow graphs (some of which may be isomorphic), depending on the order in which the variables x_i's are written on the right-hand side, say, in Eqs. (15-2).

Now, given a set of algebraic equations

$$\mathbf{Cx} = \mathbf{y}, \qquad (15\text{-}3)$$

how do we obtain the weight matrix of the signal-flow graph without first having to draw the digraph? (Like any weighted digraph, the signal-flow graph is completely described by its weight matrix.)

THEOREM 15-1

The weight matrix $\mathbf{W} = [w_{ij}]$ of the signal-flow graph corresponding to Eqs. (15-3) is given by

$$\mathbf{W} = \begin{bmatrix} \mathbf{C} + \mathbf{I} & -\mathbf{I} \\ \mathbf{0} & \mathbf{0} \end{bmatrix}^T, \qquad (15\text{-}4)$$

where $\mathbf{I}$ is the identity matrix of the same order as $\mathbf{C}$, and the superscript T denotes the transposed matrix.

The theorem is not difficult to prove and is left as an exercise. Note that the columns of all zeros in $\mathbf{W}$ correspond to the y vertices (i.e., independent variables).

Although signal-flow graphs can always be constructed from a set of equations, in many physical problems, particularly in electrical systems, signal-flow graphs are drawn directly without first writing the equations. Usually, a signal-flow graph can be drawn as easily as the equations are formulated. Also, writing equations from a signal-flow graph is a simple matter, because each vertex x_k represents one equation of the system in which

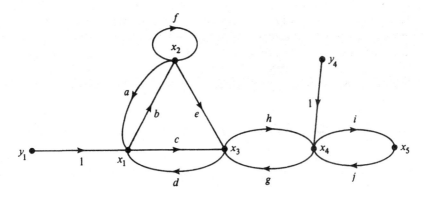

Fig. 15-2 Signal-flow graph.

x_k is equal to the sum of the products of weights of all incoming edges and the labels of the initial vertices of these edges. For example, the system of equations for the signal-flow graph of Fig. 15-2 can be immediately written down as

$$x_1 = y_1 + ax_2 + dx_3,$$
$$x_2 = bx_1 + fx_2,$$
$$x_3 = cx_1 + ex_2 + gx_4,$$
$$x_4 = hx_3 + jx_5 + y_4,$$
$$x_5 = ix_4.$$

These can be rewritten in the same form as Eq. (15-3), where

$$\mathbf{C} = \begin{bmatrix} 1 & -a & -d & 0 & 0 \\ -b & 1-f & 0 & 0 & 0 \\ -c & -e & 1 & -g & 0 \\ 0 & 0 & -h & 1 & -j \\ 0 & 0 & 0 & -i & 1 \end{bmatrix}, \qquad (15\text{-}5)$$

$$\mathbf{x} = \begin{bmatrix} x_1 \\ x_2 \\ x_3 \\ x_4 \\ x_5 \end{bmatrix}, \text{ and } \mathbf{y} = \begin{bmatrix} y_1 \\ 0 \\ 0 \\ y_4 \\ 0 \end{bmatrix}$$

Reduction of Signal-Flow Graphs

The signal-flow graph method of analysis is most useful when we want to solve for only one unknown variable, say x_j, as a function of one independent variable, say y_k. We solve by eliminating all other vertices one by one, taking care that this elimination process does not alter the net product of the edge weights of directed paths from y_k to x_j. This graph reduction corresponds exactly to the algebraic method of eliminating all other variables by systematic substitution. Some elementary reductions of a signal-flow graph are shown

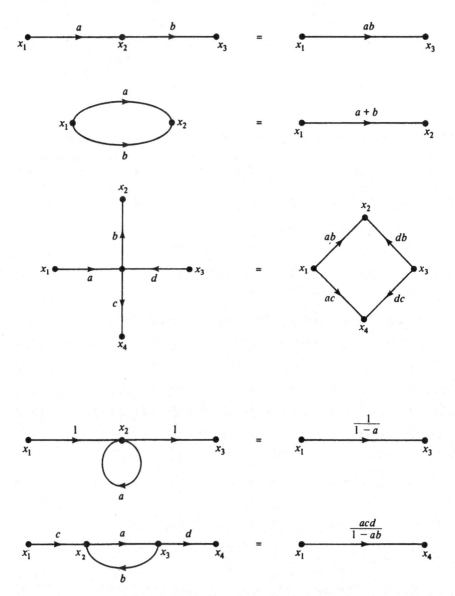

Fig. 15-3 Reductions of signal-flow graphs.

in Fig. 15-3. Repeated application of such reduction steps, selected visually, will eventually lead to elimination of all intermediate vertices. (Apply these reduction steps successively to Fig. 15-2 to eliminate all vertices except y_1 and x_3.)

Although our ability to reduce the digraph by simple inspection adds much to the power and flexibility of signal-flow graphs, it is often better to use a more methodical technique that does not depend on visual inspection. And such a method is provided by *Mason's gain formula*.

Mason's Gain Formula: Let p be a directed path from a vertex a to b in a signal-flow graph G. Then the product of the weights of all edges in this path is called the *path gain* of p (same as path product defined in Section 12-2 for undirected graphs). Similarly, the product of the weights of all edges in a directed circuit (or cycle) Γ is called the *cycle gain* of Γ. For example, a list of all directed circuits and their gains in Fig. 15-2 is

Directed Circuit	Cycle Gain
$x_2 x_2$ (self-loop at x_2)	f
$x_1 x_2 x_1$	ab
$x_1 x_3 x_1$	cd
$x_3 x_4 x_3$	gh
$x_4 x_5 x_4$	ij
$x_1 x_2 x_3 x_1$	bed

Furthermore, for a given signal-flow graph G let us define the following:

$t_1 =$ sum of cycle gains of all directed circuits

$t_2 =$ sum of products of cycle gains of all vertex-disjoint directed circuits taken two at a time

$t_3 =$ sum of products of cycle gains of all vertex-disjoint directed circuits taken three at a time

$\cdots$

$t_k =$ sum of products of cycle gains of all vertex-disjoint directed circuits taken k at a time.

Thus, for the signal-flow graph of Fig. 15-2 we have

$t_1 = f + ab + cd + gh + ij + bed,$

$t_2 = fcd + fgh + fij + abgh + abij + cdij + bedij,$

$t_3 = fcdij,$

and

$t_4 = t_5 = \cdots = 0,$

as the maximum number of vertex-disjoint directed circuits in Fig. 15-2 is only three.

Now we are ready to state and illustrate two theorems, which together constitute the most important result concerning signal-flow graphs.

THEOREM 15-2

Let a signal-flow graph G characterize a set of equations $Cx = y$; then Δ, the determinant of matrix C, is given by

$$\Delta = 1 - t_1 + t_2 - t_3 + \cdots + (-1)^q t_q, \tag{15-6}$$

where q is the maximum number of vertex-disjoint directed circuits in G.

THEOREM 15-3

Let a signal-flow graph G characterize a set of equations $Cx = y$; then the ijth cofactor of C, C_{ij}, is given by

$$C_{ij} = \sum_k P_k \Delta_k, \tag{15-7}$$

where P_k is the path gain of the kth directed path from vertex i to j, Δ_k is the value of Δ in Eq. (15-6) for that part of the digraph having no vertices in common with the kth directed path, and the summation is over all directed paths from i to j.

Combining Theorems 15-2 and 15-3, we get Mason's gain formula, which gives the response x due to the forcing function y_l as

$$\left.\frac{x_j}{y_l}\right|_{\text{all other } y\text{'s}=0} = \frac{C_{ij}}{\Delta},$$

where C_{ij} and Δ are computed from the signal-flow graph G, using Eqs. (15-6) and (15-7).

Besides the original proofs given by S. J. Mason, many different proofs of Theorems 15-2 and 15-3 have been published. All are involved—some more than others. A particularly elegant proof, due to R. Ash, is given in [13-5], pages 102-109. We shall simply illustrate the application of Mason's gain formula by means of an example.

For Fig. 15-2,

$$\Delta = 1 - t_1 + t_2 - t_3$$
$$= 1 - f - ab - cd - gh - ij - bed + fcd + fgh$$
$$+ fij + abgh + abij + cdij + bedij - fcdij.$$

That this indeed is the determinant of the corresponding matrix C as given in Eq. (15-5) can be easily verified by direct computation.

There are two directed paths from y_1 to x_3:

$$p_1 = y_1 x_1 x_3 \quad \text{with path weight } p_1 = c$$
and
$$p_2 = y_1 x_1 x_2 x_3 \quad \text{with path weight } p_2 = be.$$

Correspondingly,

$$\Delta_1 = 1 - f - ij + fij$$
and
$$\Delta_2 = 1 - ij.$$

Therefore, according to Theorem (15-3), the (1, 3)th cofactor is

$$C_{13} = \Delta_1 P_1 + \Delta_2 P_2$$
$$= (1 - f - ij + fij)c + (1 - ij)be.$$

This too can easily be verified by directly computing the (1, 3)th cofactor of the matrix in Eq. (15-5). Thus the gain

$$\left.\frac{x_3}{y_1}\right|_{y_2=0} = \frac{C_{13}}{\Delta}$$

is obtained purely by graph-theoretic computation. The result can be easily verified by inverting the corresponding matrix C as given in Eq. (15-5).

Remarks and References

The chief advantage of using signal-flow graphs (over substitution method or matrix method) lies in their ability to highlight the cause–effect relationships in the system. For example, the feedback edges shown are indeed the feedbacks in the actual system. A signal-flow graph can also be used very effectively for simplifying the system of equations before solving through matrix methods. The simplification is accomplished by flow-graph reductions. As can be seen from the literature cited in the next section, signal-flow graphs have been widely applied in the study of Markov systems.

For a computerized solution of a general problem, perhaps matrix methods would be faster. For in using Mason's gain formula, an important step is the generation of all directed circuits in the signal-flow graph; and as we saw in Chapter 11, the algorithms (suitable for computers) available for this are not very efficient.

Several variations of signal-flow graphs have been proposed and studied in the literature, following Mason's pioneering papers, [15-4] and [15-5]. For a survey of different variations, see [15-1]. Some of these variations may offer slight computational advantage, but they do so by sacrificing the cause–effect relationship, which is so nicely brought out in signal-flow graphs, as presented here. Consequently, not much has come out of these other types of graph representations, and Mason's original graphs are widely used in control theory, electrical network analysis, electrical machine theory, heat transfer, and analysis of mechanical structures.

An elementary but thorough treatment of signal-flow graphs with many applications can be found in any of the following three monographs: [15-1], [15-3], and [15-6]. All three use the original signal-flow graphs as proposed by Mason, without the later variations.

15-1. ABRAHAMS, J. R. and G. P. COVERLEY, *Signal Flow Analysis*, Pergamon Press, Inc., Elmsford, N.Y., 1965.
15-2. GHOSH, S. N., and P. K. GHOSH, "Flow Graphs and Linear Systems," *Intern. J. Control*, Vol. 14 No. 5, Nov. 1971, 961–975.
15-3. LORENS, C. S., *Flowgraphs: For the Modeling and Analysis of Linear Systems*, McGraw-Hill Book Company, New York, 1964.
15-4. MASON, S. J., "Feedback Theory: Some Properties of Signal Flow Graphs," *Proc. I.R.E.*, Vol. 41, No. 9, Sept. 1953, 1144–1156.
15-5. MASON, S. J., "Feedback Theory: Further Properties of Signal Flow Graphs," *Proc. I.R.E.*, Vol. 44, No. 7, July 1956, 920–926.
15-6. ROBICHAUD, L. P. A., M. BOISVERT, and J. ROBERT, *Signal Flow Graphs and Applications*, Prentice-Hall, Inc., Englewood Cliffs, N.J., 1962.

15-2. GRAPHS IN MARKOV PROCESSES

The simplest random process is one in which the outcomes of successive trials are independent of each other. In a coin-tossing experiment, for example, the outcome of the kth tossing is independent of the outcome of all previous tossings. Many phenomena, however, cannot be described by this simple model. There are random processes in nature in which the outcome depends on the outcome of previous trials. For example, the probability of an offspring inheriting a genetic feature does depend on the presence (or absence) of this feature in his ancestors.

A Markov process is the simplest generalization that permits the outcome of any trial to be dependent on the outcome of the trial immediately preceding it, and on no other.† This simple but powerful generalization gives the Markov process an ability to describe random processes in such diverse areas as statistical information theory, control theory, genetics, inventory control, analysis of computer proprams, and the study of social mobility of different classes, to name a few.

†This model originated by Andrei Andreivich Markov (in 1907) is a landmark in probability theory. Unlike previous mathematicians, who had modeled a random process as a sequence of independent trials, Markov saw the advantage of introducing dependence of each trial on the outcome of its predecessor. Attempts have of course been made to study models with more involved dependency of the present trial on the outcome of the past trials, but such studies generally have led to intractable results.

A Markov process is a stochastic system capable of assuming one of n states $s_1, s_2, \ldots, s_n$, and the states change only at discrete points in time. The state at the kth instant depends only on the state of the $(k-1)$th instant and not on any of the previous states. In other words, in a successive sequence of trials the outcome of the kth trial depends only on the outcome of the $(k-1)$th trial, and not on any of the preceding ones.†

Transition Probabilities: To describe a Markov process, we must specify for each state, s_i, the probability of making the next transition to each of the n states. The *transition probability* p_{ij} is the probability that if the present state of the process is s_i, the next state will be s_j. These probabilities, p_{ij}, must satisfy

$$0 \leq p_{ij} \leq 1 \quad \text{and} \quad \sum_{j=1}^{n} p_{ij} = 1, \qquad (15\text{-}8)$$

the latter because the sum of probabilities of transitions to all possible states from a given state must be unity. (Note that we have assumed that the transition probabilities are constants, and do not vary with time. Such a process is called a *stationary process* or a *time-invariant process*.†).

Transition Matrix: The n^2 transition probabilities describing a Markov process can most conveniently be given in the form of an n by n *transition matrix* $\mathsf{P} = [p_{ij}]$, subject, of course, to the two conditions in Eq. (15-8). Any square matrix with real, nonnegative elements in which the sum of each row is 1 is called a *stochastic matrix*. Thus every stochastic matrix is the transition matrix of some Markov process, and vice versa. Let us look at some properties of a stochastic matrix P.

1. If P is a stochastic matrix, its kth power P^k is also a stochastic matrix, for $k = 0, 1, 2, \ldots$ (matrix $\mathsf{P}^0 = \mathsf{I}$, the identity matrix).

2. If all rows of P are identical, then

$$\mathsf{P} = \mathsf{P}^2 = \mathsf{P}^3 = \mathsf{P}^4 = \cdots.$$

3. Since each row of a stochastic matrix adds up to 1, only $n-1$ columns need be given: the remaining column can be derived from them.

In addition to the transition matrix, we also need to know the initial probabilities

$$\pi(0) = [\pi_1(0), \pi_2(0), \ldots, \pi_n(0)],$$

†A Markov process is often called a *Markov chain* if the number of states is countable. A *finite Markov chain* is one in which the number of states is finite. In this book we are considering only stationary, finite Markov chains in which the time also changes in discrete steps (and not continuously). As there is no possibility of confusion, such a process will simply be referred to as a Markov process hereafterwards.

where $\pi_j(0)$ is the probability of the Markov process being in state s_j at the start, that is, at the time instant 0. Clearly, the probabilities in $\pi(0)$ must satisfy the following conditions:

For $i = 1, 2, \ldots, n$,

$$0 \leq \pi_i(0) \leq 1 \quad \text{and} \quad \sum_{i=0}^{n} \pi_i(0) = 1. \tag{15-9}$$

Any real-valued vector whose components satisfy the conditions in Eq. (15-9) is called a *probability vector*. The *initial probability vector* $\pi(0)$ and the transition matrix completely determine a Markov process. They are sufficient to predict the probability of the process being in any state at any time instant k.

Stochastic Graph: An alternative means of describing an n-state Markov process is an n-vertex, weighted, connected digraph G. The vertices of G correspond to the states, and an edge (s_i, s_j) with a nonzero weight p_{ij} represents the nonzero transition probability from state s_i to s_j. Such a digraph, called a *transition graph*, is not only of great value in visualizing a Markov process, but is also a powerful analytic tool in studying the process. Clearly, the weights of the edges in the transition graph G must satisfy the conditions in Eqs. (15-8). A digraph in which the edge weights are positive quantities and the sum of weights of edges emanating from a vertex is unity is called a *stochastic graph*.

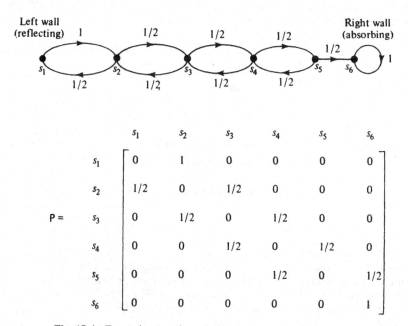

Fig. 15-4 Transition matrix and diagram for a Markov process.

As an example of a Markov process, let us consider the following version of the classic problem of random walk:

Suppose that a particle moves, according to a probabilistic mechanism, along a straight line $n = 2m$ units long between two walls. At each transition the particle moves either one unit left or one unit right, each with the probability $\frac{1}{2}$. If the particle hits the left wall, it gets reflected; but if it hits the right wall, it is absorbed into the wall.

At a given time the process is in one of $n = 2m$ states (i.e., the particle is at one of the n points along the line). Only the present state is relevant to what the next state may be and not any of the previous states. This is an n-state Markov process. For $n = 6$, the transition matrix and the transition diagram for this process are given in Fig. 15-4.

Note the similarities between a Markov process and a sequential machine, discussed in Section 12-8. Both have a finite number of states, and in both the state transitions occur at discrete points in time. The main difference between the two is that in Markov processes we deal with probabilities, which are real-valued quantities instead of a set of symbols, as in the case of sequential machines. Transitions in a Markov process do not depend on any (externally controlled) inputs, but are governed by probability distributions. Moreover, there are no outputs associated with a Markov process.

Multistep Transition Probabilities

An important question regarding a Markov process is the following: Given that the process was initially in state s_i, what is the probability of its being in state s_j after exactly k transitions? This probability $\phi_{ij}(k)$, called the *k-step transition probability* from state s_i to s_j, is given by

THEOREM 15-4

In a Markov process the k-step transition probability $\phi_{ij}(k)$ from state s_i to s_j is equal to the ijth entry in matrix P^k, the kth power of the transition matrix P.

Proof: Consider the transition digraph. The weight p_{ij} of edge (s_i, s_j) is the probability of going from vertex s_i to s_j in one step. Since the transition probabilities at each step are stochastically independent, the product

$$p_{ir} \cdot p_{rj}$$

gives the probability that the process will go from state s_i to s_r in the first step and then from s_r to s_j in the second step. Continuing this argument, the probability of going from vertex s_i to s_j along a directed edge sequence of length k is given by the product of the weights of these edges. But the probability of going from s_i to s_j in exactly k steps is the sum of the probabilities of going from s_i to s_j along all directed edge sequences of length k in the transition digraph. That this sum is given by the ijth entry in matrix P^k can be easily seen by arguments used in Theorem 9-10. ∎

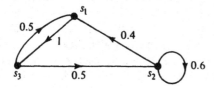

Fig. 15-5 Three-state Markov Process.

To illustrate Theorem 15-4, let us consider a three-state Markov process whose transition digraph is given in Fig. 15-5. The transition matrix and some of its powers are

$$P = \begin{bmatrix} 0 & 0 & 1 \\ .4 & .6 & 0 \\ .5 & .5 & 0 \end{bmatrix}, \quad P^2 = \begin{bmatrix} .5 & .5 & 0 \\ .24 & .36 & .4 \\ .2 & .3 & .5 \end{bmatrix},$$

$$P^3 = \begin{bmatrix} .2 & .3 & .5 \\ .344 & .416 & .24 \\ .37 & .43 & .2 \end{bmatrix}, \quad P^4 = \begin{bmatrix} .3700 & .4300 & .2000 \\ .2864 & .3696 & .3440 \\ .2720 & .3580 & .3700 \end{bmatrix},$$

$$P^{10} = \begin{bmatrix} .3099 & .3863 & .3038 \\ .3069 & .3841 & .3090 \\ .3064 & .3837 & .3099 \end{bmatrix}, \quad P^{15} = \begin{bmatrix} .3076 & .3848 & .3076 \\ .3076 & .3846 & .3078 \\ .3076 & .3845 & .3079 \end{bmatrix}$$

$$P^{17} = P^{18} = P^{19} = P^{20} = \begin{bmatrix} .3077 & .3846 & .3077 \\ .3077 & .3846 & .3077 \\ .3077 & .3846 & .3077 \end{bmatrix}.$$

The ijth entry in P^4, for example, is the probability that the process will go from state s_i to s_j in exactly four steps. Let us, for instance, examine all directed edge sequences of length four from s_3 to s_2 in Fig. 15-5. These are

$s_3 s_2 s_2 s_2 s_2$ with probability of traversing $(.5)(.6)^3 = .108$,

$s_3 s_2 s_1 s_3 s_2$ with probability of traversing $(.5)^2(.4)(1) = .1$,

$s_3 s_1 s_3 s_2 s_2$ with probability of traversing $(.5)(1)(.5)(.6) = .15$.

The sum of their probabilities, .358, is exactly the entry in the (3, 2) position of P^4.

For this example, let us make some further observations on the properties of matrix P^k.

1. Beyond a certain value of k, P^k contains only nonzero entries. This implies that there is at least one directed edge sequence of length k (and therefore a directed path of length k or less) from every vertex to every other vertex in Fig. 15-5.

2. All rows of P^k tend to become identical as k increases. This means that the k-step transition probability $\phi_{ij}(k)$ becomes independent of i for large k. This result should not come as a surprise, because the effect of the starting state s_i should wear off after sufficiently many transitions.
3. As a direct consequence of item 2, the higher powers of P become identical; that is,

$$P^k = P \cdot P^k = P \cdot P^{k+1} = \cdots$$

because P is a stochastic matrix and P^k has identical rows.

Does P^k of every Markov process exhibit these properties, or is this example a special case? To answer this question, let us take a closer look at stochastic digraphs, and try to classify them.

Classification of States

A set S of states is said to be *closed*, *trapping*, or *absorbing* if no state outside S can be reached from any state s_i in S. In other words, there is no directed edge from any vertex s_i in S to any vertex outside S. For example, $\{s_4, s_5, s_3\}$ in Fig. 15-6 is a closed set of states. So is $\{s_4\}$. A single state s_k is an absorbing or *trapping state* if and only if it has a self-loop with weight one, such as s_4 in Fig. 15-6. Clearly, the entire set of states in a Markov process trivially constitutes a closed set. If there exists no other closed set of states except the entire set of states of the Markov process, the process is called *ergodic* or *irreducible*. In other words, a process is ergodic if and only if its

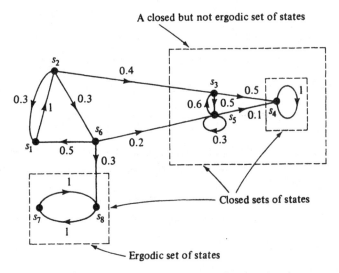

Fig. 15-6 Nonergodic Markov process showing closed sets.

transition graph is strongly connected; that is, there is a nonzero probability of going from any state to any other state. For example, in Fig. 15-5 the process is ergodic, but the process shown in Fig. 15-6 is not. A strongly connected closed set of states is called an ergodic or irreducible set.

Regular Process: Of special interest among ergodic Markov processes are those in which there exists a directed edge sequence exactly of length k (for some positive integer k) from every vertex to every other vertex in G. Such a process is called a *regular Markov process*. Clearly, every regular process is ergodic, but the converse is not true. For example, in Fig. 15-7(a) the process is ergodic but not regular. For there is no directed edge sequence of even length from s_1 to s_2 and no directed sequence of odd length from s_1 to s_3; thus there exists no k for which there is a directed edge sequence of k edges from every vertex to every other vertex.

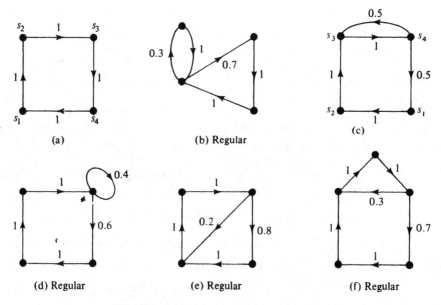

Fig. 15-7 Some ergodic stochastic digraphs.

The significance of a regular Markov process lies in the fact that for some integer k there is a nonzero probability of going from each state to every other state in exactly k steps. In terms of the k-step transition matrix $P^k = \Phi(k)$, we can make the following statement: A Markov process is regular if and only if for some integer k every entry in the k-step transition matrix $\Phi(k)$ is positive.

Given the transition digraph G of an ergodic process (i.e., G is strongly connected), how can we tell if the process is regular? To answer this question, let us introduce the following definition: A strongly connected subdigraph g in a digraph G is said to be a *minimal* if g has no proper subdigraph of two or more vertices that is strongly connected.

THEOREM 15-5

Let G be a strongly connected stochastic digraph, and let $g_1, g_2, \ldots, g_r$ be its minimal strongly connected subdigraphs, having $n_1, n_2, \ldots, n_r$ vertices, respectively. Then G represents a regular Markov process if and only if the greatest common divisor (g.c.d.) of $n_1, n_2, \ldots, n_r$ is 1.

For a proof of this theorem, see Rosenblatt's paper [15-14]. Let us simply illustrate it with some examples. The digraph in Fig. 15-7(c) has two minimal strongly connected subdigraphs, with vertex sets $\{s_3, s_4\}$ and $\{s_1, s_2, s_3, s_4\}$. Since the g.c.d. $(4, 2) = 2$, the process is not regular. On the other hand, Fig. 15-7(f) also has two minimal, strongly connected subdigraphs, one with five vertices and the other with three. Since the g.c.d. $(5, 3) = 1$, the process is regular.

Clearly, if a strongly connected transition digraph contains a self-loop, the g.c.d. is 1; therefore, the process is regular [e.g., Fig. 15-7(d)]. The reader is encouraged to write down the transition matrix P for each of the six digraphs in Fig. 15-7, and verify Theorem 15-5 by directly computing P^k (for appropriate k).

Periodic Markov Process: An ergodic process is said to be *periodic* if every state can only be entered at certain periodic intervals. The simplest example of a cyclic process is one with two states s_1 and s_2 in which only the transitions $s_1 \to s_2 \to s_1 \to s_2 \ldots$ are possible. The transition matrix P and its powers for this process are

$$\mathsf{P} = \begin{bmatrix} 0 & 1 \\ 1 & 0 \end{bmatrix}, \quad \mathsf{P}^2 = \begin{bmatrix} 1 & 0 \\ 0 & 1 \end{bmatrix}, \quad \mathsf{P}^3 = \begin{bmatrix} 0 & 1 \\ 1 & 0 \end{bmatrix}, \ldots.$$

As another example, consider the process in Fig. 15-7(c). Its transition matrix is

$$\mathsf{P} = \begin{bmatrix} 0 & 1 & 0 & 0 \\ 0 & 0 & 1 & 0 \\ 0 & 0 & 0 & 1 \\ \tfrac{1}{2} & 0 & \tfrac{1}{2} & 0 \end{bmatrix}, \quad \mathsf{P}^k = \begin{bmatrix} \tfrac{1}{3} & 0 & \tfrac{2}{3} & 0 \\ 0 & \tfrac{1}{3} & 0 & \tfrac{2}{3} \\ \tfrac{1}{3} & 0 & \tfrac{2}{3} & 0 \\ 0 & \tfrac{1}{3} & 0 & \tfrac{2}{3} \end{bmatrix},$$

for k odd and very large.

$$\mathsf{P}^k = \begin{bmatrix} 0 & \frac{1}{3} & 0 & \frac{2}{3} \\ \frac{1}{3} & 0 & \frac{2}{3} & 0 \\ 0 & \frac{1}{3} & 0 & \frac{2}{3} \\ \frac{1}{3} & 0 & \frac{2}{3} & 0 \end{bmatrix}, \quad \text{for } k \text{ even and very large.}$$

A Markov process is periodic if and only if its states can be partitioned into q subsets ($q > 1$) such that the process dwells in each of these q subsets in q consecutive transitions.

It can be shown that every ergodic process is either regular or periodic [15-11].

Markov Processes with Transient States: So far we have been considering ergodic Markov processes (i.e., those in which the transition digraphs are strongly connected). Let us now examine the processes for which the transition digraph is weakly connected. A weakly connected digraph G consists of two or more fragments (i.e., maximal strongly connected subdigraphs).

Let us consider a fragment g of G (remember fragment g could be a single vertex). If there is no edge directed out of g, then g has to have at least one edge going into it, and the vertices in g are closed. In that case, we can delete all edges going into g, and then study g independently as an ergodic process [the edge weights will satisfy Eqs. (15-8)]. On the other hand, if there is an edge going out of g, the vertices of g cannot constitute an ergodic process, because (g being maximal strongly connected) once exited, g cannot be reentered. Such a set of states, which once left cannot be entered and which among themselves are accessible from each other, is called a *transient set* of states. Sets $\{s_1, s_2, s_6\}$ and $\{s_3, s_5\}$ in Fig. 15-6 are transient sets.

The vertices of a weakly connected stochastic digraph can be uniquely partitioned into sets $T, V_1, V_2, \ldots, V_q$ such that T is the set of all transient states and each V_i is irreducible, that is, there is no edge (a, b) for $a \in V_i$, $b \notin V_i$, and the vertex set V_i is strongly connected. For example, in Fig. 15-6

$$T = \{s_1, s_2, s_6, s_3, s_5\}, \quad V_1 = \{s_7, s_8\}, \quad V_2 = \{s_4\}.$$

Clearly, a (weakly connected) stochastic digraph cannot consist of set T alone—there must be at least one set V_i. In other words, it is not possible that all states of a Markov process are transient (remember we are only considering finite Markov processes).

After a large number of transitions, a Markov process with transient states will eventually settle down into one of its irreducible subsets. Such a system has two types of distinct behavior, and one may be interested in either or both: (1) the behavior of the system before it enters an irreducible set of states, and (2) the behavior of the system after it enters an irreducible set. Behavior 2 is no different from that of an ergodic system. For once the system enters an irreducible set, it can never leave it, and thus the existence of states outside this set is immaterial. Behavior 1 will be studied briefly in Section 15-3 while analyzing computer programs.

Thus as far as the asymptotic behavior of Markov processes is concerned, we need to study only ergodic processes. Among ergodic processes, also, only the regular processes are of importance.† Therefore, in the rest of this section we shall study the asymptotic behavior of a regular Markov process.

Asymptotic Behavior of a Regular Markov Process

One of the most important questions about a Markov process is what happens to it after many many transitions? That is, after the transients die down, does the system reach a steady state, independent of the initial probabilities? If so, what is the steady-state probability vector $\pi(\infty)$ and how do we compute it? The answer lies in the behavior of P^k as k tends to infinity, and Theorem 15-6 provides it for a regular Markov process:

THEOREM 15-6

If P is a transition matrix of a regular Markov process, then its powers, P^k, as k tends to infinity, approach a stochastic matrix $\Phi = [\phi_{ij}]$ having identical rows, and each row w of Φ is a probability vector.

Outline of the Proof: We first note that since the process is regular there exists some positive integer r such that $P^r = M$ contains only positive entries. Second, we observe that premultiplying any column vector y by a stochastic matrix having only positive entries has an averaging effect on the elements of y. This averaging effect applied again and again would eventually smooth out differences that may have existed among the elements of y. That is, all components of vector $M^h y = P^{rh} y = P^k y$ will have identical elements, as h (and therefore k) becomes very large. Finally, let us observe that this condition is equivalent to P^k approaching a limit Φ as k tends to infinity, and the rows of matrix Φ are identical—each a probability vector. The reader is encouraged to fill in the details. ■

Theorem 15-6 is perhaps the most important result in the theory of Markov processes. Many interesting and useful results for a regular Markov process depend on the existence of this limit for P^k, as $k \to \infty$.

Since P^∞ (which is a shorthand notation for $\lim P^k$ as $k \to \infty$) exists, we can express the steady-state probability

$$\pi(\infty) = \pi(0)P^\infty = \pi(0)\Phi. \tag{15-10}$$

---

†To quote Feller, "The classification into persistent and transient states is fundamental, whereas the classification into periodic and aperiodic states concerns a technical detail. It represents a nuisance in that it requires constant reference to trivialities;" W. Feller, *An Introduction to Probability Theory and Its Applications,* Volume I, 3rd ed., John Wiley & Sons, Inc., New York, 1968, 387.

Writing this equation in terms of its elements,

$$\pi_j(\infty) = \sum_{i=1}^{n} \pi_i(0)\phi_{ij}; \quad j = 1, 2, \ldots, n. \quad (15\text{-}11)$$

Now, since all rows of Φ are identical, each element ϕ_{ij} is equal to a value w_j that depends only on the column index j. Thus

$$\begin{aligned}
\pi_j(\infty) &= \sum_{i=1}^{n} \pi_i(0) w_j \\
&= w_j \sum_{i=1}^{n} \pi_i(0) \\
&= w_j.
\end{aligned} \quad (15\text{-}12)$$

Thus a regular Markov process approaches the same limiting probability distribution w regardless of where it started. Moreover, this final probability vector, $\pi(\infty) = $ w, is the one that appears as rows of matrix Φ.

For a given regular Markov process, how does one compute this fixed vector w, that is, the vector that makes up all the rows of Φ and is also equal to $\pi(\infty)$? Several methods are available. Raising the transition matrix P to higher and higher powers is one, but it is obviously not a good method. A most frequently used method is the following:

Successive state probability vectors must satisfy

$$\pi(k + 1) = \pi(k)\text{P}. \quad (15\text{-}13)$$

If the state probability vector has attained its limiting value $\pi(\infty)$, it must then satisfy

$$\pi(\infty) = \pi(\infty)\text{P}, \quad (15\text{-}14)$$

and since $\pi(\infty) = $ w, according to Eq. (15-12), it can be rewritten as

$$\text{w} = \text{wP}. \quad (15\text{-}15)$$

Equation (15-15) implies n simultaneous equations, which can be rewritten as

$$\text{w}(\text{I} - \text{P}) = 0. \quad (15\text{-}16)$$

But since P is a stochastic matrix, the sum of rows of matrix I $-$ P is zero. Therefore, I $-$ P is a singular matrix; that is, the n equations in Eq. (15-16) are not linearly independent. It can be shown, however, that any $n - 1$ of these equations are linearly independent, and thus we need only one more equation to solve for the n unknowns in the row vector w. This is readily provided by the relation

$$w_1 + w_2 + \cdots + w_n = 1. \quad (15\text{-}17)$$

Thus in Eq. (15-16) if we replace any one column, say the jth, of the matrix $I - P$ on the left-hand side and change the jth entry from 0 to 1 in the right-hand side, we would incorporate Eq. (15-17) into Eq. (15-16). Let this new equation be denoted by

$$\mathbf{w}(I - P)_j^* = \mathbf{v}_j, \qquad (15\text{-}18)$$

where $\mathbf{v}_j$ is a row vector of length n with all zero entries except the jth, which is 1. Equation (15-18) can be solved by either directly inverting the matrix $(I - P)_j^*$ or by using signal-flow graphs, as shown in Section 15-1. For illustration, let us consider the three-state regular Markov process given in Fig. 15-5. Applying Eqs. (15-15) and (15-17) directly, we get

$$(w_1 \quad w_2 \quad w_3) \begin{bmatrix} 0 & 0 & 1 \\ .4 & .6 & 0 \\ .5 & .5 & 0 \end{bmatrix} = (w_1 \quad w_2 \quad w_3),$$

which is

$$w_1 = .4w_2 + .5w_3,$$
$$w_2 = .6w_2 + .5w_3,$$
$$w_3 = w_1.$$

Hence

$$w_1 = w_3 = .8w_2,$$

which combined with

$$w_1 + w_2 + w_3 = 1,$$

immediately yields the limiting probability vector $w = (\frac{4}{13} \quad \frac{5}{13} \quad \frac{4}{13})$. This indeed is the result we had obtained earlier by trying 17th and higher powers of P.

Solving the same problem using the form of Eq. (15-18) gives

$$I - P = \begin{bmatrix} 1 & 0 & -1 \\ -.4 & .4 & 0 \\ -.5 & -.5 & 1 \end{bmatrix}.$$

Replacing any one of the columns, say the second, with all 1's we get

$$(I - P)_2^* = \begin{bmatrix} 1 & 1 & -1 \\ -.4 & 1 & 0 \\ -.5 & 1 & 1 \end{bmatrix},$$

which on inverting yields

$$(I - P)_2^{*-1} = \frac{1}{1.3} \begin{bmatrix} 1 & -2 & 1 \\ .4 & .5 & .4 \\ .1 & -1.5 & 1.4 \end{bmatrix}.$$

Finally,

$$w = (I - P)_2^{*-1}(0 \; 1 \; 0)$$
$$= (\tfrac{4}{13} \; \tfrac{5}{13} \; \tfrac{4}{13})$$

which checks with the value of w obtained earlier.

Instead of solving Eq. (15-18) by algebraic inversion of matrix $(I - P)_j^*$, often it is more efficient to use a signal-flow graph to solve Eq. (15-20), particularly when matrix P is relatively sparse.

The signal-flow graph corresponding to Eq. (15-18) is directly obtained from the transition graph of the process by a simple modification:

1. Replace the vertex labels s_i's with w_i's.
2. Remove all edges incoming into the specified vertex s_j.
3. Put edges of weight -1 from every other vertex to s_j.
4. Reverse the direction of every edge. [Inverting the edge direction corresponds to transposing the matrix—as required by Eq. (15-4).]
5. Add a new vertex with label 1 to the digraph and draw an edge of unit weight from this new vertex to s_j.

In this signal-flow graph we can obtain the gain from v to every vertex, which will give the elements of w.

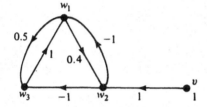

Fig. 15-8 Signal-flow graph for a Markov system.

Once again, modifying Fig. 15-5 we get the appropriate digraph, as shown in Fig. 15-8. For this signal-flow graph [using Eq. (15-5)] the determinant is given by

$$\Delta = 1 - t_1 + t_2 - t_3 + \cdots$$
$$= 1 - t_1$$
$$= 1 - (.5 - .4 - .4) = 1.3,$$

and the three relevant cofactors needed in Eq. (15-6) are

$$\Delta_{12} = .4,$$
$$\Delta_{22} = .5,$$
$$\Delta_{32} = .4,$$

which again checks with the results obtained earlier.

Medvedev [15-12] has a different method of obtaining vector w from the signal-flow graph, and through some examples he has shown that there are cases when graph-theoretic methods are superior to algebraic ones. Another computational formula using transition digraphs is given in [15-7]. For more on computing vector w via transition graphs, see [15-8].

Transient Analysis of a Markov Process

Matrix $\Phi = P^\infty$ for a regular Markov process gives us the steady-state distribution, but it does not reveal the transient behavior of the process. It does not tell us how fast P^k converges to the limit Φ, nor does it give the frequencies with which various states in the system were visited before the steady-state condition was reached. The answer to these questions lies in the behavior of the sequence $P, P^2, P^3, P^4, \ldots$. The problem is then to find a closed-form expression for matrix P^k (as a function of k). This can be obtained using z-transforms:

From Eq. (15-13) we know that

$$\pi(k+1) = \pi(k)P.$$

Taking the z-transform of both sides of this equation, we get

$$\frac{\Pi(z) - \pi(0)}{z} = \Pi(z)P, \tag{15-19}$$

where $\Pi(z)$ is the z-transform of $\pi(k)$. Rearranging Eq. (15-19), we get

$$\Pi(z) = \pi(0)(I - zP)^{-1}. \tag{15-20}$$

Application of inverse z-transform to both sides of Eq. (15-20) gives

$$\pi(k) = \pi(0) \ [\text{inverse } z\text{-transform of matrix } (I - zP)^{-1}]. \tag{15-21}$$

Thus

$$P^k = \text{inverse } z\text{-transform of matrix } (I - zP)^{-1}. \tag{15-22}$$

Equation (15-21) gives the probability vector after k transitions have taken place, for $k = 1, 2, \ldots$, and Eq. (15-22) provides the value of P^k, as a function of k, in a closed form for any Markov process (not necessarily regular).

Matrix $I - zP$ can be inverted by direct matrix methods, but the flow-graph method, as used for inverting $(I - P)^*$ in Eq. (15-18), is often found to be more efficient. The weight of each edge p_{ij} is now multiplied by z. Informally speaking, z is the transform of the unit delay, and multiplying each transition probability by z corresponds to the delay associated with each transition. For a thorough treatment of transient analysis of a Markov process via signal-flow graphs, see Chapters 3 and 4 of [15-9].

For illustration, let us consider the Markov process depicted in Fig. 15-5. we have

$$I - zP = \begin{bmatrix} 1 & 0 & -z \\ -.4z & 1-.6z & 0 \\ -.5z & -.5z & 1 \end{bmatrix},$$

$$(I - zP)^{-1} = \frac{1}{(1-z)(1+.4z-.1z^2)} \begin{bmatrix} 1-.6z & .5z^2 & z(1-.6z) \\ .4z & 1-.5z^2 & .4z^2 \\ z(.5-.1z) & .5z & 1-.6z \end{bmatrix}$$

$$= \frac{1}{(1-z)(1-.174z)(1+.574z)} \left\{ \begin{bmatrix} 1 & 0 & 0 \\ 0 & 1 & 0 \\ 0 & 0 & 1 \end{bmatrix} \right. \quad (15\text{-}23)$$

$$+ z \begin{bmatrix} -.6 & 0 & 1 \\ .4 & 0 & 0 \\ .5 & .5 & -.6 \end{bmatrix} + z^2 \begin{bmatrix} 0 & .5 & -.6 \\ 0 & -.5 & .4 \\ -.1 & 0 & 0 \end{bmatrix} \right\}.$$

Let us now compute the inverse $(I - zP)^{-1}$ by a signal-flow graph rather than algebraically. The signal-flow graph is nothing but the transition graph in which each edge is multiplied by z, and the direction of every edge is reversed. See Fig. 15-9, which is obtained directly from Fig. 15-5.

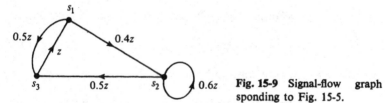

Fig. 15-9 Signal-flow graph corresponding to Fig. 15-5.

From Fig. 15-9 we immediately obtain the cycle product terms [as used in Eqs. (15-6) and (15-7)] as follows:

$$t_1 = .6z + .5z^2 + .2z^3,$$
$$t_2 = .3z^3,$$
$$\text{determinant} \quad \Delta = 1 - t_1 + t_2$$
$$= 1 - .6z - .5z^2 + .1z^3.$$

A cofactor

$$\Delta_{13} = .5z(1-.6z) + .2z^2$$
$$= z(.5-.1z),$$

which is in agreement with the 3,1 entry in Eq. (15-23).

Now taking the inverse z-transform of both sides of Eq. (15-23), we get

$$P^k = \begin{bmatrix} \frac{4}{13} & \frac{5}{13} & \frac{4}{13} \\ \frac{4}{13} & \frac{5}{13} & \frac{4}{13} \\ \frac{4}{13} & \frac{5}{13} & \frac{4}{13} \end{bmatrix} + \frac{(-.574)^{k-2}}{3.57} \begin{bmatrix} .673 & .500 & -1.174 \\ -.223 & -.170 & .400 \\ -.387 & -.287 & .673 \end{bmatrix}$$
$$+ \frac{(.174)^{k-2}}{20.35} \begin{bmatrix} .074 & -.500 & .426 \\ -.070 & .470 & .400 \\ .013 & -.087 & .074 \end{bmatrix}.$$

Remarks and References

As the brief exposition in this section shows, weighted digraphs enter very naturally into the study of finite Markov chains; in particular, the asymptotic behavior of a Markov chain can be derived swiftly from its digraph. A great deal more on Markov chains remains unsaid in this section. Recommended introductory readings on Markov chains are [15-11] and [15-13]. For z-transforms and applications of signal-flow graphs to the study of Markov chains, [15-9] is an excellent source. The following is a list of selected reading for more on the interaction of Markov processes and graph theory.

15-7. BIONDI, E., G. GUARDABASSI, and S. RINALDI, "On the Analysis of Markovian Discrete Systems by Means of Stochastic Graphs," *Automation and Remote Control*, Vol. 28, No. 2, Feb. 1967, 275–277.

15-8. GUARDABASSI, G., and S. RINALDI, "Two Problems in Markov Chains: A Topological Approach," *Operations Res.*, Vol. 18, No. 2, March–April 1970, 324–333.

15-9. HOWARD, R. A., *Dynamic Probabilistic Systems, Vol. I: Markov Models*, John Wiley & Sons, Inc., New York, 1971.

15-10. HUGGINS, W. H., "Signal Flow Graphs and Random Signals," *Proc. I.R.E.*, Vol. 45, 1957, 74–86.

15-11. KEMENY, J. G. and J. L. SNELL, *Finite Markov Chains*, Van Nostrand Reinhold Company, New York, 1960.

15-12. MEDVEDEV, G. A., "Analysis of Discrete Markov Systems by Means of Stochastic Graphs," *Automation and Remote Control*, Vol. 26, No. 3, March 1965, 481–485.

15-13. PARZEN, E., *Stochastic Processes*, Holden-Day, Inc., San Francisco, 1965, Chapter 6.

15-14. ROSENBLATT, D., "On the Graphs and Asymptotic Forms of Finite Boolean Relation Matrices and Stochastic Matrices," *Naval Research Logistics Quarterly*, Vol. 4, 1957, 151–161.

15-15. SITTLER, R. W., "System Analysis of Discrete Markov Processes," *I.R.E. Trans. Circuit Theory*, Vol. CT-3, No. 1, 1956, 257.

15-3. GRAPHS IN COMPUTER PROGRAMMING

Analysis of a given computer program has been an important problem from the early days of computer programming. The purpose of such an analysis could be to estimate the running time or storage requirement of a program, to subdivide a large program into a number of subprograms, to detect certain types of structural errors in the program, to document a program, or simply to understand a program written by someone else. For all these purposes it is very convenient to represent a program as a digraph. Each vertex represents a *program block*, that is, a sequence of computer instructions having the property that each time any instruction in the sequence is executed all are executed. Each program block has one entry point (the first instruction in the sequence) and one exit point (the last instruction in the sequence). Each edge (v_i, v_j) represents a possible transfer of control from the last instruction in the program block v_i to the first instruction in the program block v_j. Such a digraph is called a *program digraph*. A program digraph can also be thought of as an abstraction of a flow chart in which the boxes are shrunk to vertices and arrows become the edges. For example, Fig. 15-10 shows the program digraph of the flow chart in Fig. 11-9. [Ignore dashed line (v_{14}, v_1) for now.]

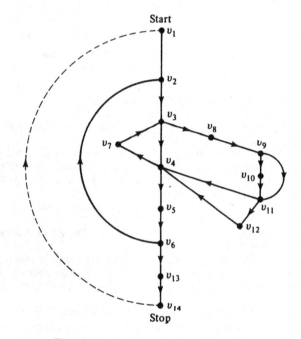

Fig. 15-10 Program digraph of Fig. 11-9.

Note that there may be more than one program digraph for a given program, because one program block may be split into several. Also observe that programs that modify their own control and processing instructions in the course of execution cannot be represented in this fashion. (Declarative statements, such as formats and dimension statements, are ignored in program digraphs.)

Some obvious but important properties of a digraph representing any valid computer program are

1. A program digraph must be connected.
2. It has precisely one vertex of zero in-degree, and this vertex corresponds to the start of the program. (If the program has several starting points, an additional *start vertex* can be introduced from which directed edges can be drawn to all of these vertices.)
3. There is precisely one vertex of zero out-degree, and this vertex corresponds to the end of the program. We shall call it the *stop vertex*. (Existence of more than one stopping point in the program can be taken care of as in item 2.)
4. Every vertex in the program digraph must be accessible from the start vertex.
5. The stop vertex must be accessible from every vertex in the program digraph.

Detection of Programming Errors

To detect and report certain types of structural errors in a source program is an essential part of a compiler's job. The most common of these errors can be checked by tracing directed paths from the start vertex to the stop vertex in the program digraph. Path tracing will detect program blocks that are never entered, or program blocks from which there is no exit leading to the stop vertex, or disjoint parts, such as a subroutine that is never called [15-33].

Estimation of Program Running Time

Given a computer program and the execution time for each of the program blocks, we are often required to estimate the running time of the program. The situation is represented by a program digraph having a weight t_i associated with each vertex v_i, where t_i is the execution time of the corresponding program block. If we can estimate the number of times each vertex is entered (i.e., the program block is executed), the running time of the program can be determined (for a particular computer, of course).

In the program digraph let each directed edge e_i be assigned a nonnegative integer f_i, where f_i is the number of times edge e_i is traversed. The number of times a vertex is entered must equal the number of times it is exited, except for the start vertex and the stop vertex. These two exceptions can also be taken care of by adding an edge of unit weight directed from the stop vertex v_n to the start vertex v_1 (see the dashed edge in Fig. 15-10). Now the edge weights in this modified digraph satisfy the Kirchhoff current law (KCL) at every vertex. (Quantity f_i may be looked upon as a flow through the *i*th edge e_i.

Recall the electrical networks in Chapter 13 and the flow networks in Chapter 14.)

If we apply KCL to a weighted, connected digraph of n vertices and e edges, we get $n-1$ linearly independent equations in e unknowns, the unknowns being the weights of (i.e., the flows through) the edges. Thus we can choose independently only a set of $\mu = e - n + 1$ flows, corresponding to the chords with respect to any spanning tree in the digraph. The remaining $n-1$ flows through the tree branches can be expressed in terms of these μ flows. For example, in the program digraph of Fig. 15-10, $\mu = 18 - 14 + 1 = 5$. Therefore, the flow through every edge can be expressed in terms of five unknowns. Thus the iteration counts through 14 boxes in the flow chart are expressed in terms of only five unknowns. These five unknowns can also be chosen conveniently by picking a spanning tree. For example, in Fig. 11-9 we know that the iteration count of the edge (v_4, v_7) is $N-1$, where N is the number of vertices in the original graph on which Algorithm 6 was being applied. The flows through the remaining four chords are data dependent in a more involved fashion. The minimum, maximum, and average running times of the program can be estimated by assuming appropriate probable values of these four unknowns. At this point the problem of estimating the running time becomes quite difficult. For some simple examples, the reader is referred to [15-25], [15-26], and pages 95–102 and 364–369 of [11-39].

In this connection an interesting question is as follows:

Given a connected digraph G with vertices $v_1, v_2, \ldots, v_n$ and e edges, having weights $f_1, f_2, \ldots, f_e$ associated with the edges, what are necessary and sufficient conditions that G corresponds to some program digraph such that v_1 is the start vertex and v_n is the stop vertex? Some necessary conditions are obvious: Each f_i must be a nonnegative integer. The in-degree of $v_1 = 0 =$ out-degree of v_n. Also, the f_i's must satisfy KCL at each vertex except at v_1 and v_n. Moreover, the sum of weights of edges going out of v_1 should be equal to the sum of weights of edges going into v_n, both being equal to unity. Are these conditions sufficient also? The answer, as given by the following construction, is yes.

From the given, weighted digraph G let us construct an unweighted digraph H as follows: Replace every edge e_j with f_j parallel edges, where f_j is the weight of the edge e_j in the digraph G. Clearly, the digraph H will be balanced [i.e., $d^+(v) = d^-(v)$ for every vertex v in H] if and only if KCL is satisfied at every vertex in G. Now, from Theorem 9-1, a digraph is balanced if and only if it is an Euler digraph; that is, there exists a directed Euler walk from v_1 to v_n in H. This is possible if and only if there exists a directed edge sequence in G from vertex v_1 to v_n such that every edge e_k appears in it exactly f_k times, and the edge (v_n, v_1) does not appear in this directed edge sequence. The last statement is equivalent to G being a program digraph with v_1 as start vertex and v_n as stop vertex. Thus we have.

Theorem 15-7

Let G be a connected, weighted digraph with n vertices and e edges. Let all the edge weights $f_1, f_2, \ldots, f_e$ be nonnegative integers, and such that they satisfy KCL at each vertex, except v_1 and v_n. Furthermore, let the in-degree of $v_1 = 0$ = out-degree of v_n, and the sum of the weights of the edges going out of $v_1 =$ the sum of the weights of the edges going into $v_n = 1$. Then G corresponds to a program digraph in which v_1 is the start vertex, v_n is the stop vertex, and the weight f_i of the ith edge is the number of times that edge is traversed in the program.

Program Segmentation

Sometimes one comes across a program so large that it cannot be accommodated in its entirety into the working memory of the available computer. In such a case the program must be segmented before execution. Then the segments (pieces) of the program are brought from the slow bulk memory (drum, disk, or tape) and executed one at a time. The size of each segment must be small enough to be accommodated into the working memory and yet must be large enough so that there would not be too many transfers between the fast working memory and the slow bulk memory. Thus we have a problem of finding an optimal partitioning of the program digraph into subdigraphs such that the sum of weights of vertices (here the weight s_i of vertex v_i is the amount of storage space required by the ith program block) does not exceed a specified value. A similar problem arises in a multiaccess, time-sharing environment, where each user is given a burst of service of fixed duration. The program segments have to be chosen judiciously, not so large that its execution will exceed the allotted time and yet not too short to require inefficient transfers between memories.

If the program digraph is acyclic (i.e., the program has no loops, which is rare for any nontrivial computer program), the partitioning problem is solved easily. We sort the vertices in a topological order, and starting from the first vertex, we partition the sequence into largest possible (topologically sorted) subsequences such that the total vertex weight of no subsequence exceeds the specified value.

The difficulty in segmentation arises because of directed circuits in the program digraphs (i.e., loops in the program). A cut made across a directed circuit implies interchanges between the two segments, and hence between fast and slow memory. Thus one would like to avoid a segmentation that causes vertices of one directed circuit to belong to more than one segment. The simplest method to accomplish this is to identify all fragments (i.e., maximal strongly connected subdigraphs) in the program digraph. Then compute the memory requirement of each fragment by adding the weights of all its vertices. If the largest of these memory requirements does not exceed the size of the working memory, then the problem is solved. For we need not cut any directed circuit, and small fragments can always be combined to yield a segment of reasonable size.

In practice, however, it is found that the largest fragment is usually too large to fit in the working memory. For example, in Fig. 15-10 almost the entire program digraph forms a fragment. In such a case, cutting across directed circuits is unavoidable. The simplest approach suggested in the literature is to find a partition that severs the least number of directed circuits. This is a difficult problem. To enumerate all strongly connected subdigraphs in the program digraph and then consider each as a possible segment, although suggested as a solution in the literature [15-32], is a horrendous task. Even generation of all directed circuits in a digraph is extremely time consuming, as we observed in Chapter 11. Another method suggested for program segmentation is by first ordering the vertices in a certain fashion [15-20]. This method of segmentation involves more labor than finding all fragments, but less labor than generating all strongly connected subdigraphs of the program digraph.

Segmentation of a program is a very difficult problem, to say the least. We are quite far from having found a procedure for an efficient solution for this important and interesting graph-theoretic problem.

Even if we were able to obtain a partitioning that minimized the number of severed directed circuits, the solution might not be optimal. Every directed circuit is not traversed the same number of times. Obviously, cutting a directed circuit with higher iteration count is worse than cutting one with lower iteration count. We must have the iteration count of each loop—information rarely available *ab initio*, because of its data dependence. A stochastic analysis of the program, assuming that it behaves as a Markov system, is often the answer to this problem.

Stochastic Model of a Program

One method used to estimate the relative frequencies of traversal of different edges and vertices is to assume the program digraph to be a stochastic digraph, in which the weight p_{ij} of edge (v_i, v_j) is the conditional probability† that the program execution will go to program block v_j given that it has executed the program block v_i. Once the program reaches v_n, the stop vertex, the probability of its branching to any other vertex is zero. To satisfy the conditions in Eq. (15-8), we add a self-loop of weight one at vertex v_n. Thus v_n is an absorbing state, the only absorbing state in the system, and the remaining vertices correspond to transient states. A very simple stochastic program digraph and its transition matrix P are shown in Fig. 15-11.

†Note that these weights p_{ij}'s have nothing to do with weights f_k's assigned in the previous analysis of the program digraph. While f_k's obey KCL at each vertex, p_{ij}'s obey Eq. (15-8).

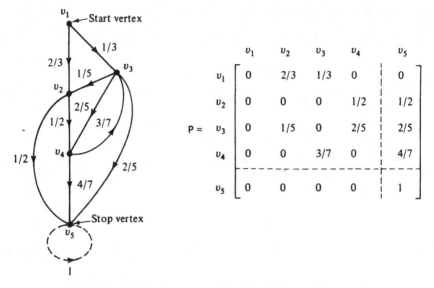

Fig. 15-11 Stochastic program digraph and its transition matrix.

The transition matrix P of any stochastic program digraph can be expressed in the form

$$P = \begin{bmatrix} Q & T \\ \hline 0 & 1 \end{bmatrix}, \qquad (15\text{-}24)$$

where Q is an $(n-1)$ by $(n-1)$ submatrix corresponding to the transient states, T is an $(n-1)$ by 1 column vector, and 0 is the row vector of $n-1$ zeros.

Let us look at matrix P^k, which represents the k-step transition probabilities. Clearly,

$$P^k = \begin{bmatrix} Q^k & T' \\ \hline 0 & 1 \end{bmatrix}. \qquad (15\text{-}25)$$

(T' is a matrix that we need not compute here.)

The ijth entry in Q^k is the probability of being in transient state v_j after exactly k transitions from the starting state v_i (also transient). Let us first show that Q^k becomes 0 as k becomes large.

In the stochastic program digraph G with n vertices, let p_i be the probability that starting from vertex v_i the program will not reach v_n stop vertex in n steps (or less). Since there exists at least one directed path (of length $n-1$ or less) with a nonzero path product from v_i to v_n, quantity $p_i < 1$. Let p be the largest of all p_i's. The probability of not reaching v_n in n steps is less than p, in $2n$ steps it is less than p^2, and so on. Since $p < 1$, these probabilities tend to zero. That is, the sum of the entries in the ith row of Q^k as $k \to \infty$ becomes zero, for $i = 1, 2, \ldots, n - 1$. Now since $Q^k = 0$ for some large k, we can write

$$I + Q + Q^2 + \cdots + Q^{k-1} = (I - Q)^{-1}, \qquad (15\text{-}26)$$

which can be easily verified by multiplying both sides with $I - Q$. Equation (15-26) says that matrix $I - Q$ is nonsingular. For brevity, let us denote matrix $(I - Q)^{-1}$ by $R = [r_{ij}]$.

The sum of probabilities of reaching v_j from v_i in 1 step, 2 steps, 3 steps, $\ldots$, and $k - 1$ steps is equal to r_{ij}, the ijth entry in $(I - Q)^{-1}$, according to Eq. (15-26). This is precisely the average number of times vertex v_j appears in the random paths starting from v_i. In a program, since we always start from v_1 and end at v_n, the first row of R gives us the average iteration counts of all $n - 1$ transient vertices. That is,

r_{1j} = the average number of times program block v_j will be executed in a typical run.

Matrix $R = (I - Q)^{-1}$, besides giving the average number of occurences of different vertices, is a storehouse of a lot of other useful information about the transient behavior of the stochastic program digraph. For example, let h_{ij} denote the probability that the program will ever execute v_j having executed v_i. Clearly,

$r_{ij} = h_{ij} \cdot$ (average number of times v_j occurs, given that the system started in v_j)

$\qquad = h_{ij} \cdot r_{jj}.$

Therefore,

$$h_{ij} = \frac{r_{ij}}{r_{jj}}. \qquad (15\text{-}27)$$

To extract another piece of information from matrix R, let β_j denote the probability that the program starting from v_j will never return to v_j. Now, to pass r_{jj} times through a vertex v_j, the process must reach v_j once and then return there $r_{jj} - 1$ times. Therefore, the probability of returning to v_j after leaving it once is given by

$$\frac{r_{jj} - 1}{r_{jj}} = 1 - \frac{1}{r_{jj}}.$$

Hence the probability, β_j, of never returning to v_j after leaving it once is

$$\beta_j = \frac{1}{r_{jj}}. \qquad (15\text{-}28)$$

Finally, let α_j denote the probability that v_j is executed exactly k times (starting from the start vertex v_1). Then

$$\alpha_j = (h_{1j})(1 - \beta_j)^{k-1}(\beta_j). \tag{15-29}$$

For illustration let us continue with the example of Fig. 15-11. The Q matrix is

$$Q = \begin{bmatrix} 0 & \frac{2}{3} & \frac{1}{3} & 0 \\ 0 & 0 & 0 & \frac{1}{2} \\ 0 & \frac{1}{5} & 0 & \frac{2}{5} \\ 0 & 0 & \frac{3}{7} & 0 \end{bmatrix},$$

and matrix $R = (I - Q)^{-1}$ comes out to be

$$R = (I - Q)^{-1} = \begin{bmatrix} 1 & \frac{26}{33} & \frac{20}{33} & \frac{7}{11} \\ 0 & \frac{58}{33} & \frac{3}{11} & \frac{7}{11} \\ 0 & \frac{14}{33} & \frac{14}{11} & \frac{7}{11} \\ 0 & \frac{6}{55} & \frac{6}{11} & \frac{14}{11} \end{bmatrix}.$$

Therefore,

$r_{12} = \frac{26}{33}$ is the average number of times vertex v_2 gets executed,

$h_{14} = \frac{\frac{7}{11}}{\frac{14}{11}} = \frac{1}{2}$ is the probability that v_4 will be executed,

$\beta_3 = \frac{11}{14}$ is the probability that v_3 will never be executed given that it has just been executed.

And so on.

Having computed the average number of executions of each vertex v_j, one can immediately get the expected execution time for the entire program as

$$\tau = \sum_{j=1}^{n} t_j r_{1j}, \tag{15-30}$$

where t_j is the execution time of the program block v_j. Equation (15-30) assumes that there is no parallel processing (i.e., no two program blocks are executed simultaneously).

For the purpose of segmenting a fragment g in the program digraph, we can compute the least frequently used edge in g, delete it from the fragment, and check if the resulting digraph can be partitioned into appropriate size subfragments. If not, we remove the least frequently used edge in the remaining digraph. This process is continued till subfragment g is segmented into the required size subfragments. This is the stochastic segmentation procedure suggested in [15-33].

The most difficult part of stochastic analysis of a program is the labor and inaccuracies involved in the evaluation of the transition matrix **P**, because the branching probabilities are data dependent. They can, however, be estimated by simulation methods using sample input data [15-33]. Another difficulty is that for many programs the assumption about the weights p_{ij}'s being statistically independent is not valid.

Remarks and References

In analysis and design of application programs and system software you are likely to encounter graph theory more often than any other branch of mathematics. As we have just seen, a weighted digraph is a natural and useful representation of a computer program, and is of immense aid in timing analysis, segmentation, and in detecting certain common types of structural errors. In addition, there are other programming applications that were not discussed here. Some of these are

1. Program optimization, [15-16].
2. Automatic flow charting, [15-19] and [11-4], page 245.
3. Graphs as data structures, [15-34].
4. Parallel-processing design and evaluation, [15-23].
5. In proving equivalence of two programs, or proving validity of a program by transforming the program digraph into canonical forms.

The following list of papers is a sample of the growing literature on utilization of graphs in the art of computer programming.

15-16. ALLEN, F. E., "Program Optimization," *Ann. Rev. Automatic Programming*, Vol. 5, 1969, 239–307. M. I. HALPERN and C. J. SHAW (eds.), Pergamon Press, Inc., Elmsford, N.Y., 1969.
15-17. BAER, J. L., and R. CAUGHEY, "Segmentation and Optimization of Programs from Cyclic Structure Analysis," *Proc. AFIPS Conf.*, Vol. 40, 1972, SJCC, 23–35.
15-18. BEIZER, B., "Analytic Techniques for the Statistical Evaluation of Program Running Time," *AFIPS Conf.*, Vol. 37, 1970, FJCC, 519–524.
15-19. BERZTISS, A. T., and R. P. WATKINS, "Directed Graphs and Automatic Flowcharting," *Proc. 4th Austral. Comput. Conf. Adelaide*, 1969, 495–499.
15-20. EARNEST, C. P., K. G. BALK, and J. ANDERSON, "Analysis of Graphs by Ordering of Nodes," *J. ACM*, Vol. 19, No. 1, Jan. 1972, 23–42.
15-21. HAMBURGER, P., "On an Automated Method of Symbolically Analyzing Times of Computer Programs," *Proceedings of the 21st ACM National Conference*, Thompson Book Co., Washington, D.C., 1966, 321–330.
15-22. KARP, R. M., "A Note on Application of Graph Theory to Digital Computer Programming," *Information and Control*, Vol. 3, 1960, 179–190.
15-23. KARP, R. M., and R. E. MILLER, "Properties of a Model of Parallel Computations: Determinancy, Termination, Queueing," *SIAM J. Appl. Math.*, Vol. 14, 1966, 1390–1411.

15-24. KERNIGHAN, B. W., "Optimal Sequential Partitions of Graphs," *J. ACM*, Vol. 18, No. 1, 1971, 34–40.
15-25. KNUTH, D. E., "Mathematical Analysis of Algorithms," *Proc. IFIP Congress 71*, Ljubljana, Aug. 1971, 1135–1143.
15-26. KNUTH, D. E., "The Analysis of Algorithms," *Proc. International Congress of Mathematics*, Nice, Sept. 1970.
15-27. KRAL, J., "One Way of Estimating Frequencies of Jumps in a Program," *Comm. ACM*, Vol. 11, 1968, 475–480.
15-28. KREIDER, L., "A Flow Analysis Algorithm," *J. ACM*, Vol. 11, No. 4, Oct. 1964, 429–436.
15-29. LOWE, T. C., "Automatic Segmentation of Cyclic Program Structures Based on Connectivity and Processor Timing," *Comm. ACM*, Vol. 13, No. 1, Jan. 1970, 3–9.
15-30. MARIMONT, R. B., "Applications of Graphs and Boolean Matrices to Computer Programming," *SIAM Rev.*, Vol. 2, 1960, 259–268.
15-31. MARTIN, D., and G. ESTRIN, "Models of Computations and Systems—Evaluation of Vertex Probabilities in Graph Models of Computations," *J. ACM*, Vol. 14, No. 2, April 1967, 281–299.
15-32. PROSSER, R. T., "Applications of Boolean Matrices to the Analysis of Flow Diagrams," *Proceedings of the Eastern Joint Computer Conference*, Dec. 1959, Spartan Books, New York, 133–138.
15-33. RAMAMOORTHY, C. V., "The Analytic Design of a Dynamic Look Ahead and Program Segmenting Scheme for Multiprogrammed Computers," *Proceedings of the 21st ACM National Conference*, Thompson Book Co., Washington, D.C., 1966, 229–239.
15-34. ROSENBERG, A. L., "Data Graphs and Addressing Schemes," *J. Computer and System Sci.*, Vol. 5, No. 3, June 1971, 193–238.
15-35. SCHURMANN, A., "The Application of Graphs to the Analysis of Distribution of Loops in a Program," *Information and Control*, Vol. 7, 1964, 275–282.
15-36. VERHOFF, E. W., "Automatic Program Segmentation Based on Boolean Connectivity," *SJCC Proc.*, Vol. 38, 1971, 491–495.

15-4. GRAPHS IN CHEMISTRY

Although Arthur Cayley used trees to represent the structures of organic molecules 100 years ago (and, indeed, much of the early interest in the study of trees was motivated by this representation), it is only recently that graph-theoretic techniques are coming into use for characterization and identification of chemical compounds. This is due to (1) the advent of the electronic computer with its ability to handle graphs, and (2) the ever-intensifying need of the chemist to have a mechanized information retrieval system capable of dealing with the millions of organic compounds known today.

Given a chemical substance and some of its properties (such as molecular weight, chemical composition, mass spectrum, etc.), the chemist would like to find out if this substance is a known compound. If he is able to identify this compound, he may like to know some additional properties of the compound, or if the compound is "new" he would like to know its structure, and then include it in the dictionary of known compounds. It is essential, therefore, to have a standard representation for a compound, and the representation must be compact, unambiguous, and amenable to classification.

It was shown in Section 3-6 how a chemical compound can be represented by means of a connected graph, with the atoms as the vertices and the bonds between them as edges. For compactness the hydrogen atoms are omitted from the representation, as they are implied by every unused valence of the other atoms. For example, the structural graph of aminoacetone C_3H_7NO, with its H atoms stripped off, is shown in Fig. 15-12. [Recall that the valence for carbon (C) is 4, for nitrogen (N) it is 3, and for oxygen (O) it is 2.]

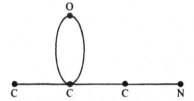

Fig. 15-12 Structural graph of aminoacetone.

The structural graph of a chemical compound, in general, contains much more information than the molecular formula does. For example, the molecular formula $C_{10}H_{22}$ can denote any of its 75 isomers (75 being the number of unlabeled trees with 10 vertices and with no vertex of degree five or more), while the graph specifies the exact isomer. It must be kept in mind, however, that a structural graph does not contain all the information contained in the three-dimensional model of the chemical compound. The structural graph does not specify the bond distances or the bond angles of the molecule. Since these are known only for a small number of organic molecules anyway, this is not much of a handicap. A slightly more serious shortcoming of a graph is its inability to distinguish between stereoisomers [15-40]. Thus, except for stereochemistry purposes, the structural graph gives a reasonably adequate description of a chemical compound.

Canonical Representation of a Molecule

As pointed out above, a standard representation of chemical structures is a precondition for a computerized information retrieval system. From the structural point of view, organic molecules can be divided into two classes: (1) the aliphatic compounds, and (2) the ring compounds. The structural graph of an aliphatic compound has no circuit, except possibly circuits of length two arising out of multiple bonds, which are represented by parallel edges, as shown in Fig. 15-12. The graph of a ring compound contains at least one circuit of length three or more.

Since the graph of an aliphatic compound is a tree (if we ignore parallel edges), it can easily be given a canonical representation as follows: Every tree has a unique centroid or a pair of centroids (parallel edges are considered as single edges for the purposes of locating the centroid). The centroid can be used as the root of the tree (recall Section 10-3), and each subtree attached to the root is a radical. The subtrees can be ordered by the number of vertices they contain in a nondecreasing order. Each radical is further subdivided into subradicals, which are ordered in the same fashion. This process produces a unique linear code for each tree—a string. For example, the code for the tree in Fig. 15-12 is C(C) (=O)(C(N)). For more details on coding of aliphatic compounds, see [15-42].

Cyclic compounds are less tractable, because no unique centroid (or pair of centroids) can be defined in a graph with circuits. Fortunately, the chemist need not be concerned with the general problem of coding a graph (a very difficult problem, as discussed in Chapter 11). The structural graph of almost every ring compound is (1) planar, (2) a regular graph of degree three, and (3) contains a Hamiltonian circuit. It is not very difficult to find a unique linear code for such a graph. There exists an n-sided polygon in such a graph of n vertices, and a description of the graph requires only some notation for the remaining $n/2$ edges. These edges may be represented by a sequence of n numbers consisting of their end vertices. For details on this coding scheme for most ring compounds, see [15-42].

Matching of Chemical Structure

The problem of determining whether or not two chemical compounds (having the same chemical formula) are identical is the same as the graph isomorphism problem, considerably simplified by the labels of the vertices. Finding a unique code for a graph implies the solution of the isomorphism problem as well, because two graphs would be isomorphic if and only if their codes were the same. For chemical structures, however, it is generally easier to perform a direct vertex-by-vertex matching than to first find a unique code for each graph. We shall describe one such algorithm for matching of chemical structures based on Sussenguth's paper [11-59].

The underlying idea behind this algorithm is to use various properties (such as labels, degrees, adjacencies, etc.) of vertices in the two graphs to generate pairs of vertex subsets, which must match if the graphs are to be isomorphic. An increasing number of properties are used to partition vertices into smaller and smaller subsets. Eventually, either every vertex in one graph is uniquely paired off with a vertex in the other graph, or two subsets of vertices characterized by identical properties in the two graphs do not have the same number of vertices. (A third case arises when more than one isomorphism exists between the two graphs.) The process can be best explained with an example:

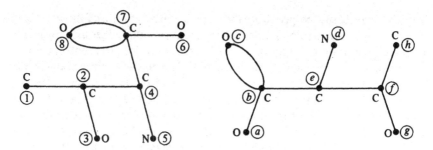

Fig. 15-13 Structural graphs of two molecules.

Let us determine whether or not the two molecules in Fig. 15-13 are identical (H atoms are not shown, as usual). The vertices are arbitrarily named (1), (2), ..., (8) in G and (a), (b), ..., (h) in J.

The process of generating matching subsets with common properties is outlined in Table 15-1. For instance, vertices representing carbon atoms in G must correspond to those representing carbon atoms in J. Of these, the pendant C vertices in G must correspond to pendant C vertices in J, and so on. (Note that Table 15-1 shows only a part of the subsets that are actually generated and matched.) From Table 15-1, we conclude that G and J are

Criteria for Subset Formation		Corresponding Subsets in		Subset Number
		G	J	
Vertex label:	C	1, 2, 4, 7	b, e, f, h	1
	O	3, 6, 8	a, c, g	2
	N	5	d	3
Degree:	One	1, 3, 5, 6	a, d, g, h	4
	Two	8	c	5
	Three	2, 4	e, f	6
	Four	7	b	7
Bond:	Single	1, 2, 3, 4, 5, 6, 7	a, b, d, e, f, g, h	8
	Double	7, 8	b, c	9
Intersection of subsets 1 and 4		1	h	10
Vertices adjacent to				
1/h		2	f	11
5/d		4	e	12
7/b		4, 6, 8	a, c, e	13
Intersection of subsets 4 and 13		6	a	14
Remaining vertex		3	g	15

Table 15-1 Matching of Chemical Structures

isomorphic and the vertex correspondence is

$$\begin{pmatrix} 1 & 2 & 3 & 4 & 5 & 6 & 7 & 8 \\ h & f & g & e & d & a & b & c \end{pmatrix}.$$

A similar procedure can be used to identify one given graph as a subgraph of another.

Computerized Chemical Identification

Given the chemical formula of a "new" substance and the valence rules, one can generate the list of all distinct chemical structures possible, using graph enumeration techniques. Computer programs have been written to perform this operation. (It is necessary, of course, to have a coding scheme that provides a unique representation for a structural graph.) This method of producing an exhaustive list of all possible isomers gets out of hand as the number of atoms in the molecule increases. For example, there are over $\frac{1}{2}$ million structures possible for $C_{20}H_{41}OH$. It is therefore necessary to provide additional chemical information (such as the type of radicals ruled out as unstable) to keep the list to a manageable size. A computer program can be written to compare each of the structures in the list against various sets of analytical data, particularly mass spectra.

As a part of the continuing effort toward a system of automated identification of chemical compounds, a computer language, called DENDRAL, has been developed at Stanford University. See [15-38] and [15-39]. One of the programs in DENDRAL generates the list of all tree-type potential isomers from an input of molecular formula and mass spectrum. The program, written in LISP language, consists of 40,000 words, and is run on a PDP-6 time-sharing system at Stanford. One of the long-term goals of such an effort is to develop a tool for automated chemical exploration of the planets [15-42].

Remarks and References

Lederberg and Feigenbaum and their team at Stanford University have done the pioneering work in computerized chemical identification via graph theory. A number of technical reports and papers (four of them referenced in [15-44]) describe various aspects of the DENDRAL program. See also [15-37]. For a very readable description of the essentials of DENDRAL, see the paper by Feigenbaum and Lederberg [15-38]. Another paper by Lederberg [15-40] is recommended as a well-written exposition of how graphs can be used for representing structures of organic molecules—both tree type and ring type.

Sussenguth in his doctoral thesis and in a subsequent paper [11-59] has given an algorithm for matching chemical structures. He reports that the computation time in his algorithm varies only as the square of the number of vertices; and that his computer program when run on an IBM 7090 took 6 to 7 seconds for matching 50-vertex graphs and only a few thousandths of a second to detect if the graphs were not matched.

A survey of computer methods in handling chemical structures is available in [15-44], which includes most of the relevant references through 1966. Other papers recommended are [15-38], [15-43], and [11-59]. The search for a good coding system is far from over. Papers proposing alternative notational systems continue to appear in the *Journal of Chemical Documentation*.

15-37. BUCHANAN, B. G., G. L. SUTHERLAND, and E. A. FEIGENBAUM, "HEURISTIC DENDRAL: A Program for Generating Explanatory Hypotheses in Organic Chemistry," *Machine Intelligence*, Vol. 4 (Meltzer and D. Michie, eds.), Edinburgh University Press, Edinburgh, 1969.

15-38. FEIGENBAUM, E. A., and J. LEDERBERG, "Mechanization of Inductive Inference in Organic Chemistry," in *Formal Representation of Human Judgement* (B. Kleinmuntz, ed.), John Wiley & Sons, Inc., New York, 1968, 187–218.

15-39. GLUCK, D. J., "A Chemical Structure Storage and Search System Developed at Du Pont," *J. Chem. Doc.*, Vol. 5, No. 1, Feb. 1965, 43–51.

15-40. LEDERBERG, J., "Topology of Molecules," in *The Mathematical Sciences*, The M.I.T. Press, Cambridge, Mass., 1969, 37–51.

15-41. LYNCH, M. F., J. M. HARRISON, W. G. TOWN, and J. E. ASH, *Computer Handling of Chemical Structure Information*, Elsevier Publishing Company, Amsterdam, 1971.

15-42. MEETHAM, A. R., "Partial Isomorphisms in Graphs and Structural Similarities in Tree-Line Organic Molecules," *Proc. IFIP Congress*, 1968.

15-43. PENNY, R. H., "A Connectivity Code for Use in Describing Chemical Structures," *J. Chem. Doc.*, Vol. 5, No. 2, May 1965, 113–117.

15-44. TATE, F. A., "Handling Chemical Compounds in Information Systems," *Ann. Rev. Inf., Sci. Tech.* Vol. 2 (C. A. Cuadra, ed.), John Wiley & Sons, Inc., New York, 1967, 285–309.

15-5. MISCELLANEOUS APPLICATIONS

There is virtually no end to the list of problems that can be solved with graph theory. In addition to applications covered in the last four chapters, many applications were mentioned in earlier chapters, for example, binary search trees for file organization (Chapter 3), design of printed-circuit board (Chapter 5), dimers problem in crystal physics (Chapter 8), teleprinter's problem (Chapter 9), and ranking problem (Chapter 9). The following are some additional examples of applications.

Information Retrieval: In a modern information retrieval system each document carries a number of *index terms* (also called *descriptors*). The index terms are represented as vertices, and if two index terms v_i and v_j are closely related (such as "graph" and "tree"), they are joined with an edge (v_i, v_j).

The simple, undirected (and possibly disconnected) large graph thus produced is called the *similarity graph*. Components (i.e., maximally connected subgraphs) of this graph produce a very natural classification of documents in the system. For retrieval, one specifies some relevant index terms, and the maximal complete subgraph that includes the corresponding vertices will give the complete list of index terms which specify the needed documents. Establishing graph isomorphism is needed in a situation such as an information retrieval system for chemical compounds. The reader is referred to Salton [15-60] for more on the subject. Reference [11-1] is also recommended.

Analysis of Lumped Physical Systems: In Chapter 13 we saw how a system consisting of two-terminal electrical components was represented (and analyzed) by means of a graph. In that case the graph looked very much like the network schematic diagram. This approach can be generalized so that a graph (called the *system graph*) is used to model any physical system built from a finite number of interconnected components, given the model of each component, of course. The system graph is a convenient tool in analysis of the entire physical system. See Trent [15-61] or Koenig, Tokad, and Kesavan [15-57] for more details.

Matrix Inversion: For inverting a large (say, 100 by 100) sparse matrix M by a computer, the straightforward application of the Gaussian elimination method is inefficient, is susceptible to poor accuracy, and causes storage problems. The following graph-theoretic method has been found to be better:

1. Replace each nonzero entry in the given matrix M with a 1, and permute the rows and the corresponding columns of the resulting binary matrix to make all diagonal entries as 1's.
2. The matrix X so obtained is now regarded as the adjacency matrix of a digraph G (the self-loops corresponding to 1's along the diagonal are deleted).
3. The resulting digraph is partitioned into its fragments.
4. A fragment, if too large, is "torn" further into smaller fragments by removing an appropriate edge.
5. The smaller matrices are inverted, and from them the inverse of the original matrix M^{-1} is obtained.

For further details see Harary [15-54] and Iyer [15-56].

Graphs of Groups: Cayley showed that every group of order n can be represented by a strongly connected digraph of n vertices, in which each vertex corresponds to a group element and edges carry the label of a generator of the group (originally, Cayley used edges of different colors to show different generators). Thus the graph of a cyclic group of order n is a directed circuit of n vertices in which every edge has the same label. The digraph of a group uniquely defines the group by specifying how every product of elements corresponds to a directed edge sequence. This digraph, known as the Cayley diagram, is useful in visualizing and studying abstract groups. For more details on graphs of groups, see [15-52].

Linguistics: Graphs have been used in linguistics to depict parsing diagrams. The vertices represent words and word strings and the edges represent certain syntactical relationships between them. A set of words (vocabulary) and a set of rules (grammar) for forming strings (sentences) characterize a language. In other words, the language then is a set of all legal strings so generated. (Natural languages, because of their complexity, have defied attempts at such complete specifications.) One problem in computational linguistics is to identify whether or not a given string belongs to a language, whose vocabulary and grammar are given. For more on graphs in computational linguistics, see [15-55].

Sociological Structures: Digraphs under the name *sociograms* have been used to represent relationships among individuals in a society (or group). Members are represented by vertices and the relationship (admiration, association, influence, etc.) by directed edges. Connectedness, separability, complete subdigraphs, size of fragments, and so forth, in a sociogram can be given immediate significance. A number of tribes have been studied by anthropologists and are classified according to their kinship structures. For more on this topic, see Flament [15-51], Harary [15-53], and Chapter 8 of Anderson [15-45].

Graph theory has also been used in economics [15-46], logistics, cybernetics, artificial intelligence, pattern recognition, genetics, reliability theory, fault diagnosis in computers, studying the structure of computer memory, and the study of Martian canals [15-63]. A mathematical model of disarmament [1-2] has been attempted with graph theory and so have the conflict in the Middle East and the structure of Mozart's opera, *Cosi fan tutte* [15-53]. And thus goes the ever-increasing list of applications of graph theory. Admittedly, in some of these applications a graph is used only as a means of visual representation, and no more than a trivial use is made of graph theory itself. There are many cases, however, where important and not-so-obvious results are obtained through a deeper use of graph theory.

Bibliographies and Further Reading: Although throughout the text we have provided selected readings, the number of published papers on graph theory is much larger (over 3000). There are several good bibliographies available, where most of the published material on the subject is listed. The best known among these are Zykov [15-64], Turner and Kautz [15-63], Turner [15-62], and Deo [15-49]. An informative article containing a systematic list of definitions and a bibliography of graph theory as applied to physics is [15-50]. Some additional books recommended are Mayeda [15-59], Chen [15-48], Marshall [15-58], and Behzad and Chartrand [15-47]. I hope your interest in graph theory has been aroused sufficiently so that you will go exploring in the cited literature on your own.

15-45. ANDERSON, S., *Graph Theory and Finite Combinatorics*, Markham Publishing Company, Chicago, 1970.

15-46. AVONDO-BODINO, G., *Economic Applications of the Theory of Graphs*, Gordon and Breach, Science Publishers, Inc., New York, 1962.

15-47. BEHZAD, M., and G. CHARTRAND, *Introduction to the Theory of Graphs*, Allyn and Bacon, Inc., Boston, 1972.

15-48. CHEN, W., *Applied Graph Theory*, North-Holland Publishing Company, Amsterdam, 1971.

15-49. DEO, N., "An Extensive English Language Bibliography on Graph Theory and Its Applications," National Aeronautics and Space Administration/JPL (California Institute of Technology) Technical Report 32-1413, October 1969; Supplement 1, April 1971.

15-50. ESSAM, J. W., and M. E. FISHER, "Some Basic Definitions in Graph Theory," *Rev. Mod. Phys.*, Vol. 42, No. 2, April 1970, 272–288.

15-51. FLAMENT, C., *Applications of Graph Theory to Group Structure*, Prentice-Hall, Inc., Englewood Cliffs, N.J., 1963.

15-52. GROSSMAN, I., and W. MAGNUS, *Groups and Their Graphs*, The Random House/Singer School Division, New York, 1964.

15-53. HARARY, F., "Graph Theory as a Structural Model in the Social Sciences," in *Graph Theory and Its Applications* (B. Harris, ed.), Academic Press, Inc., New York, 1970, 1–16.

15-54. HARARY, F., "Sparse Matrices and Graph Theory," in *Large Sparse Sets of Linear Equations* (J. K. Reid, ed.), Academic Press, Inc., New York, 1971, 139–150.

15-55. HARRIS, Z., *Mathematical Structure of Language*, John Wiley & Sons, Inc. (Interscience Division), New York, 1968.

15-56. IYER, C., "Computer Analysis of Large-Scale Systems," Ph.D. Thesis, Department of Electrical Engineering, University of Hawaii, Honolulu, May 1971.

15-57. KOENIG, H. E., Y. TOKAD, and H. K. KESAVAN, *Analysis of Discrete Physical Systems*, McGraw-Hill Book Company, New York, 1967.

15-58. MARSHALL, C. W., *Applied Graph Theory*, John Wiley & Sons, Inc., (Interscience Division), New York, 1971.

15-59. MAYEDA, W., *Graph Theory*, John Wiley & Sons, Inc., (Interscience Division), New York, 1972.

15-60. SALTON, G., *Automatic Information Organization and Retrieval*, McGraw-Hill Book Company, New York, 1968.

15-61. TRENT, H. M., "Isomorphism Between Oriented Linear Graphs and Lumped Physical Systems," *J. Acoust. Soc. Am.*, Vol. 27, 1955, 500–527.

15-62. TURNER, J., "Key-Word Indexed Bibliography of Graph Theory," in *Proof Techniques in Graph Theory* (F. Harary, ed.), Academic Press, Inc., 1969, 189–330.

15-63. TURNER, J., and W. H. KAUTZ, "A Survey of Progress in Graph Theory in the Soviet Union," *SIAM Rev.*, Supplement Issue, Vol. 12, 1970, 1–68.

15-64. ZYKOV, A. A., "Bibliography of Graph Theory," in *Theory of Graphs and Its Applications* (M. Fiedler, ed.), Academic Press, Inc., New York, 1964.

Appendix

A BINET-CAUCHY THEOREM

The following classical result, known as the Binet-Cauchy theorem, is useful in calculating the determinant of the product of two matrices: If Q and R are k by m and m by k matrices, respectively, with $k < m$, then the determinant of the product

det(QR) = the sum of the products of corresponding
major determinants of Q and R.

The term major determinant (or simply major) means the determinant of the largest square submatrix of Q (or R) formed by taking any k columns (or rows) out of m. The term corresponding implies that if columns $i_1, i_2, \ldots, i_k$ of Q are chosen for a particular major, the corresponding major of Q must consist of rows $i_1, i_2, \ldots, i_k$ of Q. Clearly, there are $\binom{m}{k}$ such product terms.

Before proving the theorem, let us illustrate with an example:
Let

$$Q = \begin{bmatrix} 4 & -3 & -2 \\ 2 & -1 & 0 \end{bmatrix},$$

and

$$R = \begin{bmatrix} 1 & -1 \\ -2 & 0 \\ 3 & -2 \end{bmatrix};$$

$$\det(QR) = \det\begin{bmatrix} 4 & -3 \\ 2 & -1 \end{bmatrix} \cdot \det\begin{bmatrix} 1 & -1 \\ -2 & 0 \end{bmatrix} + \det\begin{bmatrix} 4 & -2 \\ 2 & 0 \end{bmatrix}$$

$$\cdot \det\begin{bmatrix} 1 & -1 \\ 3 & -2 \end{bmatrix} + \det\begin{bmatrix} -3 & -2 \\ -1 & 0 \end{bmatrix} \cdot \det\begin{bmatrix} -2 & 0 \\ 3 & -2 \end{bmatrix}$$

$$= 2 \cdot (-2) + 4 \cdot 1 + (-2) \cdot 4 = -8.$$

Proof: To evaluate $\det(QR)$, let us devise and multiply two $(m+k)$ by $(m+k)$ partitioned matrices:

$$\begin{bmatrix} I_k & Q \\ 0 & I_m \end{bmatrix} \cdot \begin{bmatrix} Q & 0 \\ -I_m & R \end{bmatrix} = \begin{bmatrix} 0 & QR \\ -I_m & R \end{bmatrix},$$

where I_m and I_k are identity matrices of order m and k, respectively. Therefore,

$$\det \begin{bmatrix} Q & 0 \\ -I_m & R \end{bmatrix} = \det \begin{bmatrix} 0 & QR \\ -I_m & R \end{bmatrix}.$$

That is,

$$\det(QR) = \det \begin{bmatrix} Q & 0 \\ -I_m & R \end{bmatrix}. \tag{A-1}$$

Let us now apply Cauchy's expansion method to the right-hand side of Eq. (A-1), and observe that the only nonzero minors of any order in matrix $-I_m$ are its principal minors of that order. We thus find that the Cauchy expansion consists of these minors of order $m-k$ multiplied by their cofactors of order k in Q and R together. ∎

Appendix

B NULLITY OF A MATRIX AND SYLVESTER'S LAW

If Q is an n by n matrix, then $Qx = 0$ has a nontrivial solution $x \neq 0$ if and only if Q is singular; that is, det $Q = 0$. The set of all vectors x that satisfy $Qx = 0$ forms a vector space called the *null space* of matrix Q. The rank of the null space is called the *nullity* of Q. Furthermore, it can be shown that

$$\text{rank of Q} + \text{nullity of Q} = n. \tag{B-1}$$

These definitions and Eq. (B-1) also hold when Q is not square but a k by n matrix, $k < n$.

An important result involving nullity of matrices is Sylvester's law of nullity, which can be stated as follows:

Sylvester's Law: If Q is a k by n matrix and R is an n by p matrix, then the nullity of the product cannot exceed the sum of the nullities of the factors; that is,

$$\text{nullity of QR} \leq \text{nullity of Q} + \text{nullity of R}. \tag{B-2}$$

Proof: Since every vector x that satisfies $Rx = 0$ must certainly satisfy $QRx = 0$, we have

$$\text{nullity of QR} \geq \text{nullity of R} \geq 0. \tag{B-3}$$

Let s be the nullity of matrix R. Then there exists a set of s linearly independent vectors

$$\{x_1, x_2, \ldots, x_s\}$$

forming a basis of the null space of R. Thus

$$Rx_i = 0, \quad \text{for } i = 1, 2, \ldots, s. \tag{B-4}$$

Now let $s + t$ be the nullity of matrix QR. Then there must exist a set of t linearly independent vectors

$$\{x_{s+1}, x_{s+2}, \ldots, x_{s+t}\}$$

such that the set

$$\{x_1, x_2, \ldots, x_s, x_{s+1}, x_{s+2}, \ldots, x_{s+t}\}$$

forms a basis for the null space of matrix QR. Thus

$$QRx_i = 0, \quad \text{for } i = 1, 2, \ldots, s, s+1, s+2, \ldots, s+t. \quad \text{(B-5)}$$

In other words, of the $s + t$ vectors x_i forming a basis of the null space of QR, the first s vectors are sent to zero by matrix R and the remaining nonzero Rx_i's ($i = s + 1, s + 2, \ldots, s + t$) are sent to zero by matrix Q. Vectors

$$Rx_{s+1}, Rx_{s+2}, \ldots, Rx_{s+t}$$

are linearly independent; for if

$$0 = a_1 Rx_{s+1} + a_2 Rx_{s+2} + \cdots + a_t Rx_{s+t}$$
$$= R(a_1 x_{s+1} + a_2 x_{s+2} + \cdots + a_t x_{s+t}),$$

then vector $(a_1 x_{s+1} + a_2 x_{s+2} + \cdots + a_t x_{s+t})$ must be the null space of R, which is possible only if $a_1 = a_2 = \cdots = a_t = 0$.

Thus we have found that there are at least t linearly independent vectors which are sent to zero by matrix Q, and therefore

$$\text{nullity of Q} \geq t.$$

But since

$$t = (s + t) - s$$
$$= \text{nullity of QR} - \text{nullity of R},$$

Eq. (B-2) follows.

Substituting Eq. (B-1) into Eq. (B-2), we find that

$$\text{rank of QR} \geq \text{rank of Q} + \text{rank of R} - n. \quad \text{(B-6)}$$

Furthermore, in Eq. (B-6) if the matrix product QR is zero, then

$$\text{rank of Q} + \text{rank of R} \leq n \quad \blacksquare \quad \text{(B-7)}$$

SUBJECT INDEX

A

Abelian group, 114, 116
Abelian monoid, 114
Abelian semigroup:
 definition, 113
 with identity element, 114
Absorbing state, 429
Abstract graph, 88-89
Accessible, 203
Activities:
 critical, 403
 definition, 400
 dummy, 401
 duration of, 400
Activity network, 400
Activity-vertex representation, 408
Acyclic digraph, 230, 410
Acyclic network, 400
Adjacency matrix, 157-161, 220-227
 as data structure in algorithms, 270
 powers of, 159, 222
 properties of, 158, 220
 relationship with other matrices, 161
Adjacent:
 definition, 7
 edges, 177
Algebra (*see* Algebraic system)
Algebraic system:
 definition, 113
 with one operation, 114
 with two operations, 118

Algorithms, 268-327
 bridge-finding, 323
 chromatic number, 313
 circuit generation, 287
 connectedness and components, 274
 cut-vertices and blocks, 284
 definition, 269
 efficiency of, 269
 feedback edge-set, 313
 feedback vertex-set, 313
 fundamental circuits, 280
 fragment-finding, 304
 generating all directed circuits, 287
 Hamiltonian circuit, 313
 isomorphism, 310
 minimal cut, 312
 minimal edge cover, 313
 minimal spanning tree (*see* Algorithms, shortest spanning tree)
 maximal clique, 312
 maximal matching, 312
 planarity-testing, 304
 shortest-path, 290
 shortest path between all pairs of vertices, 297
 shortest path from specified vertex to another vertex, 292
 shortest path from specified vertex to all others, 292
 shortest spanning tree, 62, 279
 smallest dominating set, 313
 spanning-tree, 277

Steiner tree, 313
transitive closure, 300
traveling salesman problem, 313
topological sorting, 403
AMBIT/G, 317
Arbitrarily traceable graphs, 29
Arborescence, 206
  number of, 223, 238
  root of, 206
Articulation point (*see* Cut-vertex)
Assignment problem, 178, 396
Automata (*see* Sequential machines)
Automatic flow charting, 448
Automorphism, 267

B

Backtrack, 288, 301
Balanced digraph, 197
Bases of circuit subspace, 126
Bases of cut-set subspace, 127
Basic cut-set (*see* Fundamental cut-sets)
Basis vectors, 124
BCD code, 344
Bicenters, 47
Bicentroidal trees, 248
Bichromatic graph, 166
Binary operation, 113
Binary group code, 352
Binary matrix, 138
Binary relation, 198
Binary tree, 49
Binet-cauchy theorem, 219, 366, 373, 458
Bipartite, complete, 192
Bipartite graph, 168
Block, 80, 284
Block-diagonal form, 274
Boolean addition, 330
Boolean algebra, 328
Boolean arithmetic, 170, 173
Boolean function, 329
Boolean multiplication, 330
Branch, 3 (*see also* Edge)
Branch of tree, 56
Breadth-first search, 302
Bridge, 286

C

Canonic form of switching function, 350
Canonic form of program digraph, 448
Canonical representation of molecules, 450
Cayley diagram, 456
Cayley's theorem, 54, 164
Center of tree, 45
Central tree, 60
Centroidal tree, 248
Chain (*see* Walk)
Characteristic polynomial, 311
Chemical identification, 451, 453
Chord, 56, 212, 278
Chord-set, 56
Chromatic number, 166, 171, 313
Chromatic partitioning, 169, 171
Chromatic polynomial, 174, 177
Circuit, 20
  directed, 202
  fundamental, 57
  Hamiltonian, 30
  subspace, 126, 130
Circuit correspondence between graphs, 84
Circuit-generation algorithms, 284
Circuit matrix, 141-145, 216-217, 337, 359, 380
Circuit-path decomposition, 306
Circuit vector, 125
Classification of graphs according to connectivity, 85
Clique, 32, 312
Closed state, 429
Cocycle (*see* Cut-set)
Coding a graph, 311
Coefficient of internal stability (*see* Independence number)
Coloring problem, 165
Combinatorial dual, 104, 106
Combinatorial graph (*see* Abstract graph)
Combinatorial optimization problem, 396
Commutative field, 117
Commutative group, 114
Commutative ring, 117
Commutative semigroup, 113
Complete graph, 32
Complete matching, 178
Completely regular graph, 111
Completely specified machine, 342

SUBJECT INDEX 465

Component, 21, 55, 202, 274, 275, 278
Computation time of algorithms, 270
Computer logic, partitioning of, 165
Computer programs as digraphs, 194, 439
Condensation, 203, 230
Configurations:
  counting series, 257
  definition, 257
Connectedness:
  definition, 21
  in digraph, 202, 221
  minimal, 42
  strong, 202
  weak, 202
Connectedness and components algorithm, 274
Connection matrix (*see* Adjacency matrix)
Contact network, 329
Cook-Karp class of algorithms, 316
Cotree (*see* Chord-set)
Counting series, 243, 257
Counting trees (*see* Enumeration, of trees)
Covering (*see* Edge-covering)
Covering subgraph (*see* Edge-covering)
Covering number, 183
Critical path method (CPM), 400-408
Cross variable, 357
Cut, 387
Cut-node (*see* Cut-vertex)
Cut-set, 68-71
  capacity, 80
  minimal, 68
  proper, 68
  properties of, 69-71
  simple, 68
  subspace, 130
Cut-set matrix, 151, 153, 220, 380
Cut-set vector, 125
Cut-vertex, 76, 284
Cycle (*see* Circuit, Directed circuits)
Cycle gain, 421
Cycle index, 253-254
Cycle structure, 252
Cyclic code, 351
Cyclic exchange, 59
Cyclic interchange (*see* Cyclic exchange)
Cyclic representation of permutation, 251
Cyclomatic number (*see* Nullity)

D

Data structure in graph algorithms, 270-273
Decanting problem, 13
Decision tree, 41
Decyclization, 232
Deficiency, 181
Degree:
  matrix, 139, 164
  of a permutation, 251
  of a vertex, 7
Degree-constrained shortest spanning tree, 63
Deletion of edge, 27
Deletion of vertex, 27
DENDRAL, 453
Depth-first search, 301-304
Deterministic sequential machine, 342
Diameter:
  of a graph, 163
  of a tree, 48
Digraph, 194-237
  acyclic, 301, 410
  adjacency matrix of, 220
  asymmetric, 197
  balanced, 197
  complete, 197
  definition, 194
  disconnected, 202
  edge, 236
  Euler, 203
  game, 410
  irreflexive, 199
  kernel of, 411
  pseudosymmetric, 197
  reflexive, 199
  regular, 197
  representation of permutations, 251
  simple, 197
  strongly connected, 202
  symmetric, 197, 199
  transitive, 200
  weakly connected, 202
  weighted, 400
Dihedral group, 266
Dimension of vector space, 124
Dimer problem, 185-186
DIP, 317

Directed circuits, 202, 212, 230, 232, 287, 291, 421, 443
Directed graph (see Digraph)
Directed Hamiltonian circuits, 312
Directed path, 231, 288, 403, 423
Disconnected graph, 21, 139, 159, 161
Distance matrix, 61, 273 (see also Weight matrix)
Distance between two spanning trees, 59
Distinct representatives, 179
Division ring, 117
Dominating set, 172-173
Domination number, 173
Dummy activity, 401
Dual of graph, 103, 105, 190

E

Eccentricity of vertex, 46
Edge, 1
  adjacent, 177
  backward, 389
  capacity of, 79, 385
  covering, 182-183
  current, 357
  directed, 195
  edge-current vector, 358
  edge digraph, 236, 408
  forward, 389
  gain, 418
  incident into a vertex, 195
  incident out of a vertex, 195
  initial vertex of, 195
  isomorphism, 87
  listing, 271
  parallel, 2
  pendant, 183
  sequences of, 160, 222, 333
  series, 9
  terminal vertex of, 195
  train, 20 (see also Walk)
  variables, 357
  voltage, 357
  weight of, 61

Edge, connectivity, 75
Edge covering, 182-183
Edge current, 357
Edge digraph, 236, 408
Edge gain, 418

Edge isomorphism, 87
Edge listing, 271
Edge train, 20 (see also Walk)
Edge variables, 357
Edge voltage, 357
Edge-current vector, 358
Edge-disjoint Hamiltonian circuits, 32
Edge-disjoint subgraphs, 17
Edge-disjoint union of circuits, 115, 212
Edge-disjoint union of cut-sets, 71
Edge-voltage vector, 358
Electrical network:
  application of graph theory to, 356-383
  as flow problems, 399
  use of computers in, 268
  use of incidence matrices in, 271
Elementary reduction, 99
Elementary tree transformation (see Cyclic exchange)
Embedding:
  of graph, 90
  on sphere, 94
Enumeration:
  of digraphs, 263
  of multigraphs, 262
  of simple graphs, 260
  of trees, 52, 240-250
Enumerator, 243
Equivalence classes, 201
Equivalence relation, 239
Ergodic process, 429
Error-checking code, 352, 354
Euler graph, 23, 115, 210
Euler lines:
  directed, 203, 210, 225, 227
  number of, 205, 226, 238
  in spanning arborescence, 210
Euler's formula, 96
Events:
  critical, 403
  in projects, 400
Event-vertex representation, 408
Execution time (see Computation time)
Exterior region, 94

F

Faces (see Regions)
False vertices, 350

Fary's theorem, 93
Field, 117
Figure counting series, 258
Finite fields, 119
Finite-state machines (*see* Sequential machines)
First Betti number (*see* Nullity)
Five-color theorem, 188
Float of activity, 406
Flow, 385
Flow chart, 269
Flow network, 384, 398
Flow problem, 384-399
  matching problem, as, 182, 396
  use of computers in, 268
Forest, 55
Fortran Extended Graph Theoretic Language (FGRAAL), 317
Forward calculation, 405
Four-color conjecture, 10, 187-190
Fragments:
  definition, 202
  finding all, 312
  in program segmentation, 443
Free trees:
  definition, 48
  number of unlabeled, 248
Fronds, 303
Full symmetric group, 253
Function:
  definition, 256
  equivalence classes of, 256
Fundamental circuits:
  algorithm, 280-284
  and cut-sets, 73
  definition, 57, 71
  for digraph, 212
  matrix, 144
    application in electrical networks, 359
    deriving of, 144
    for digraph, 219
Fundamental cut-sets:
  definition, 71
  for digraph, 212
  matrix, 153
    relationship with other matrices, 153
    in synthesis of contact networks, 336-339
Fusion of vertices, 28, 274

G

Galois field:
  modulo m, 118-119
  modulo 2, 138
Game, 409-413
  comparison with puzzle, 409
  digraph of, 410
  finite, 409
  perfect-information, 409
  states in, 410
  two-person, 409
Generating functions, 241
Geometric dual of a graph, 103
Geometric representation of a graph, 89
Graph Information Retrieval Language (GIRL), 317
Graph Algorithm Software Package (GASP), 316
Graph Extended Algol (GEA), 317
Graphs:
  arbitrarily traceable, 29
  as data structures, 448
  bipartite, 168
  bichromatic, 166
  circuit-free, 55
  in coding theory, 351-353
  complement of, 56, 76
  complete, 32
  complete bipartite, 192
  connected, 21
  decomposition of, 26
  definition, 1
  directed (*see* Digraph)
  disconnected, 21
  drawing of, 2
  equivalence, 200
  Euler, 23, 28
  finite, 7
  in game theory, 409-413
  general, 2
  infinite, 7
  isomorphic, 14, 139
  Kuratowski, 90, 93
  labeled, 53
  linear, 1
  nonplanar, 90
  nonseparable, 151
  null, 9
  nullity of, 57, 60

SUBJECT INDEX

operations on, 26
oriented, 195 (see also Digraph)
planar, 90
rank of, 57
regular, 8
"rigid", 209
ring sum of, 26
separable, 142
self dual, 107
signal-flow, 416-423
similarity, 455
simple, 2
stochastic, 426
subspaces of, 133
transition, 426
tree, 60
two-connected, 83
union of, 26
unicursal, 24
uniquely colorable, 172
universal, 32
unlabeled, 53
  number of, 239
vertex, 9
weighted, 34, 61-63
Graph-theoretic algorithms, 269-316
  performance, 270
Graph-theoretic languages, 316-317
GRASPE, 317
Gray codes, 351
Group:
  abelian, 114
  definition, 113
  permutation, 250
  of subgraphs, 115
Graph Theoretic Programming Language
  (GTPL), 316

H

Hamiltonian circuit, 30-34
  number of, 268
  origin of, 10
Hamiltonian path, 30-34
  finding, in a graph, 312
  shortest, 63 (see also Traveling salesman problem)
Hamming distance, 349
Height of a tree, 50
Heuristic procedure, 310

HINT, 317
Homeomorphic graphs, 100
Huffman graph-theoretic codes, 352

I

Identification of chemical compounds, 449
Identity element, 113, 116
Identity permutation, 252
Immediate successors, 272
Impedance matrix, 371
Incidence, 7
Incidence matrix, 137-140
  for digraph, 214
  in electrical networks, 359
  as input in algorithms, 271
  rank of, 140
  reduced, 141, 214
  relationship with other matrices, 161
  in synthesis of contact networks, 336
In-degree, 195, 287, 400
Independence number, 170-171
Independent circuits, 144
Independent set of vertices, 169-170
Independent set of edges, 193
Infinite graph, 7
Infinite region, 94
Information retrieval, 449, 454
Instant Insanity, 18
Intermediate vertices, 386
Internal states, 342
Internal vertices, 49
Internally stable set, 169
Intersection of graph, 26
Intersection of subspaces, 131
In-tree, 207 (see also Arborescence)
In-valence (see In-degree)
Invariant of a graph, 311
Inward demidegree (see In-degree)
Isograph (see Balanced digraph)
Isolated vertex, 8
Isomorphic graphs, 14, 139
Isomorphic digraphs, 196
Isomorphism, 14, 53, 159, 209, 239, 274, 284, 310, 451

J

Join, 132
Jordan curve theorem, 91, 189
Jordan's method of elimination, 337

SUBJECT INDEX

K

$\mathcal{K}$-chromatic graph, 166
k-connected graph, 78
Kernel, 411
Kirchhoff matrix, 223
Kirchhoff's current law, 358, 441
Kirchhoff's voltage law, 359
Königsberg bridge problem, 3, 23
Kruskal's algorithm, 62, 280
Kuratowski graphs, 90, 93, 341
Kuratowski's theorem, 100

L

Labeled graph, 53
Labeled trees, 240
Latin square, 193
Level of vertex, 49
Line digraph, 236
Line:
  Euler, 23
  unicursal, 24
  (*See also* Edge)
Linear combination, 123
Linear complex, 3 (*see also* Graphs)
Linear dependence, 123
Linear programming, 216, 386
Linearly independent, 123, 216
Linguistics, graphs in, 456
Link (*see* Chord)
Longest-path analysis, 301
Loop:
  definition, 1, 21
  in electrical networks, 360
Loop impedance matrix, 371
Lossy networks, 392
Lower bound on edge capacity, 392
Lumped physical systems, 455

M

Map coloring, 187
Map-construction approach, 304
Markov chain, 425
Markov process, 424-439
  asymptotic behavior, 433
  definition, 424
  periodic, 431

transient analysis, 437
with transient states, 432
Marriage problem, 180
Mason's gain formula, 421-422
Matching, 177-182
  in bipartite graphs, 182
  definition, 177
  maximal, 178
  perfect, 186
Matching number, 178
Matching problem (*see* Assignment problem)
Matrix:
  adjacency, 157-159
  circuit, 142-143
  cut-set, 151, 153
  incidence, 137, 139
  inversion, 455
  relation, 201
  representing a graph, 137
  stochastic, 425
  transition, 425
  transmission, 332
  weight, 61
Max-flow min-cut theorem, 86, 387-388
Maximal complete subgraph, 312
Maximal flow, 312, 385-386
Maximal matching, 178, 312
Maximal planar graph, 111
Maximal strongly connected subgraphs (*see* Fragments)
Maxwell's formula, 366
Meshes (*see* Regions)
Method of paired comparisons, 227
Minimal cost flow, 393-395
Minimal covering, 184, 328
Minimal decyclization, 232, 313
Minimal spanning tree, 61, 277-279
Minimum-feedback arcs, 232
Monoid, 113
Multicommodity flow, 395-396
Multiple sources and sinks, 390

N

Network analysis problem, 305
Network:
  activity, 400-409
  contact, 329-341
  electrical, 5, 356-381
  in planning and scheduling, 400

synthesis of, 334
transport, 384-389
Network flows, 79
Network functions, 370
Nim, 410
Nodal analysis, 362
Node (*see* Vertex)
Node admittance matrix, 363
Node-removal method, 334
Node voltages, 361, 370
Nonplanar graph, 90, 306, 341
Nonpolynomial algorithms, 315
Nonseparable graph, 76, 284
Null graph, 9, 122
Nullity of a graph, 57, 60
Number of different arborescences, 238
Number of different directed Euler lines, 238
Number of free unlabeled trees, 248
Number of labeled graphs, 239
Number of labeled trees, 240
Number of rooted labeled trees, 241
Number of rooted unlabeled trees, 243
Number of unlabeled graphs, 239

O

1-connected graph, 78
1-factor (*see* Matching, perfect)
1-isomorphic, 81
One simplex, 3 (*see also* Edge)
Operation on graphs, 26
Operations research, graphs in, 384-414
Optimal-policy matrix, 298
Ordered trees, 209
Orientation of graph, 195-196
Orthogonal complements, 132
Orthogonal vectors, 130
Otter's formula, 249
Out-degree, 195, 287, 343, 400
Outer region (*see* Infinite region)
Out-tree, 207 (*see also* Arborescence)
Out-valence (*see* Out-degree)
Outward demidegree (*see* Out-degree)

P

Pair group, 255
Palm tree, 303

Parallel edges, 2, 271, 401
Parallel processing design, 448
Parenthesis-free notation (*see* Polish notation)
p-partite, 168
Partitioning algorithm, 313
Partitioning problem, 165
Partitions, 243
Passive edges, 363
Path:
  critical, 403
  compared with walk and circuit, 21
  directed, 201
  Hamiltonian, 31
  length of, 20
Path length, 51
Path matrix, 156, 336
Path product, 330
Path-finding algorithm, 273
Paton's algorithm, 281
Pendant vertex, 9, 43, 196
Performance of graph-theoretic algorithms, 314
Permutation, 250-255
  degree of, 253
Permutation group, 250, 253, 256
Persistent state, 344
PERT, 232, 268, 400-409
Planar graph, 90, 108, 165
Planarity testing algorithm, 99, 304-310
Plane representation, 90, 95, 97, 273
Planning and scheduling of networks, 400-409
Point (*see* Vertex)
Polish notation, 208
Pólya's counting theorem, 238, 250, 257-264
Polynomial-bounded algorithms, 314
Precedence matrix, 220
Precedence relationship, 400
Predecessor matrix, 220
Preference graph, 227
Primitive connection matrix, 332
Prim's algorithm, 62, 279
Probability vector, 426
Program:
  error detection in, 441
  optimization of, 448
  segmentation of, 443
Program block, 440
Program digraph, 440, 445
Project cost curve, 406

Proper coloring, 165-168
 definition, 165
 of edges, 177
 of regions, 186

Q

Quadratic flow-cost function, 399

R

Radius of a tree, 48
Random digraph, 296
Random graph, 278, 321
Random processes, 424
Random walk, 427
Randomly generated graph (*see* Random graph)
Rank:
 of graph, 57, 60
 of incidence matrix, 214
Ranking by Hamiltonian path, 228
Ranking by score, 228
Ranking with minimum violations, 229
Reachability algorithm, 300
Reachability matrix, 235
Reachable vertex, 203
Realizability:
 of a circuit matrix, 341
 of matrices, 162
 of a single-contact function, 335, 340
Reduced incidence matrix, 153, 339
Reference vertex, 214
Reflected binary code, 351
Regions, 93
 adjacent, 187
 coloring, 187
Regular graph, 92
Regular Markov process, 430
Regularization of planar graph, 189
Relation, 198-201
 digraph of, 220
 equivalence, 200
 matrix of, 201
 reflexive, 199
 symmetric, 199
 transitive, 200
Relay contact, 329
Ring, 117

Ring sum, 26
 of circuits, 115, 212
 of cut-sets, 72
RLC network, 362
Rooted tree, 48, 241, 243
 number of unlabeled, 243
Running time, 439, 441

S

s-field (*see* Skew field)
Scaffolding (*see* Spanning tree)
Search techniques, 271
Seating problem, 6, 32
Second-shortest path, 301
Self-dual graphs, 107
Self-loop, 1, 195, 271
Semicircuits, 202, 212
Semigroup, 113
Semipath, 201
Semiwalk, 201
Separable graph, 76, 284
Sequential circuits, 342
Sequential machines, 165, 194, 342, 344
Series edges, 99
Set:
 of basic circuits, 107
 definition, 112
 empty, 112
 null, 112
 with one operation, 112-116
 with two operations, 116-119
Shift register, 205
Shortest-distance arborescence, 294
Shortest-distance tree, 294
Shortest-path algorithms, 290
Signal-flow graph, 194, 416-423, 436
Signal transmission network, 418
Single-contact network, 335
Sink, 385
Skeleton of graph (*see* Spanning tree)
Skew field, 117
Slack:
 free, 406
 total, 406
Snake-in-the-box code, 352
Sociograms, 456
Source, 385
Source vertices, 418
Spanning arborescence, 209, 303
Spanning forest, 55, 146, 277

Spanning tree:
  algorithm for, 277-279
  all, in a graph, 55, 58, 238, 277, 280, 376
  application to electrical networks, 356, 359
  computer running time, use in estimating, 442
  definition, 55, 73, 209, 277
  degree-constrained shortest, 63
  minimal, 61, 279
  number of, 218
  root of, 281
  shortest (see Minimal spanning tree)
  sign of, 218, 376
  weight of, 61
Spanning tree matrix, 164
Sparse graph, 300
Sparse matrix, 271
Star graph, 184
Start vertex, 441
Starting state, 344
State:
  absorbing (see Closed state)
  closed, 429
  ergodic, 429
  persistent, 344
  transient, 432
  trapping (see Closed state)
State assignment problem, 346
State diagram (see State graph)
State equivalence, 345
State graph:
  definition, 342
  properties, 343
  reduction of, 347
State table, 342
Static flow, 385
Stationary process, 425
Steady-state probabilities, 434
Steiner tree, 313
Stereographic projection, 95
Stochastic graph, 426
Stochastic matrix, 425
Stochastic program-digraph, 445
Stochastic system, 425
Stochastically independent transition probabilities, 427
Stop vertex, 441
Storage requirement of program, 439
Strongly connected, 202, 203, 222, 312
Structural isomers, 53

Subgraph, 16, 21, 141, 273
Submatrix, 141
Subnetworks, 408
Subsequence, largest monotonically increasing, 44
Subset, 112
Subspace, 125
Successor listing, 272
Supersink, 390
Supersource, 390
Switching function, 184, 329
Switching network, 146, 271, 328
Sylvester's law, 146, 152, 460
System graph, 380, 455

T

Teleprinter's problem, 204-205
Terminal vertex of path, 20
Thickness, 109
Three-terminal devices, 373
Through variable, 357
Tie (see Chord-set)
Time-invariant process, 425
Topological order, 402, 443
Topological sorting, 231, 313, 402
Tournament, 197, 227-230
Transient vertices, 446
Transition function, 342
Transition matrix, 220, 425, 444
Transition probabilities, 425, 427
  multistep, 427
Transitive closure of digraph, 300
Transitivity (see Relation)
Transmission, 331
Transmission matrix, 332
Transport network, 384-389
Transportation problem, 393
Trapping state (see Closed state)
Traveling salesman problem, 34, 280, 313
Tree, 39-54
  binary, 48
  centers in a, 45
  central, 60
  decision, 41
  diameter of, 48
  in digraphs, 206-211
  external path length of, 51
  family, 41
  free, 48

SUBJECT INDEX 473

height of, 50
labeled, 54
null, 39
number of, 238
ordered, 209
path length of, 51
radius of, 48
rooted, 48
sorting (*see* Decision tree)
shortest-distance, 294
spanning, 55
unlabeled, 54
Tree admittance product, 366
Tree graph, 60
Tree pairs, 379
Tree-felling procedure, 280
True vertices, 350
Tutte's map-construction method, 108
Two-connected graphs, 83
Two-chromatic graph, 166-167
Two-isomorphic graphs, 104, 143, 336
Two-person games, 409
Two-terminal contact network, 334
Two-tree, 368

U

Unicursal graph, 24
Unicursal line, 24
Unimodular matrix, 214, 380
Union of graphs, 26
Unique code for graph, 451
Unique embedding, 98
Uniquely colorable graphs, 172
Uniqueness of dual graphs, 103
Universal graph, 32
Unlabeled graphs, number of, 239
Utilities problem, 4, 88

V

Valency (*see* Degree)
Vector:
    definition, 120
    orthogonal, 130
Vector space, 120-121
    application in analysis of networks, 359
    of graph, 121-122

Vertex, 1
    closing, 410
    degree of, 7
    eccentricity of, 46
    end, 9
    even, 22
    forbidden, 288
Vertex, fusion of, 28
    intermediate, 20
    internal, 49
    isolated, 8
    label of, 15
    level of, 49
    merged (*see* Vertex, fusion of)
    odd, 22
    pendant, 9, 43, 196
    reference, 141
    starting, 410
Vertex coloring, 165-169, 187
Vertex connectivity, 75, 78
Vertex cover, 193
Vertex graph, 9 (*see also* Null graph)
Vertex-disjoint subgraphs, 17
Vertex-edge incidence matrix (*see*
    Incidence matrix)
Vertex-labeling process, 390
Violation in ranking, 229
Vulnerability, 77, 284

W

Walk, 19-21
    closed, 20
    compared with path and circuit, 21
    different types of, 35
    directed, 201
    open, 20
Weight:
    of edge, 61
    of spanning tree, 61
    of subtree, 248
    of vertex, 248
Weight matrix, 61, 273, 418
Weighted graph, 61-63
    complete, 34
Whitney's theorem, 98, 106
Windows (*see* Regions)

AUTHOR INDEX

A

van Aardenne-Ehrenfest, T., 234
Abrahams, J. R., 424
Aitken, A. C., 162
Allen, F. E., 448
Amoia, V., 64
Anderson, J., 448
Anderson, S., 457
Ash, J. E., 454
Auguston, J. G., 318
Avondo-Bodino, G., 457

B

Baer, J. L., 448
Balk, K. G., 448
Bartee, T. C., 354
Basili, V. R., 318
Battersby, A., 414
Beckenbach, E. F., 265, 414
Behzad, M., 457
Beizer, B., 448
Bellman, R., 318
Bellmore, M., 36
Benedict, C. P., 110
Berge, C., 11, 415
Berkowitz, S., 318
Berztiss, A. T., 318, 448
Biondi, E., 439
Birkhoff, G., 354

Boisvert, M., 424
Boothroyd, J., 318
Bott, R., 225
Bray, T. A., 414
Bredeson, J. G., 354, 355
Brooks, R. L., 192
deBruijn, N. G., 206, 234
Bruno, J., 110
Bryant, P. R., 381
Buchanan, B. G., 454
Busacker, R. G., 11, 414

C

Calahan, D. A., 381
Caldwell, S. H., 354
Cartwright, D., 234
Caughey, R., 448
Cayley, A., 10, 11, 52, 54, 164, 238, 248, 449, 456
Chan, S. P., 381
Chartrand, G., 457
Chase, S. M., 318
Chen, W. K., 135, 457
Chen, Y. C., 234
Chien, R. T., 86, 381
Collins, N. L., 65
Cook, S. A., 316
Cooke, K. L., 318
Corneil, D. G., 318, 319
Cottafava, G., 64

C (cont.:)

Coverley, G. P., 424
Crespi-Reghizzi, S., 318
Cummins, R. L., 65

D

Dantzig, G. B., 319
Dawson, D. F., 381
Dean, R. A., 135
Deo, N., 36, 65, 110, 457
Dickson, D. C., 319
Dijkstra, E. W., 280, 292, 319
Dimsdale, B., 415
Dirac, G. A., 36
Dreyfus, S. E., 319

E

Earnest, C. P., 448
Edmonds, J., 319
Elias, P. A., 86
Elmaghraby, S. E., 415
Essam, J. W., 457
Estrin, G., 449
Euler, L., 3, 11, 23, 96, 115, 203, 210, 226
Even, S., 319

F

Fary, I., 93, 110
Feigenbaum, E. A., 454
Feinstein, A., 86
Fisher, M. E., 457
Flament, C., 457
Flores, B., 320
Floyd, R. W., 297, 319
Frank, H., 415
Fraser, J. J., 319
Frazer, W. D., 319, 355
Friedman, D. P., 319, 320
Frisch, I. T., 415
Ford, G. W., 265
Ford, L. R., 86
Fulkerson, D. R., 86, 415

G

Gardner, M., 415
Ghosh, P. K., 424
Ghosh, S. N., 424
Ghouila-Houri, A., 415
Gibbs, N. E., 319
Gluck, D. J., 454
Goldman, J., 135
Good, I. G., 204, 234
Gotlieb, C. C., 318, 319
Gould, R., 135
Graham, G. D., 319
Grossman, I., 457
Grundig, P. M., 415
Guardabassi, G., 439

H

Hakimi, S. L., 36, 354, 355
Halmos, P. R., 135
Hamburger, P., 448
Hamilton, W. R., 10, 30, 31, 63, 165, 312
Harary, F., 11, 36, 162, 234, 265, 457
Harris, Z., 457
Harrison, J. M., 454
Harrison, M. A., 355
Hart, R., 319
Held, M., 319
Henderson, D. A., Jr., 320
Herstein, I. N., 135
Hill, F. J., 355
Hobbs, A. M., 110
Hohn, F. E., 162
Holt, R. C., 319
Hopcroft, J. E., 319
Howard, R. A., 439
Hu, T. C., 65, 319, 415
Huffman, D. A., 65, 352, 354, 355
Huggins, W. H., 439

I

Iri, M., 415
Isaacson, J. D., 320
Iyer, C., 457

K

Karl, J., 449
Karp, R. M., 316, 319, 448
Kasteleyn, P. W., 192
Kaufmann, A., 234
Kautz, W. H., 457
Kemeny, J. G., 439
Kendall, M. G., 234
Kernighan, B. W., 319, 320, 449
Kesavan, H. K., 457
Kim, W. H., 86, 192, 381
King, C. A., 320
Kirchhoff, G., 10, 11, 58, 223, 356, 358, 359, 381, 383, 441
Klee, V., 355
Klein, M. M., 415
Knoedel, W., 320
Knuth, D. E., 65, 320, 449
Koenig, H. E., 457
König, D., 11
Kreider, L., 449
Kruskal, J. B. Jr., 62, 65, 280
Kuo, F. F., 381
Kuratowski, K., 90, 93, 100, 304, 341

L

Lederberg, J., 454
Lehmer, D. H., 320
Lempel, A., 234
Lietzmann, W., 11
Lin, S., 36, 320
Liu, C. L., 192
Lockett, J. A., 318
Lorens, C. S., 424
Lowe, T. C., 449
Lukasiewicz, J., 208
Lynch, M. F., 454

M

MacLane, S., 107, 110, 304
Magnus, W., 457
Marble, G., 320
Marimont, R. B., 449
Markov, A. A., 423, 424, 425, 431, 432, 433, 437
Marshall, C. W., 457
Martin, D., 449
Mason, S. J., 421, 422, 424
Matula, D. W., 320
Maxwell, J. C., 356, 366, 382
Maxwell, L. M., 9, 135
Mayberry, J. P., 225
Mayeda, W., 355, 457
McIlroy, M. D., 320
Medvedev, G. A., 439
Meetham, A. R., 454
Menger, K., 86
Mesztenyi, C. K., 318
Michie, D., 65
Miller, K. S., 135
Miller, R. E., 355, 448
Minker, J., 318
Minty, G. J., 280, 320, 355
Mirsky, L., 192
Moder, J. J., 415
Montalbano, M., 415
Moon, J. W., 65, 234
Morpurgo, R., 318
Mulligan, G. D., 320
Munro, J. I., 320

N

Nemhauser, G. L., 36
Norman, R. Z., 234, 320

O

Onaga, K., 415
Ore, O., 11, 36
Otter, R., 249

P

Palmer, E. M., 265
Parson, T. D., 110
Parzen, E., 439
Paton, K., 281, 287, 320
Penny, R. H., 454
Percival, W. S., 356
Perfect, H., 192
Peterson, G. R., 355
Peterson, W. W., 355
Phillips, C. R., 415
Plisch, D. C., 415
Pohl, I., 320

P *(cont.:)*

Pólya, G., 238, 248, 250, 257
Prabhaker, M., 320
Pratt, T. W., 319, 320
Prim, R. C., 62, 65, 280
Prosser, R. T., 449

R

Rabin, M. O., 320
Ramamoorthy, C. V., 449
Read, R. C., 192, 265, 318, 320
Reed, M. B., 9, 12, 381
Reinboldt, W. C., 318
Reingold, E. M., 319, 320
Rinaldi, S., 439
Riordan, J., 65
Robert, J., 424
Roberts, S. M., 288, 320
Robichaud, L. P. A., 424
Robinson, R. W., 265
Rosenberg, A. L., 449
Rosenblatt, D., 439
Rota, G. C., 135
Rouse Ball, W., 12
Rovner, P. D., 320

S

Saaty, T. L., 11
Sakarovitch, M., 415
Salton, G., 457
Saltzer, C., 354, 355
Schurmann, A., 415, 449
Seppänen, J. J., 320
Seshu, S., 12, 381
Shannon, C. E., 86, 328
Shirey, R. W., 110
Sittler, R. W., 439
Smith, C. A. B., 36, 415
Snell, J. L., 439
Steiglitz, K., 110
Steiner, J., 313
Sussenguth, E. H. Jr., 321
Sutherland, G. L., 454

T

Tabourier, Y., 321
Talbot, A., 381

Tarjan, R. 287, 301, 319, 321
Tate, F. A., 454
Tiernan, J. C., 288, 321
Tokad, Y., 457
Torres, W. T., 319
Town, W. G., 454
Trent, H. M., 457
Tucker, A. C., 65
Turner, J., 457
Tutte, W. T., 36, 108, 110, 225, 234, 341, 355

U

Uhlenbeck, G. E., 265
Unger, S. H., 321

V

Verhoff, E. W., 449

W

Wang, H., 415
Warshall, S., 297, 321
Watkins, R. P., 448
Weinberg, L., 110, 321
Weinblatt, H., 321
Welch, J. T. Jr., 321
Wells, M. B., 321
Welsh, D. J. A., 355
Whitney, H., 86, 98, 106, 110, 304
Whitney, V. K. M., 321
Wilcox, G. W., 36
Wilkov, R. S., 192
Williams, T. W., 135
Wing, O., 234
Witzgall, C., 414
Wolfberg, M. S., 321

Y

Yen, J. Y., 321

Z

Zykov, A. A., 457